T0074338

NEUROPHYSIOLOGY

MONOGRAPHS IN MODERN NEUROBIOLOGY
Walter B. Essman, Series Editor

PSYCHOPHARMACOLOGY:
An Introduction to Experimental and Clinical Principles
By Luigi Valzelli

NEUROCHEMISTRY OF CEREBRAL ELECTROSHOCK
By Walter B. Essman

CURRENT BIOCHEMICAL APPROACHES
TO LEARNING AND MEMORY
Edited by Walter B. Essman and Shinshu Nakajima

FUNCTIONAL CHEMISTRY OF THE BRAIN
By Adrian Dunn and Stephen C. Bondy

DEPRESSION AND SCHIZOPHRENIA:
A Contribution on their Chemical Pathologies
By H.M. van Praag

NEUROTRANSMITTERS, RECEPTORS & DRUG ACTION
Edited by Walter B. Essman

NEUROPHYSIOLOGY
By P.P. Newman

NEUROPHYSIOLOGY

P. P. Newman, M.D.

Department of Physiology,
School of Medicine,
University of Leeds
Leeds, England

ISBN 978-94-011-6683-6 ISBN 978-94-011-6681-2 (eBook)
DOI 10.1007/978-94-011-6681-2

Published in the UK and Europe by
MTP Press Limited
Falcon House
Lancaster, England

Published in the US by
SPECTRUM PUBLICATIONS, INC.
175-20 Wexford Terrace
Jamaica, N.Y. 11432

Softcover reprint of the hardcover 1st edition 1980

ISBN 978-94-011-6683-6

To the memory of
JOHN F. FULTON
whose illustrious writings are a
fountain of wisdom and knowledge,
an inspiration for all time.

Preface

This book was developed from a course of lectures and practicals given to first- and second-year medical students at the University of Leeds. My aim has been to provide a comprehensive account of the nervous system and its functions, which I hope will help the student to attain a better understanding of clinical neurology. For this reason a good deal of attention has been paid to the study of control systems, and emphasis laid on those mechanisms that are frequently deranged by injury or disease. In particular, a useful coverage has been given to disturbances of the motor and sensory systems that commonly occur in human beings.

Throughout the text numerous references have been made to the great pioneers of the past and to present-day investigators whose contributions have added enormously to our knowledge of the subject or who have pointed the way to important advances. Perhaps the most striking change in recent years has been the application of new techniques in neurophysiology for more precise measurement and analysis of experimental results. A biophysical approach is now mandatory and some of its broad outlines have been included. This should present no difficulty to individuals interested in biological methods, especially those who wish to pursue careers in science or to become professional physiologists. With this idea in mind, I have included references to laboratory procedures and many illustrations from original research. The reader need not be reminded that neurophysiology is essentially a practical subject built on years of experimental work. Concepts that appear to be firmly established may have to be revised from time to time in the light of new evidence and further developments.

I am indebted to certain authors for permission to reproduce a number of illustrations. Grateful thanks are due to Mrs. Jacqueline Hill, who typed the entire manuscript, Mr. Grayshon Lumby for his art work, and Mr. Peter Hargreaves for photographic assistance. Finally, I should like to express my thanks for the help and patience and constant encouragement of the publisher.

P.P.N.

Introduction

The nervous system is the means by which we communicate with others and also with ourselves. It is essentially a network of communications with impulses coursing at different speeds controlled like traffic at check-points on the roads, obeying a code of instructions and generally avoiding collisions. Primitive nervous systems deal fairly efficiently with this kind of motor system. In more advanced forms of life, the brain exerts greater control over its communication system by introducing voluntary effort together with feelings and thoughts; in human beings, there is the capacity for creative ability and for many other functions of intelligence.

In order to study all these activities of the nervous system it is useful to describe three operating parts. The first part to be considered is the output. This is usually in the form of a physical change such as a muscular contraction or chemical secretion, but any combination of such changes may occur to form a recognizable pattern of behavior. The second part may be called the input. This comprises all the sense organs of the body, including the eyes and ears, and thus provides information about the state of our environment and any changes that are occurring in it. The third part is the control unit, or central nervous system. It receives, selects, and stores each event and issues instructions that enable us to respond in a sensible and often predictable way; at the same time it is capable of modifying a response in accordance with what has been learned from previous experience.

The output of the nervous system is the means for action. In anatomical terms the output may be somatic or visceral. The somatic output is concerned principally with the muscular system—movements, tone, and posture; the visceral output is a function of the autonomic and endocrine systems. The distinction is no doubt far too rigid, for patterns of behavior are more often the result of extensive somato-visceral interactions organized with great precision and dependent upon the integrated activity of wide regions of the nervous system.

The input to the nervous system generally begins at the receptors, which are devices of high sensitivity. Each receptor is a little biological unit reacting to a particular change by discharging nerve impulses. A single impulse soon becomes many impulses, as the signal passes through the ramifications of repeated stages until the final destination is reached. When a large number of receptors are discharging simultaneously, the mass of impulse traffic could be overwhelming and even harmful if allowed to proceed unchecked. Consequently, the activity of any group of receptors must be related to activity elsewhere, the problem of the nervous system being to ensure that all its lines of communication are properly coordinated. To this end, effective control of the sensory input is imposed at almost every anatomical level. This often starts at the receptor itself, a good example being the feedback loops of muscle spindles. In the spinal cord impulses from different receptors often converge on the same afferent pathways or share the same neurons with the result that competition is intense, and the weaker or delayed discharges are blocked. In the brain itself, the problem of maintaining a desirable level of activity takes many forms. If the brain is to achieve stability it must be able to correct displacements in any direction; it is just as important to correct for a decreased level of excitability as it is to prevent excessive activity. The physiological term for increasing neuronal activity is "facilitation"; the reduction of neuronal activity is called "inhibition." These are two important mechanisms which tend to maintain within limits the entire sensory inflow to the cerebral cortex.

The brain and the spinal cord together operate as a control unit on two different levels. One level is reflex behavior. A reflex may be regarded as an inborn reaction with relatively fixed properties, for the modifications which can be imposed on reflexes are strictly limited. The second level of control involves the higher centers of the nervous system and includes the complex process of learning. With repetition by training or by drawing on experience, the nervous system is capable of improving its performance. Initial clumsiness or bewilderment soon transforms into skill and speed. Obviously, the brain possesses a storage mechanism, which is essential for learning and which has almost unlimited potentialities for rapid self-analysis.

How, then, does the brain work and where is the seat of its intelligence? The brain is a complex machine operating with incredible speed on a perfectly ordered plan. It has at its service a variety of sense organs scanning the environment like antennae and providing a continuous stream of information. A network of communications transmits the signals along well-defined pathways to a number of action stations. Here the best use is made of all the details available for making appropriate decisions, corrections or commands. Neurophysiology deals with all aspects of this plan, and because it is a medical science based on experimental investigation, many questions about the functions of the brain can be answered with a fair amount of certainty. In the following pages, an account will be given of the general principles of nervous activity and of the mechanisms that keep that activity under constant control. Also shown will be how the brain makes extensive use of quite simple reactions to build more complex systems consistent with the behavioral needs of the whole organism. Since normal functioning is dependent on the integrity of the component parts, due reference will be made to disturbances of function that may result from localized injury or disease. Thus a study of neurophysiology provides the necessary background for sound diagnosis and accurate understanding in the treatment of neurological disorders. There remains, however, the problems of human creative intelligence and adaptiveness to which the contributions of neurophysiology are still somewhat academic. This is not surprising, for it has long been recognized that subjective elements, such as those of the mind, require a different sort of discipline.

Contents

PART I

Nervous Activity and Muscular Contraction

CHAPTER I

Neurophysiological Concepts and Methods

NEUROPHYSIOLOGICAL CONCEPTS

The neuron

This term is used to describe the conducting elements of the nervous system. The essential parts of a neuron are the cell body or soma, the cell processes or dendrites, the axon or nerve fiber, and the terminal filaments or nerve endings (Fig. 1).

Nerve cells vary considerably in size, shape, and population density. The average size of a cell is about 50 μm. Under the microscope the commonest types of cells appear to be round, pyramidal, or flask-shaped. The membrane surrounding the cell and its processes consists of a central core of lipid material with outer layers of protein. The membrane has important electrical and metabolic properties that present a barrier to the free movement of ions. Consequently, the composition of the interior of the cell is very different from that of the external medium. The ease with which the membrane can change its properties, especially its permeability for diffusible ions, accounts for the excitability of the neuron.

Within the cytoplasm are found the nucleus and other sub-cellular organelles concerned with the vital processes of synthesis and secretion. These include the Nissl granules, composed of ribose nucleoproteins and distributed around a system of tubules and cavities to form the endoplasmic reticulum of the cell; the Golgi apparatus, containing a heavy concentration of ribosomes; the mitochondria, active in most of the chemical reactions of the cell; and the neurofibrils, a collection of microtubules that are involved in the transport of materials down the axon.

Dendrites are branching processes radiating from the cell body for short distances. Their arrangement varies in complexity and pattern from the simple bipolar and unipolar forms found in ganglia to a profuse arborescence like that of the Purkinje's cells in the cerebellum. The dendrites of spinal motoneurons are intermediate in type, while a specialized form exists in the cerebral cortex where pyramidal cells have short basal and long apical dendrites. Their main function is to make contact with other neurons, serving as receptors by conducting impulses to cells. The sites of impulse transmission are on the soma-dendritic membrane, which provides for the excitatory and inhibitory actions of synapses.

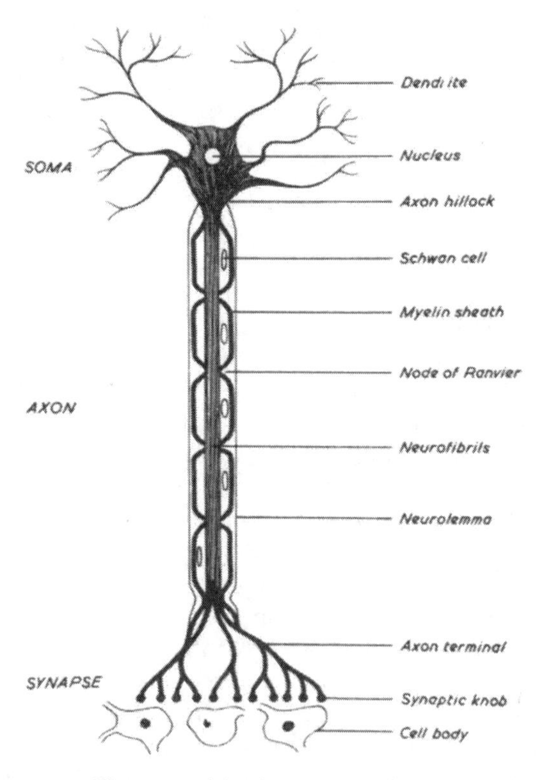

Fig. 1. Diagram of a neuron. The axon, arising from the axon hillock and dividing into a number of branches, contains a central core of neurofibrils and is surrounded by a myelin sheath developed from the Schwann cells. The Ranvier's nodes interrupt the myelin sheath at intervals. The region where the axon terminal contacts another neuron is called a synapse.

The axon is essentially the long process of a nerve cell. It arises from a part of the cell body known as the axon hillock and generally gives off a number of collaterals before proceeding to the next stage of transmission. It carries information from one region to another within the central nervous system or it may serve as a channel of communication between the central nervous system and the periphery. Axons carrying information from the sense organs travel together in bundles known as sensory, or afferent, nerves, while those traveling out of the central nervous system are called efferent nerves. Axons vary in diameter as well as in length. Some assume a myelin sheath arising from an investment of Schwann's cells at intervals along the length of the axon, while others remain unmyelinated. The myelin sheath is important for rapid conduction of information along the axon and for the process of regeneration in nerves that have been damaged.

Axon terminals are specialized structures where the chemicals used in transmission are released. Since there is no direct continuity between one neuron and the next or between a nerve and the organ which it supplies, the information must be carried over by other means. On reaching its terminal each axon divides into a number of branches before expanding into an "end-button" or synaptic knob. Within each knob there is a large number of vesicles containing a transmitter substance, which is released on arrival of a nerve impulse at the point of contact between the axon terminal and the underlying cell membrane of another neuron. In the case of a peripheral nerve supplying a muscle, a comparable transmission process occurs at the motor end plate.

Nerve impulses

The body possesses two systems of communication for sending and receiving instructions–hormonal and neural. Hormones, or "chemical messengers," are produced by endocrine glands and are distributed to their target organs by the circulating blood. Although the brain can exert a controlling influence on the endocrine system, it remains a relatively slow method of communication. The nervous system, on the other hand, employs fast electrical signals called nerve impulses. These may be generated at sensory nerve endings when a receptor is stimulated or generated as a consequence of the events occurring at a synaptic junction. At any moment of time a neuron may be (1) reacting to the continuous bombardment of both excitatory and inhibitory influences converging upon it; (2) discharging impulses down its own axon to other neurons; or (3) remaining in a state of quiescence. There is great variation in the excitability of individual neurons, so that different patterns of nerve impulses may be discharged by the same neuron on different occasions. Further, impulses do not all travel at the same rate; movement depends largely upon the type of nerve fiber and whether it is myelinated. A third point is that the number of impulses traveling in succession along the same fiber is limited by the need for a rest period or refractory state. This, however, is relatively short, lasting only about 1 msec, and it follows that the maximum frequency possible could be as high as 1000 impulses per second. This fact alone will indicate how neuronal activity can be varied over a wide range of frequencies. In general, the instructions carried by nerve impulses are arranged in the form of a code based upon the firing pattern and frequency of their discharge.

Conduction

When a peripheral nerve, for example, the sciatic nerve of the frog, is dissected away from the body, it can be easily demonstrated that impulses are conducted equally well in either direction. In the intact body conduction is in only one direction, depending upon the origin of the impulses. Thus, afferent fibers arising from receptors normally conduct impulses towards the central nervous system, and motor fibers supplying the muscles and glands conduct impulses away from the central nervous system. This functional direction is called orthodromic conduction. Under experimental conditions in the intact body, it is possible to drive the impulse in the wrong direction, and this is called antidromic conduction. The sciatic nerve, like most peripheral nerves, contains a mixed population of sensory and motor fibers bound up together to form a nerve trunk. In the central nervous system conduction occurs along anatomical tracts, which constitute the bulk of the white matter, while the cellular elements occupy the grey matter together with the supporting tissue or neuroglia.

Returning to the rate at which an impulse travels, an approximate measure can be made by observing the time it takes for the impulse to reach a recording electrode from a fixed distance away. In practice, the conduction velocity is estimated from the difference between the latent periods of recordings made at two points on the nerve following stimulation. After measuring the distance between the two points, the result is calculated in meters per second or $m.s.^{-1}$, the units in which conduction velocities are now expressed.

Transmission

Nerve impulses do not pass directly from one neuron to another or from nerve ending to effector organ; the information carried by nerve impulses is transmitted by the release from the axon terminal of chemical substances that may be either excitatory or inhibitory in nature. Since the process of transmission occurs at synaptic junctions, the whole pattern of nervous activity is really determined by prevailing conditions at different synapses. Excitatory action is evident when a neuron which was previously silent discharges at least one nerve impulse. Inhibitory action may be regarded as a reduction in firing frequency up to the point of complete quiescence. Transmitter substances are released from vesicles in the presynaptic fiber or axon terminal soon after the arrival of a nerve impulse. A potential change is then generated at the synaptic junction in the region of the closest contact between the terminal and subsynaptic element. Thus, transmission is achieved by both chemical and electrical means in most vertebrates, although it should be mentioned that purely electrical synapses have been described in crustacea and in some fish.

Convergence

There are relatively few independent lines of transmission in the central nervous system, the majority of them being shared like party lines common to a number of users. Individual cells will, therefore, receive synaptic inputs from many different sources, as can be seen from the clusters of synaptic knobs covering much of the surface of the soma and its dendrites (Fig. 2).

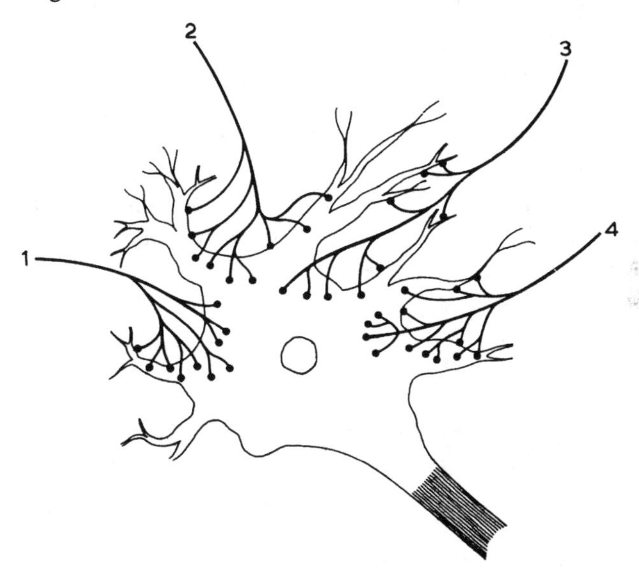

Fig. 2. Convergence of synaptic knobs on soma and dendrites. Four different presynaptic fibers are shown terminating on a single cell. The synaptic endings are most dense on the basal regions of the dendrites and on the surface of the soma.

This coming together of presynaptic axonal terminals is known as convergence. Apart from the obvious economy achieved by neuronal convergence, its effect on the activity of the cell may be considerable. In the first place, the quantity of transmitter substance released by one presynaptic fiber may not be sufficient to produce a significant change; the potential build-up is then said to be below threshold. One way in which a threshold level is attained is through the arrival of a succession of impulses in a short space of time, allowing the necessary build-up. This method is known as temporal summation. To be effective the transmitting substance has to persist beyond the period of the refractory phase. Second, when impulses arrive simultaneously at various points on the soma-dendritic membrane, the build-up of transmitter to threshold level can be quickly achieved. This method of synaptic transmission is know as spatial summation. If, for example, a sufficient number of synaptic knobs is excited, the cell will be caused to discharge. Third, neuronal convergence brings to the cell various excitatory and inhibitory influences and provides for their integration and subsequent behavior of the cell.

Divergence

A presynaptic fiber usually divides many times near its termination to contact a number of other neurons. A good example of this is the mossy fiber input to granule cells in the cerebellum where divergence is very great. This arrangement gives excellent opportunities for bringing a large proportion of cells under the influence of a single afferent source. In the spinal cord, divergence of impulse pathways is responsible for the spread of reflex effects from one segment to another, with the result that widespread activity may follow a strictly localized input.

Excitatory actions

The electrical activity of a neuron depends upon the physico-chemical properties of its surface membrane. This is an extremely important structure investing the entire neuron–cell body, dendrites, and axon. All neuronal membranes have certain properties in common, the most important being the membrane's capacity to alter its permeability to the movements of ions and therefore to permit the flow of electric current across it. Nervous activity may be regarded as a breakdown of the resting conditions in which the membrane is held in a polarized state–that is, the membrane acts as a barrier to the free diffusion of ions, which would lead to equal concentrations of ions on the two sides. Under resting conditions, the polarized membrane maintains a measurable concentration difference.

Excitatory actions evoke a discharge or depolarization of the surface membrane. When this occurs at a receptor or within the axon a nerve impulse is generated. At synaptic junctions, excitatory action is mediated by a chemical transmitter, which acts on the subsynaptic element, causing increased permeability of the postsynaptic membrane for certain ions. The resultant flow of ions causes an electric current to be generated between the synaptic knob and the cell, producing an excitatory postsynaptic potential (EPSP). The cell will discharge an impulse when depolarization of the postsynaptic membrane reaches a critical level.

Inhibitory actions

These also depend upon a change in the permeability of the resting cell membrane with consequent movement of ions, but owing to the nature of the change, the flow of current is in the reverse direction. At the synaptic junctions inhibitory action is mediated by a specific transmitter substance, which changes the ionic permeability of the subsynaptic membrane. As a result, the membrane becomes hyperpolarized, and an inhibitory postsynaptic potential (IPSP) is produced. The change in membrane potential depresses the excitability of the cell below the threshold level for impulse discharge.

Excitatory-inhibitory interactions

It can be envisaged that a single neuron has contact with a large number of synaptic knobs or terminal axons of other neurons, some of which are excitatory and some inhibitory. What happens to the output of the neuron depends upon the relative number of opposing synapses that become active at a given moment of time. Excitatory synapses cause a depolarization of the postsynaptic membrane, which causes the cell to discharge an impulse if sufficient transmitter is released. In this action the excitatory effects of many such synapses can summate in order to reach a threshold level of depolarization. Inhibitory synapses depress the level of depolarization by producing an outward flow of current, which tends to hyperpolarize the postsynaptic membrane. The net result depends upon the averaging out of these two opposing mechanisms.

An impulse is generated when the excitatory synaptic action is dominant—that is, when the current flow causes a depolarization of the membrane by about 10 mv. With more intense current flow, the frequency of impulse discharge is correspondingly increased. In order to produce inhibition the extent of depolarization in the postsynaptic membrane must be kept below the firing level. In this action the more synaptic knobs that can contribute to the inhibition, the more difficult it becomes to shift the membrane potential toward threshold; when inhibitory synaptic action becomes effective, the frequency of discharge is correspondingly reduced. Finally, a pronounced change of membrane potential in the direction of hyperpolarization may cause the neuron to cease to discharge altogether. Under normal circumstances the activity of a single neuron can be considered as the outcome of numerous interactions between excitatory and inhibitory synapses in which fluctuations in the permeability of the postsynaptic membrane play the dominant role.

Neuronal circuits

One of the problems in studying the organization of the nervous system arises from the complexity of the connections between individual neurons. Even some of the basic patterns are complicated by neuronal convergence, which causes fields of activity to overlap. Nevertheless, the basic patterns of organization are sufficiently known and appear over and over again to imply that they have a specific functional role to play. This is true of the stretch or monosynaptic reflex, which employs a direct route of conduction from the sensory receptor to the motoneuron it activates. A more common arrangement consists of a chain of short links or interneurons placed between the input and output systems, while in higher orders of organization, the circuitry can be so involved as to defy identification by ordinary anatomical means.

Two fundamental types of interneuronal activity have been described:

1. The multiple chain. This applies to the division of an axon into branches, repeated at several synaptic stages; the result is that many alternative routes become available. The conduction pathway is often called polysynaptic and is typical of the flexor reflex, whereby impulses can spread over several internuncial relays before reaching their destination in the motoneurons.

2. The closed chain. This applies to the bifurcation of an axon so that one of the branches completes a circuit by looping back through one or more relays to supply the original neuron. The effect of a closed chain is that a neuron can be re-excited to give an increase in the total response or to prolong that response. Alternatively, when one of the relays is an inhibitory interneuron, the closed chain acts as a negative feedback mechanism, reducing the size of the response or shortening its action. Negative feedback is valuable for suppressing weakly excited or unwanted signals.

Facilitation

This term was first applied to features of reflex activity in the spinal cord, but is now used to describe any response in the nervous system which is increased by intervention from another source. The facilitatory effect of one set of neurons on another is a consequence of convergence in which the combined response of two synchronous volleys is greater than the simple sum of their separate responses. Facilitation applied to the activity of a single neuron refers to an increase in the frequency of neuronal discharge.

Occlusion

This feature of synaptic transmission arises directly from neuronal sharing. The term indicates a reduction in response of a pool of neurons supplied by two or more afferent pathways. The reduction is a consequence of overlapping fields of activity in which the combined response of two synchronous volleys is less than the simple sum of their separate responses. Occlusion is an important property of the nervous system, for it often determines whether a given neuron will respond to a second stimulus.

Potentiation

When the same presynaptic fiber is excited by two stimuli separated by a short interval of time, the postsynaptic response to the second stimulus is often increased in size. The first stimulus is called a conditioning stimulus, the second a testing stimulus. When repetitive stimulation is employed for conditioning, a much larger test response may be observed because of greater mobilization of synaptic vesicles and increased liberation of transmitter substance. This is known as post-tetanic potentiation, and it may be responsible for several features of nervous activity, including the phenomenon of rebound in spinal reflexes, and for certain functions of the memory process.

Storage

The brain is capable of storing the information it receives for an unlimited time in the form of memory traces. Many different theories have been proposed to account for the storage mechanism, the most acceptable being some physico-chemical alteration in synaptic structure. It is necessary to distinguish between short-term memory, in which the entire process is erased unless reinforced by repetition, and long-term memory, which requires a consolidation period during which a permanent trace is laid down.

Plasticity

This term implies that the synaptic element is not a rigid structure, fixed very early in life, but one that can alter its functional efficiency according to use. It is actually an extension of the well-known property of muscle that undergoes hypertrophy with training and atrophy with disuse. There is some difficulty, however, in applying such a concept to the central nervous system where growth processes are extremely limited, at least in human beings. Nevertheless, changes at the molecular level, depending upon brain protein synthesis, underline the important relationship between cellular metabolism and function.

METHODS OF INVESTIGATION

Materials

Isolated tissues

1. Nerve fibers are used to study excitability and conduction properties and to measure the changes induced by the passage of nerve impuses. The sciatic nerve of the frog is convenient for class experiments since it will last for several hours when kept moist in Ringer's solution. For demonstrating electrical activity the nerve is placed over silver wires in a recording bath and covered with mineral oil to prevent drying. Studies of membrane properties are facilitated by the use of large-diameter axons, such as the unmyelinated giant fibers of squid and sepia kept in cold sea water. Microelectrodes can be inserted directly into the axon, or test solutions can be injected with a microsyringe to study their effects on ionic movements. The transport of materials along nerve fibers has been demonstrated by mounting a nerve in a microscope slide groove, using a darkfield condenser, and labeling proteins with radioactive tracer elements.
2. Muscle fibers can be kept alive long enough for investigation of their electrical and contractile properties or for studying the reactions occurring at the motor end-plate. Muscles commonly used are the frog sartorius and rat diaphragm. After dissection, the muscle is mounted on a block in a chamber molded from paraffin wax, and carried on a mechanical stage under a binocular dissecting microscope. In measuring membrane poten-

tials, the tip of a microelectrode is moved from the surface into the interior of the muscle fiber. To locate the neuromuscular junction, a second micropipette, filled with acetylcholine, delivers small quantities of the drug with direct current pulses from the stimulator. When the pipette is close to the junction, each pulse produces a localized twitch. Recording the end-plate potential becomes relatively simple when neuromuscular transmission is blocked by curare. The effect of this drug is to reduce the size of the potential below the threshold of the muscle fiber so that no contractions occur. The preparation can be used to study the events of impulse transmission from the end-plate to other parts of the muscle fiber and to show the effects of adding various agents like calcium and magnesium to the fluid in the muscle chamber.

3. Nerve-muscle preparations. The frog sciatic-gastrocnemius preparation is the best known. After pithing the frog and destroying the brain, the sciatic nerve is dissected from its origin in the spinal cord down to the knee. Next, the gastrocnemius muscle is isolated from the tibia and a thread tied round the cut end of the Achilles tendon. The preparation is transferred to a recording board, fixed by a strong pin through the knee joint, while the nerve is placed on a pair of stimulating electrodes. Finally, the thread is attached to a strain gauge for converting mechanical forces into electrical signals, which can be displayed on a polygraph. Many of the basic properties of skeletal muscle contraction can be demonstrated with this preparation. For more advanced studies, using the rat phrenic nerve-diaphragm or the frog nerve-sartorius, the muscle is held in position in one compartment of a bath and the nerve in another compartment where the bathing fluids can be renewed or kept at a constant value. Some baths are designed to allow mechanical stretch of the muscle before stimulation.

4. Smooth muscle isolated from the body will contract spontaneously if kept in a warm, oxygenated medium of suitable composition. A length of rabbit ileum is frequently used, one end being attached to the base of a muscle bath and the other end to a writing lever. The contractions may be influenced by introducing a chemical substance into the medium; adrenaline, for example, causes relaxation while acetylcholine increases contractions. The rhythmic activity of smooth muscle is entirely independent of any extrinsic nerve supply.

Animal preparations

1. A decapitate or spinal preparation allows the experimenter to study the isolated spinal cord free from all influences descending from the brain. The initial operation is usually performed under ether anesthesia, which is blown off during the next hour while the animal undergoes the effects of

spinal shock. A transection of the cord can be made at any spinal level, but if the cervical cord is chosen, artificial respiration will be required because of paralysis of the respiratory muscles. All spinal animals have relatively low blood pressure and respond poorly to fluctuating external temperatures.

2. Decerebrate preparations have the advantage that the remaining parts of the nervous system can be studied without anesthesia. The classical method of decerebration is to remove both cerebral hemispheres through a hole made by trephining the skull after clipping the common carotid arteries. When decerebration is accomplished by making a section through the brain stem (at any desired level), the connections to the cerebellum remain intact. Decerebration by the anemic method requires exposure of the basilar artery on the ventral surface of the medulla and pons. Both common carotid arteries and the rostral end of the basilar artery are ligated. The blood supply through the circle of Willis is thus effectively cut off, depriving the cerebral hemispheres of their circulation, while the vertebral arteries maintain the flow to the brain stem and cerebellum.

3. The isolated encephalon preparation is useful for studying the functions of the cerebral cortex when general anesthesia is undesirable. The animal is prepared under ether, the spinal cord and vagi are transected in the neck, and breathing is maintained artificially by a respiratory pump.

Intact animals

1. Acute experiments. The most frequently used method of general anesthesia is the adminstration of barbiturates. Among these, sodium pentobarbital has a long-lasting effect with good muscular relaxation; sodium thiopental provides a shorter-lasting effect and is useful when minimal anesthetic action is desired. Initial doses of the drug are injected into the peritoneum, and a venous cannula is implanted in one of the limbs for additional doses in order to maintain the required state of anesthesia. This can often be assessed from the size of the pupil or by monitoring the electrical activity of the cerebral cortex. Since most barbiturates have a depressant action on the central nervous system, other general anesthetics are commonly employed. These include Dial, surital, urethane, and chloralose. During the preparation of the animal, the skin areas, in particular, are infiltrated with a local anesthetic. A check is kept on the arterial blood pressure throughout the experiment, and care is taken to maintain adequate ventilation and a constant room temperature.

2. Chronic experiments. The functions of the brain may be studied in the unanesthetized animal, provided steps are taken to avoid frighting or hurt-

ing the animal. The advantages are considerable since the responses with the animal remaining free and unrestricted, resemble those it would have in its natural state. Electrical stimulation and recording are carried out by implanted electrodes stereotaxically positioned in a preliminary operation with the animal under anesthesia. The electrode terminals are wired to a harness on the back of the animal's neck.

Various methods are used for stimulating the brain. The *direct* method employs spring-loaded leads, which allow the animal to move about a soundproof cage equipped with a television camera. The *remote control* method requires the use of special apparatus. In one type, the electromagnetic field is set up by a primary coil surrounding the cage; induction shocks are delivered through a secondary coil buried beneath the scalp. Another type provides for two-way communication by radio transmitters. The *chemitrode* method consists of a large number of tubes permanently implanted in the brain through which substances can be injected, perfusates measured, and electrical activity recorded, while the animal takes part in the normal life of the colony.

Human subjects

1. Clinical observations. Since the results of animal experimentation cannot always be applied to human beings, a good deal of information about the functions of the nervous system has come from clinical sources. A neurological examination is important in considering any medical problem. It is necessary to observe the presence or absence of normal functions and, if a lesion is detected, to establish as far as possible the extent and location of the disturbance. A lesion of the nervous system may be the only evidence of a clinical disorder, but in many patients a neurological disturbance may appear as a complication of some other disease.
2. Clinical tests. Electroencephalography is a method of recording the electrical activity of the cerebral cortex by means of electrodes placed in contact with the scalp. Normal subjects show characteristic patterns or brain rhythms, which may undergo a change in the presence of a disturbed functional state. Thus, local abnormalities in waveform and frequency may indicate the site of an epileptic focus.

Electromyography is a method of recording the electrical activity of skeletal muscle by the insertion of recording needle electrodes. In muscle disease there is usually a loss of active muscle fibers, so that the motor unit potentials are reduced in number and amplitude during voluntary contraction.

Electro-diagnostic tests on peripheral nerves give a measure of the conduction times and functional state of the nerves and of their reflex capabilities. Delayed or slowed conduction can be demonstrated as a prolongation of the latent period when the nerve is stimulated electrically.

3. Surgical operations. Only a few examples will be given here of neurophysiological methods employed in open brain surgery. Explorations for epileptic foci and their accurate localization are greatly facilitated by recordings from different sites within the brain at the time of operation. Microelectrode recordings of single neuron activity have been attempted in order to define the limits of abnormal brain tissue. Surgical lesions by electro-coagulation for the relief of brain disorders, for example, tremor and rigidity in Parkinson's disease, are now carried out on the principle of stereotaxis. The apparatus used fits over the patient's head and is designed to enable an instrument to reach any point within the cranium with absolute precision.

Techniques

Histological

1. Cytoarchitecture pertains to the micro-structure of nerve cells and their processes. The tissue to be studied is first kept in a fixative for a few days (usually a solution of formalin), and then dehydrated with alcohol before being embedded in celloidin or paraffin wax. Thin sections are cut on a sliding or rotary microtome before the process of staining. Some of the various stains that act on nervous tissue are more reliable than others; hematoxylin, toluidine blue O, and thionine are general purpose stains. Appropriate techniques for using dyes with special staining properties are required to reveal the organelles within the cytoplasm.

2. Nerve fibers are treated histologically in a number of ways. Osmic acid stains the myelin sheath black, leaving the axon unstained. The Weigert-Pal method requires a solution containing potassium bichromate, which gives the myelin sheath an affinity for hematoxylin. Degenerated fibers do not take up the stain so that areas of sclerosis, where the nerve fibers have been destroyed, remain colorless. Marchi's method for demonstrating nerve degeneration relies upon the disintegration of myelin into fatty droplets, which are stained black with osmic acid. The method of nerve degeneration is used extensively for mapping myelinated tracts after making lesions at selected sites. When the animal is sacrificed some weeks later, the pathways of individual axons can be traced accurately to their terminal destinations.

3. The identification of structures explored in microelectrode penetrations is routinely investigated after completion of an experiment. The tract

of the microelectrode is made visible by staining with cresyl violet or thionine after the brain has been frozen or fixed in 100% formalin for a few days. Serial sections are then cut in a block of tissue parallel to the axis of penetration. For marking the tip of the microelectrode, a 10 μA DC current, 20 sec in duration, is passed through the electrode as anode to produce an electrolytic lesion. Another method, using a metal microelectrode, deposits iron by electrolysis, and the section is stained blue when the tissue is perfused with potassium ferrocyanide. The ejection of pontamine blue from a dye-filled micropipette also gives satisfactory markings for subsequent examination.

Electrical techniques

The standard apparatus used in most neurophysiological laboratories consists of a stimulator for delivering electrical pulses, an amplification system for increasing the size of the signals, a recording device for audio-visual display, and a signal processor for analysis of the data.

1. Stimulator. This instrument is designed to deliver single pulses, twin pulses, or a series of pulses at a specified frequency for any length of time. The *electronic stimulator* has four basic controls–frequency, the number of pulses per second; delay, the time of delivery in relation to the sweep time; duration, the interval of time for current flow or pulse width; and amplitude, the size of the pulse or stimulus intensity. In order to observe an event initiated by a stimulus, the sweep of an oscilloscope is synchronized to the output of the stimulator. This synchronizing lead is called a *trigger*. It ensures that the sweep displayed on the oscilloscope will follow the repetition rate set by the frequency control of the stimulator.

The *digital stimulator* delivers pulses over a wide range of settings and at precise times as required by the experimenter. It also provides a calibrated time scale to feed the sweep of an oscilloscope. It operates by counting pulses from a crystal controlled generator to produce various outputs determined by the settings of decade switches. The provision of gates and dividing circuits enables square waves and pulse trains of precise duration and number to be selected in each cycle period, and the process is repeated for as many cycles as required.

2. Amplification and recording. The size of biological signals is very small, normally only a few millivolts in amplitude. Consequently, high amplification is necessary to display the signals and to distinguish them from the noise level of recording instruments.

The *cathode-ray oscilloscope* is a very sensitive instrument suitable for displaying, measuring, and storing the electrical events associated with nervous activity. Despite the versatility of technological design to meet specific laboratory requirements, most of the instruments share certain features:

• The display unit consists of a beam of electrons generated at the back of the cathode-ray tube and focused on a fluorescent screen to produce a luminous spot. Additional elements providing low-velocity electrons are required to make storage operation possible. The position of the spot on the screen is controlled by voltages applied to two pairs of parallel plates within the tube. The horizontal, or X plates, when suitably charged, move the spot to the left or right; the vertical, or Y plates likewise move the spot up or down. Multiple traces will not deteriorate and can be viewed or erased when they are no longer desired.

• The time base unit allows the spot to sweep across the screen at different rates. Sweep rates are selected according to the nature of the event being measured. A transparent scale permits direct reading of the time for the spot to travel a given distance across the screen, usually the distance between two divisions. Owing to persistence of luminosity, a rapidly moving spot appears on the screen as a continuous beam of light. If it is desired to display any part of the event with greater accuracy, use can be made of a delay circuit introduced into the sweep. An alternative method is to magnify a portion of the sweep by increasing the gain of the horizontal amplifier and positioning the display until the desired part is on view.

• The amplifier or gain control increases the size of the signal which is applied to the Y plates; the signal is displayed as a vertical deflection of the beam, the amplitude of which can be read directly on a graduated scale. The standard oscilloscope has two independent electron beams, each controlled by its own amplifying system and time base. This feature enables it to display two different input signals simultaneously on the same or on different time bases. When one beam is used for recording an event, the other beam can be used for displaying a time scale or event marker.

• Multiple input amplifiers are an extension of the dual-beam principle for the display of several input channels. The display is achieved by electronic switches, which either chop the trace into a number of segments during the course of a sweep or alternate between channels during sweep retrace intervals.

Paper-driven recorders. Ink-writing pens give permanent records on continuously moving paper and provide simultaneous tracings for two or more input channels within a limited range of paper speeds. Although originally developed to meet the requirements of clinical electroencephalography, the recorders are eminently suitable for displaying low-frequency signals such as strychnine spikes and muscle action potentials. The inertia of the pens, however, is too great for satisfactory recording of high-frequency signals.

Ultraviolet recorders can accommodate a large number of channels and faithfully reproduce data at very high frequencies. They utilize sensitive galvanometers and an optical system to give visible traces over a wide range of paper speeds.

3. Signal processing. This is a field of electronic engineering which is now of considerable interest to neurophysiologists for analysis and computation of laboratory data. In the past, the only way to study the patterns and time relationships of nervous activity was with a combination of oscillograph tracings and photography. The method is still of value for answering simple questions like the latency of an evoked response. With the progression from vacuum tubes to transistors and integrated circuits, new types of recording instruments are available for processing and displaying potentials as they emerge on line or after storage. Among these, two kinds of laboratory computer have been developed for data analysis: In one kind, the modes of operation are built in to provide a means of averaging signals and a range of histograms. In the other kind, the instrument is far more versatile, but must be supplied with programs compiled for each operation. Both kinds of computer accept experimental data directly, sample events at selected intervals of time, convert the samples into a code of binary digits, and store the result in memory registers. The output system permits the content of the store to be displayed on a cathode-ray tube, X-Y plotter, or teletypewriter.

Evoked potentials

The electrical changes induced in the brain by stimulation of a sensory pathway are known as evoked potentials. They represent the summation of action potentials causing a flow of current in a localized region where excitation has occurred. Since the response is reproducible on repeating the stimulus, the procedure has become a useful tool for investigating functional pathways between one part of the brain and another as well as for mapping the cortical representation of the sensory projection systems. The following important characteristics may be noted:

- The *form* of an evoked potential consists of a primary wave followed by a series of secondary waves. The primary wave is resistant to anesthetics. It has two components, a presynaptic potential, initiated by the arrival of afferent impulses, and a postsynaptic potential, derived from the discharge of the evoked neurons. The secondary waves are depressed by deep anesthesia; they are mainly attributable to spread of activity in neighboring neuronal circuits, for example, between the apical dendrites of pyramidal cells and interneurons in the superficial layers of the cerebral cortex.
- The *polarity* of a primary wave recorded from the surface of the cerebral cortex is initially positive-going, because current flows from the surface to

the deeper layers of the cortex where depolarization occurs after the arrival of the afferent inflow. The site of maximum negativity is located in layer IV, the position of the basal dendrites of the pyramidal cells. The polarity is reversed in sign when the recording electrode penetrates the pial surface to a depth where the thalamocortical afferent fibers terminate. The positive component of the potential gradient is termed the "source" of the current flow, while the negative component is termed the "sink."

• The *latency* of the evoked potential is determined mainly by the conduction distance between the point of stimulation and the point of recording. Other factors, however, may have to be taken into account such as the conduction velocity of the afferent pathway and the number of intervening synapses. Furthermore, as conduction times may be altered by increasing the stimulus strength or by recruitment of additional pathways, considerable variations of latency may be observed with repetitive nerve stimulation. Accordingly, estimations of the mean latency of an evoked response are likely to be more consistent.

• The *localization* of an evoked potential is generally determined by underlying activity in specific afferent terminals. This activity depends upon the distribution of anatomical connections between the site of recording and the site of stimulation. The evoked potential, however, may spread to remoter regions where no afferent fibers terminate directly if the excitability of the tissues is sufficiently intense.

Strychnine neuronography

This technique was introduced by Dusser de Barenne as an alternative to lesion experiments for mapping pathways. When a small piece of filter paper soaked in strychnine sulfate is applied to the pial surface in close proximity to a recording electrode, cortical spiking occurs after a few minutes. If a second electrode is placed elsewhere in the brain, spiking may be observed more or less synchronously in a simultaneous recording of the two traces (Fig. 3). Since spike activity evoked by strychnine does not involve synaptic transmission, the existence of a direct link between two electrode points can be successfully demonstrated.

Microelectrode technique

The electrical changes recorded by means of gross electrodes cannot give much information about the properties of the individual units making up a population of neurons. In order to study the activity of single neurons, glass

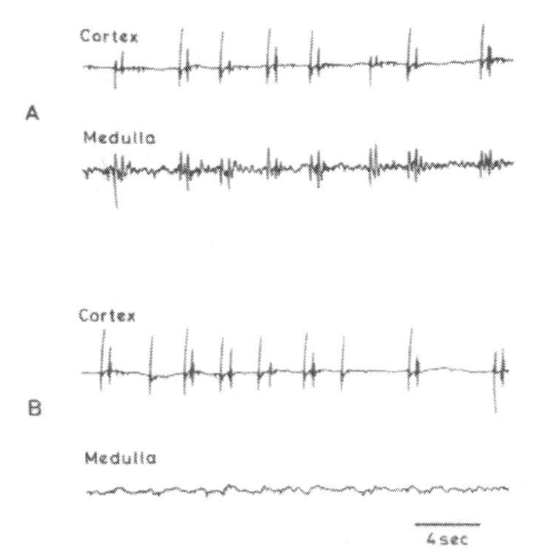

Fig. 3. Method of strychnine neuronography. The upper trace shows a recording of strychnine spikes from the pial surface of a cerebral cortex. A simultaneous recording of spikes from a medulla is seen in the lower trace. In B, the medullary electrode is outside the responsive region.

or metal microelectrodes are employed with tip diameters measuring 0.5 μm to 5.0 μm. The resistance of electrodes with such fine tips is relatively high in comparison with the input resistance of the amplifiers used for recording and, to compensate for this, a cathode follower or source follower circuit is required in the first stage. Recordings of the electrical activity generated by single neurons have been approached in two ways: Extracellular recordings measure the potentials between the immediate vicinity of the neuron and a reference electrode; intracellular recordings measure the potentials generated between the interior of a nerve cell and the media outside it.

1. Extracellular recordings give information on the discharge properties of the units being sampled as the microelectrode is moved from the surface into the interior of the brain or spinal cord. A large number of units display spontaneous activity, but their discharge patterns may be modified by an appropriate peripheral stimulus. Other units, which are silent in the absence of stimulation, respond to a stimulus by discharging one or more spikes. When the microelectrode tip lies close to the cell membrane, the polarity of the evoked spike is initially positive and is followed by a negative deflection. On the other hand, when the recording tip lies at a distance from the center of an active field, the polarity of the evoked spike is often initially negative. These changes in sign reflect the direction of current flow from soma and dendrites to the recording site.

2. Intracellular recordings give information on the electrical properties of the resting and active membrane—for example, resistance and capacitance

as well as the nature of the potentials generated as a result of synaptic transmission. Excitatory synapses generate a potential so that net current flow is inward across the subsynaptic membrane, thus producing a localized depolarization (EPSP)—i.e., the membrane potential becomes less negative than it is at rest. Conversely, inhibitory synapses generate a potential so that net current flow is outward at the subsynaptic membrane, thus producing a localized hyperpolarization (IPSP)—i.e., the membrane potential becomes more negative than it is at rest.

Pharmacological methods

Under this heading are described some of the techniques employed for studying the action of drugs on the nervous system. Many of the substances are manufactured by the body itself; they may be active or in the form of precursors requiring conversion by internal processes. Other drugs, which do not develop naturally, can cause significant changes when applied locally to the tissues or administered internally in sufficient doses.

1. Tissue extracts. Nervous tissue essentially is composed of lipid and protein structures together with the enzyme systems necessary for synthesis, storage, and utilization. The distribution of the various chemical substances found in the nervous system is by no means uniform, some tissues possessing a higher concentration of active principle than others. The preparation of tissue extracts and their separation into fractions by centrifuging are standard methods employed prior to identification and measurement of the components. The tissue to be studied is excised, weighed, and frozen in liquid air. An extract is prepared by adding reagents to preserve the metabolic characteristics before fixing and to prolong the experimental period unless rapid analysis is feasible. Cell-free extracts are prepared by grinding the tissue in hypotonic solutions.

2. Biological assays. Tests on the whole animal or on biological preparations are used for the assay of a substance when no adequate chemical methods are available. The results are then compared with a similar test made on known concentrations of the substance. Among biological preparations that were commonly used at one time are the arterial blood pressure of the cat, the frog's rectus abdominis muscle, the rabbit's uterus, and the guinea pig's ileum. The introduction of newer techniques using automated measuring instruments has largely displaced the biological method because of greater speed and accuracy.

3. Neurochemical analysis. The application of refined analytical methods to the chemistry of the single cell has generated a growing interest in behavioral processes occurring at the molecular level. Since it is established that nervous activity depends upon the presence of neurotransmitter sub-

stances, research is being directed toward biochemical mechanisms and the modifications imposed on neurotransmission at membrane sites within a synaptic unit. Instruments used for this kind of work need to measure the concentration levels and turnover rates of these substances with a high degree of precision.

Chromatographic methods are employed for identification of molecular particles, for separating the components of a complex mixture, and for determining the effects of drugs on metabolic pathways. Several versions of chromatography have been developed to meet specific requirements. These include absorbent beads or paper sheets, gas, liquid, and ion-exchange columns. The general principle of separating chemical components is based on the rate of migration of the sample through a porous bed or column; particles that are held loosely move rapidly through the column, and those held tightly move more slowly. Substances of high molecular weight are best separated by ion-exchange chromatography–e.g., the separation of individual aminoacids from their original protein compound. Sephadex ion exchanges are extensively used for isolation and purification of a wide variety of molecular substances. In other methods of analysis radioactive labels are used for counting levels of unknown samples after absorption on a gel slicer. The counts for each slice are usually displayed by automatic printout equipment.

4. Intraventricular injection. This method was introduced to localize the site of action of a drug within the brain in order to avoid the peripheral effects of drugs injected into the blood stream. A cannula, covered by a rubber cap, is inserted through a burr hole in the skull, with the tip of the cannula resting in the lateral ventricle. It can be left in place for months and the drug injected at any time, painlessly and without an anesthetic. The outflow is collected from a second cannula positioned in the fourth ventricle. To limit the site of action still further, any part of the ventricular system can be perfused by inserting additional cannulae, one acting as outflow and one delivering the drug. After the end of each experiment, the site reached by the drug is checked by substituting a dye to ascertain which regions have been stained. The method is of value in studying the central action of drugs that cannot reach the nervous structures in the brain when injected into the blood stream because of the blood-brain barrier.

Surgical procedures

1. Peripheral nerves. During the course of an experiment it is sometimes useful to block a conducting pathway in order to establish a particular finding or to confirm an observation. Temporary blocking can be achieved in one of several ways–e.g., by using a local anesthetic, by freezing, or by

applying a direct positive-going current to reduce the excitability of the nerve membrane. The effects of these methods are all reversible. Severe loss of conduction is seen when the nerve is damaged from drying or crushing, and permanent loss ensues after nerve section.

2. Ablation. In order to study the action of certain parts of the nervous system, the method of regional ablation has been widely adopted. The grey matter on the surface of the brain can be isolated by undercutting or else selected regions can be removed altogether by suction, diathermy or the blade of a knife. These operations require careful hemostasis to prevent shock and undue fall of systemic blood pressure. The extent of the lesion must be verified afterwards by histological examination. Ablation experiments have the disadvantage that they may cause functional disturbances in neighboring structures which cannot always be recognized. Furthermore, the results may be misleading since the lesion may include fibers of passage or pathways that belong strictly to another part of the brain.

3. Discrete lesions. The use of stereotaxic surgery is of value for the placement of discrete lesions in a localized region of the brain. Relatively small areas can be destroyed by electrocoagulation or currents may be passed through a micropipette to produce a more restricted electrolytic lesion. When the structures destroyed are mainly excitatory in function, the effects observed are either reduced or totally absent. On the other hand, if a lesion is made through an inhibitory pathway, the normal response may become exaggerated or responses which were previously suppressed may suddenly become revealed.

All the techniques described above have been used to gain useful information about the nervous system and knowledge of the mechanisms that make it work. It is hoped that they will lead to a better understanding of the problems of nervous function in human beings and also of the abnormalities and disturbances found in human disease.

SELECTED BIBLIOGRAPHY

Brazier, M.A.B. *Electrical activity of the Nervous System.* New York: Macmillan, 1960.

Burns, B.D. (Ed.). Electrical recording from the nervous system. In *Methods in Medical Research,* vol. 9, Chicago: Year Book Med. Pub., 1961.

Carmichael, E.A., Feldberg, W., and Fleischhauer, K. Methods for perfusing different parts of the cat's cerebral ventricles with drugs. *J. Physiol.* 173:354-367, 1964.

Curtis, D.R., and Eccles, R.M. The excitation of Renshaw cells by pharmacological agents applied electrophoretically. *J. Physiol.* 141:435-445, 1958.

Hubel, D.H. Tungsten microelectrode for recording from single units. *Science* 125:549-550, 1957.

Knight, B.W. Dynamics of encoding in a population of neurones. *J. Gen. Physiol.* 59:734-766, 1972.

Minckler, J. *Introduction to Neuroscience.* Saint Louis: Mosby Press, 1972.

Moore, G.P., Perkel, D.H., and Segundo, J.P. Statistical analysis and functional interpretation of neuronal spike data. *Ann. Rev. Physiol.* 28:493-518, 1966.

Nastuk, W.L. Physical techniques in biological research. In *Electrophysiological Methods,* vol. 5, part A. New York: Academic Press, 1964.

Perkel, D.H. Spike trains as carriers of information. In *The Neurosciences: second study program,* 587-596, New York: Rockefeller Univ. Press, 1970.

Somjen, G. *Sensory Coding in the Mammalian Nervous System.* London: Plenum Press, 1976.

CHAPTER II

Nerve Impulses

Toward the end of the last century many scientists were investigating the electrical properties of nerve and muscle and making measurements of the ionic changes associated with excitation and conduction. It was discovered that the cell membrane acted as an effective barrier, limiting the movements of certain ions, thereby producing a concentration difference between the inside and the outside of the cell. However, it was Bernstein who in 1902 first proposed the theory that resting nerve and muscle had a polarized membrane and that during activity the membrane became selectively permeable to potassium ions. The inequality of potassium between intracellular and extracellular compartments was considered to be the ionic basis for maintaining the resting potential.

The theory withstood well the test of time until 1952 when Hodgkin, Huxley, and Katz published their findings on the giant axon of the squid. These authors showed that the cell membrane was also permeable to chloride and sodium ions under resting as well as under active conditions. They found that sodium ions were not distributed on the two sides of the membrane in the same proportion as the other ions, for sodium appeared to be relatively impermeable despite the electrochemical forces in favor of its entry into the cell. They believed that the intracellular concentration of sodium ions was kept low by the operation of a metabolic pump. They also explained how the action potential was initiated by rapid movement inward of sodium ions when the resting membrane potential was suddenly reduced, These changes of membrane potential and their associated current flows represent a stage in the excitation process leading to the development of the nerve impulse.

The technique of recording the electrical changes in nerves was first demonstrated on the cathode ray oscilloscope by Erlanger and Gasser in 1922. Their pioneer experiments laid the foundation for recording events in single nerve fibers during the passage of an impulse. The successful development of electrophysiology has greatly aided our understanding of the processes that allow for the restoration of the membrane to its polarized state.

RESTING STATE

Electrical properties of the cell membrane

The nerve cell and its processes are bounded by a physiological membrane that regulates the interchange of materials between the inside of the cell and the outside. Such materials can pass through the membrane in either direction, but the rate at which the particles move and their ability to penetrate the membrane vary considerably. As a consequence, the intracellular and extracellular fluids have quite different compositions; the former has a high concentration of potassium salts, and the latter is largely a solution of sodium chloride. The respective solutions are termed electrolytes, because the particles they contain are electrically charged with positive or negative ions. Since particles of opposite charge are attracted to each other, there is always a tendency for positively charged ions (cations) to travel toward any point where there is a concentration of negatively charged ions (anions). Conversely, anions move in the opposite direction toward a positive pole. The movements of ions through the electrolytes set up a flow of current in both directions. If the currents generated are equal and opposite, the net current flow will be zero. These simple facts endow the cell membrane with important electrical properties:

1. Permeability. The membrane is selectively permeable to different ions, usually less permeable to large molecules and more permeable to smaller molecules.

2. Resistance. The membrane does not permit the free movement of ions across it. The barrier to free movement is regarded electrically as a resistance.

3. Conductance. This is the inverse of resistance. It means that the membrane is a medium for current flow when charged particles are carried through it. The membrane can be visualized as a sieve with pores of varying size regulating the number and velocity of the particles moving into and out of the cell.

4. Capacitance. The lipid component of the membrane structure has insulating properties. Without the existence of pores or protein channels for diffusion, a membrane would behave as a perfect insulator, separating the two conducting solutions and, for all practical purposes, preventing permeation by ions. The semi-permeable nature of the resting nerve membrane means that certain ions are able to diffuse through it, while others are held back. The membrane thus exhibits the features of a capacitor in which positively charged ions are held on one side and negatively charged ions on the other side.

5. Potential difference. Any two points at the surface of a resting cell are at the same potential, or electrically neutral. It follows that no current flows between the two points. If one of the points is in contact with the interior of the cell, a potential difference associated with a flow of current is obtained. The potential difference is caused by unequal concentration of ions between the two points, which had previously been kept apart by the insulating power of the membrane. In the resting state the membrane is therefore said to be polarized.

Movement of ions

The transport of ions across surface membranes may occur passively by diffusion or actively by an electrochemical force. Ions *diffuse* from a region of high concentration to a region of low concentration—i.e., in a "downhill" direction. When the concentrations are equal on the two sides of the membrane, the dissolved substance is in equilibrium and uniformly distributed. When the concentrations are unequal, a gradient exists across the membrane.

An *electrochemical force* derived from the metabolism of the cell actively transports ions from a region of low concentration to a region of high concentration—i.e., in an "uphill" direction. Active transport is made possible when ions are coupled with other molecules and carried across the membrane by means of a pumping process. The energy required for operating the pump may be provided by the conversion of glucose into glycogen or from the breakdown of adenosine triphosphate (ATP).

Distribution of ions

The principal ions of the extracellular fluid are Na^+ and Cl^- which are derived from the plasma and exist in high concentrations; the principal ion of the intracellular fluid is K^+. From Table 1 it can be seen that there is a difference in concentration for Na^+, K^+ and Cl^- ions at the normal resting state of the nerve membrane. Measurements of these concentrations show considerable variations between the different species. With large neurons, such as the giant axon of the squid, measurements can be made directly. With most other neurons, the preferred method is an indirect one, which is based on observations of ionic relationships first reported by Gibbs, and later demonstrated in a series of experiments by Donnan.

TABLE I. Ionic Concentrations in Extracellular and Intracellular Fluids

Ion species	Outside mM/L	Inside mM/L	Ratio $\frac{o}{i}$
	Invertebrates		
Na^+	460	65	7:1
K^+	10	344	1:34
Cl^-	540	80	7:1
	Vertebrates		
Na^+	150	15	10:1
K^+	5	150	1:30
Cl^-	125	9	14:1

Gibbs-Donnan Equilibrium. In order to investigate how water and ions are distributed across a semi-permeable membrane, Donnan used a vessel divided into two compartments by a colloidon partition. When one compartment contains a saline solution and the other distilled water, the sodium and chloride ions diffuse through the membrane until there is equal concentration of the salt on both sides. If an impermeable anion, such as congo red, is then added to one compartment, the distribution of sodium and chloride concentrations will change. Congo red is a sodium salt and, therefore, the concentration of sodium in that compartment is raised. Sodium ions will tend to diffuse down the concentration gradient, taking chloride ions with them across the membrane owing to electrostatic attraction. Thus there will be a loss of chloride in the compartment containing congo red and a gain of chloride in the opposite compartment; the loss of chloride is replaced by the congo red anion. A steady state is reached and diffusion stops when the concentration of chloride is exactly balanced by an equal amount of anion on the other side of the membrane. The result is known as the Gibbs-Donnan Equilibrium and is expressed as follows:

$$[Na^+]_1 [Cl^-]_1 = [Na^+]_2 [Cl^-]_2$$

It is seen that despite the unequal concentration of chloride and redistribution of sodium, the product of their concentrations in the two compartments is the same. Equilibriums of the Donnan type are found across nerve membranes since impermeable protein anions are part of the membrane structure. To some extent they contribute to the net negative charge of the intracellular fluid.

Membrane potentials

Since the nerve cell draws upon its store of charged ions for impulse activity, it is important to determine how a potential difference is maintained across the cell membrane. The answer is explained partly by the permeability characteristics of the membrane itself and partly by the interaction of forces that govern the movements of ions across it. As previously mentioned, passive movements of ions occur down the concentration gradient whenever there is an unequal distribution on the two sides. If this were allowed to proceed unopposed, an equilibrium state would be reached and the potential difference reduced to zero. The movements of ions, however, are governed also by their electrical gradient–i.e., from a positive to negative or negative to positive direction. When the two forces are taken together (the sum of the concentration and electrical gradients), the net force affecting the distribution of ions is known as the electrochemical potential.

Equilibrium potentials

The equilibrium potential for any ion can be defined as the electrical force required to balance the ionic movements caused by diffusion. In other words, if the equilibrium potential is the same as the membrane potential, there will be no net movement of the ion into or out of the cell. If the equilibrium potential is not the same as the membrane potential, the ratio of the external and internal concentrations of the ion will change unless other forces like metabolic pumps are exerted to maintain the concentration difference. The electrical force required to maintain a concentration difference for any ion species can be expressed in the form of an equation.

Nernst Equation. The equation relates the concentration of an ion to the voltage resulting from its concentration gradient. As applied to nerve cells, it expresses the value of the membrane potential at which movements of an ion in and out of a cell are exactly balanced–i.e., when the ion has equal fluxes in both directions. The equation has the following form:

$$E_v = \frac{RT}{FZ} \log \frac{[I]_o}{[I]_i}$$

E is the potential difference in volts, R is the gas content (8.316 joules per degree), T is the absolute temperature (293° at 20°C), F is the Faraday (96,500 coulombs per mole of charge), Z is the valency of the ion, $[I]_o$ is the concentration of the ion outside the cell, and $[I]_i$ is the concentration of the ion inside the cell. A simplified form of the Nernst equation is obtained by converting the natural logarithms to base 10 logarithms. The equation then becomes

$$E_{mv} = 58 \log_{10} \frac{[I]_o}{[I]_i}$$

In this form it can be appreciated that if the external concentration of the ion is increased tenfold, the membrane potential will be increased by 58 mv. The equilibrium potential for any ion can be calculated from the Nernst equation by substituting the appropriate concentration values for that ion:

Chloride ions. From Table I the external concentration of Cl^- ions is about 14 times greater than the concentration inside. Therefore, ions will tend to diffuse inward by virtue of the concentration gradient. Because the cell contains fixed protein anions, the negatively charged chloride ions will be repelled as they travel toward the negatively charged interior. A steady state is reached when the inward movement down the concentration gradient is exactly balanced by the outward movement along the electrical gradient. From the Nernst equation

$$E_{cl} = 61.5 \log 10 \frac{[Cl]_l}{[Cl]_o} = -70 \text{ mv at } 38°C$$

The equilibrium potential for chloride is therefore –70 mv, which is also the normal resting potential of the nerve membrane. At this potential difference, the number of chloride ions crossing the membrane in the two directions is exactly equal despite the fourteen-fold difference in concentration. If the potential difference were less than the equilibrium potential (i.e., less than –70 mv), the concentration difference could not be maintained.

Potassium ions. From Table I the internal concentration of K^+ ions is about 30 times greater than the concentration outside. Therefore, ions will tend to diffuse outward by virtue of the concentration gradient. The movement outward will be opposed by a driving force inward caused by the electrical gradient of the membrane. From the Nernst equation

$$E_K = 61.5 \log 10 \frac{[K]_o}{[K]_i} = -90 \text{ mv at } 38°C$$

The equilibrium potential for potassium is therefore –90 mv, which is 20 mv more negative than the normal resting potential. Since the potassium ion is not in equilibrium with the resting potential, there will be a greater outward movement than inward movement of the ion, which will lead to a steady loss of potassium from the cell. In order to maintain the high concentration difference that exists for potassium ions, there must be a further mechanism which operates to counterbalance the outward movement due to diffusion. In fact, the deficiency in inward movement is compensated for by active transport, which is linked to the sodium pumping mechanism.

Sodium ions. From Table I the external concentration of Na^+ is about 10 times greater than the concentration inside. Therefore, ions will tend to diffuse inward by virtue of the concentration gradient. The membrane potential will also favor the entry of sodium ions into the cell, and yet the intracellular sodium remains low. From the Nernst equation

$$E_{Na} = 61.5 \log 10 \frac{[Na]_o}{[Na]_i} = +60 \text{ mv at } 38°C$$

The equilibrium potential for sodium is therefore +60 mv, which is far removed from the normal resting potential of −70 mv. This means that the interior of the cell would need to be 60 mv more positive than the outside in order to achieve equal movements of sodium ions in the two directions across the membrane. In order to maintain the concentration difference that exists for sodium ions, it is necessary to have an outward pumping mechanism to counterbalance the concentration and electrical gradients so that a steady state will be attained. Studies with radioactive Na^+ have shown that these ions penetrate the membrane spontaneously, though not as freely as K^+ and Cl^-. The process by which sodium ions are removed from the cell is called the sodium pump and to maintain the resting potential, sodium ions must be continuously ejected. The energy required to operate the pump is derived from enzyme systems capable of hydrolyzing ATP. As a consequence, the active removal of sodium ions from the interior of the cell virtually equals the influx of sodium ions due to diffusion.

Sodium-potassium exchange

If a metabolic inhibitor like ouabain is added to the bathing medium, expulsion of Na^+ ions is reduced since the pump is deprived of its source of energy. The inward movement of K^+ ions is also reduced to about the same amount. These findings suggest that the two transport processes are linked. Specifically, the movement of Na^+ ions from the inside to the outside is coupled with the movement of K^+ ions from the outside to the inside of the cell. Thus, lowering or raising the external potassium concentration results in a parallel decrease or increase in the outward movement of sodium. This concept of sodium-potassium exchange is one method by which a cell can maintain differences in the concentration of the two ions across its surface membrane (Fig. 4).

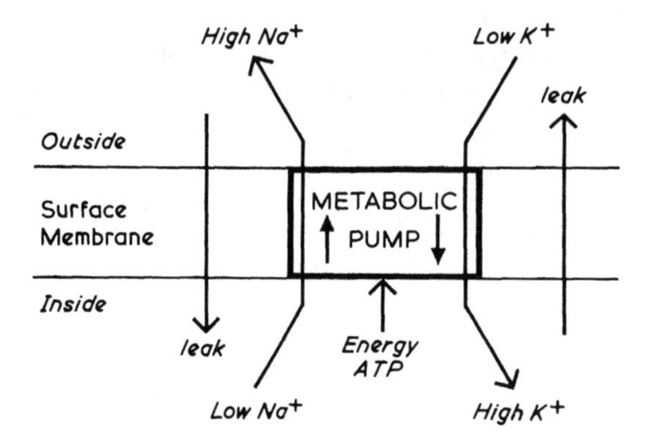

Fig. 4. Sodium-potassium exchange. Ion concentration gradients are maintained by passive leakage and active pumping of the ions across the membrane. Na+ ions diffusing through the membrane are pumped to the exterior. K+ ions leak out more rapidly than Na+ ions leak in owing to greater permeability of the membrane to potassium. The net result is that the inside of the cell, depleted of cations, becomes more negatives than the outside.

Resting membrane potential

The basic concept of how the membrane potential is generated depends upon a number of factors including the distribution of sodium, potassium, and chloride ions, the activity of the sodium-potassium exchange pump, and the internal concentration of fixed protein anions. Each of these factors contributes to maintain a steady polarized state and to restore the membrane potential when it is displaced by processes that elicit the action potential. The steady state condition occurs when outward and inward ionic fluxes are of equal magnitude.

Goldman Equation. Since the membrane potential is generated by more than one ion species, the Nernst equation cannot express all the changes resulting from passive and active ionic movements. Obviously the rate of movement of the individual ions will be a very important factor owing to differences in permeability across the membrane. For example, the permeability of K+ ions is at least 50 times greater than the permeability of Na+ ions. The Goldman equation, set out below, attempts to integrate the contributions of each ion by introducing a permeability coefficient (P) derived from the ratio between the concentration of the ion and its corresponding diffusional activity:

$$V = \frac{RT}{F} \log \frac{P_K[K]_o + P_{Na}[Na]_o + P_{Cl}[Cl]_i}{P_K[K]_i + P_{Na}[Na]_i + P_{Cl}[Cl]_o}$$

where V = the transmembrane potential (potential inside/potential outside).

This equation is an extension of the Nernst equation in which the expressions P_K, P_{Na} and P_{Cl} represent the resting permeabilities of these ions when there is no net flow of ionic current. The greater permeability of the resting membrane to potassium as compared to sodium ions means that the membrane potential really depends on the leakage of potassium ions to the exterior after the concentration gradients have been built up by the metabolic action of the sodium-potassium pump. As a consequence, the outside of the membrane becomes positively charged.

Summary

The interchange of materials between a cell and its environment is controlled by the properties of the cell membrane, which acts as a barrier to free diffusion but allows the selective transfer of dissolved particles and ionized substances.

The transport of ions across the cell membrane occurs by two dinstinct processes: *Passive* transport may be (1) along a concentration gradient from a region of high concentration to a region of low concentration, and (2) along an electrical gradient where there is a separation of positive and negative charges. *Active* transport is mediated by metabolic pumps that can work against both the concentration and electrical gradients.

The existence of a potential difference across the membrane means that there are opposite charges separated by the membrane despite the forces that allow ions to penetrate it. The potential difference of the resting nerve membrane is about –70 mv, which suffices to maintain high concentration of extracellular Na^+ and Cl^- ions and a high concentration of intracellular K^+ ions.

The equilibrium potential for any ion is the transmembrane potential at which the number of those ions crossing the membrane in the two directions is exactly equal—i.e., when a steady state is reached where net flux is zero. The potential difference at which equilibrium occurs can be calculated from the Nernst equation.

The resting membrane potential is generated by forces that maintain the unequal distribution of Na^+ K^+ and Cl^- ions; as a result, the outside of the membrane becomes positively charged. A coupled sodium-potassium exchange pump actively extrudes sodium from the interior of the cell, while the greater permeability of potassium allows diffusion of these ions back down their concentration gradient. The resting potential may therefore be regarded as a secondary effect caused by leakage of excess potassium to the exterior.

ACTIVE STATE

When the membrane potential is displaced from the resting to the active state, the sequence of electrical changes that occurs is known as the action potential. Bernstein proposed in 1902 that the resting potential was generated as a result of its specific permeability to K^+ ions. He suggested that leakage of K^+ ions to the exterior and consequent lowering of the membrane potential could account for the electrical changes observed during the initiation of an impulse. However, certain modifications of this hypothesis have become necessary since experiments with radioisotopes indicate that the membrane is also permeable to Na^+ and Cl^- ions and that Na^+ ions are actively removed against the concentration gradient by means of a metabolic pump.

Modern views on the mechanism of the action potential are derived from the work of Hodgkin, Huxley, and Katz who used the technique of voltage clamping, described below, to establish their ionic theory. Stated simply, this theory proposes that the resting membrane is depolarized by a brief and highly specific increase in permeability to Na^+ ions.

The action potential is a reversal of the resting membrane potential. It occurs when there is an increased permeability to Na^+ ions, which carry their positive charge across the membrane to the interior. A slight increase in permeability will merely displace the membrane potential without initiating an impulse. If, however, the resting potential is suddenly lowered about 15 mv (i.e., from its normal value of –70 to –55 mv), the changes observed are out of all proportion to the applied stimulus. Depolarization of the membrane below a critical level causes a high rate of sodium entry, with the result that the membrane potential changes from inside negative to inside positive. The increase of sodium permeability lasts for about 0.5 msec when the inside may become up to 60 mv positive relative to the outside–i.e., approaching the equilibrium potential for sodium. The reason for the short duration of the sodium flux is that an increase of potassium permeability occurs almost simultaneously, and the outward movement of K^+ ions tends to return the membrane potential toward its resting level. When the action potential has reached its peak, sodium permeability falls, potassium permeability rises, and the potential difference across the membrane approaches the equilibrium potential for potassium. The displacement of the membrane potential at this stage is called an after-potential.

The negative after potential is caused by an excess of K^+ ions outside the membrane; the positive after-potential represents a period of hyperpolarization caused by prolonged inward movement of K^+ ions. Finally, potassium permeability falls, and the potential difference across the membrane returns to its original value (Fig. 5).

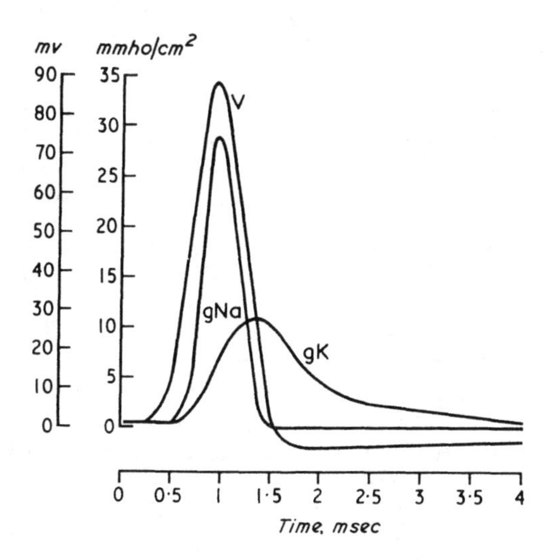

Fig. 5. Time course of a propagated action potential (v) and related sodium (gNa) and potassium (gK) conductances in a squid's giant axon. Left ordinate potential in millivolts; right ordinate conductances in millimhos/cm^2. (Modified from Hodgkin and Huxley, 1952.)

The voltage clamp

The ionic theory of the action potential was first tested in the squid axon by measuring the minimum quantity of sodium ions that must enter the axon and the equivalent amount of potassium that must leave during the rise and fall of the wave. In order to make these measurements, it was necessary to prevent the natural development of the action potential by controlling the voltage across the membrane and holding it at different levels. For example, the membrane potential could be lowered in stages from the normal resting level to zero, while the associated current flow was measured at each stage.

The diagram in Figure 6 illustrates the principle of the technique in which a feedback amplifier is used to produce the internal current needed to change the membrane potential. An internal electrode consisting of two fine silver wires is inserted into the axon. The resting membrane potential is recorded by a meter connected to one of the silver wires and to an external electrode. The other silver wire is connected to a feedback amplifier, which supplies an opposing current to the axon and thus maintains the membrane potential as constant. In other words, an equal current is generated to oppose that caused by ionic movement. Providing the gain on the amplifier is sufficient, the imposed current becomes a measure of the ionic current and, therefore, of membrane conductance.

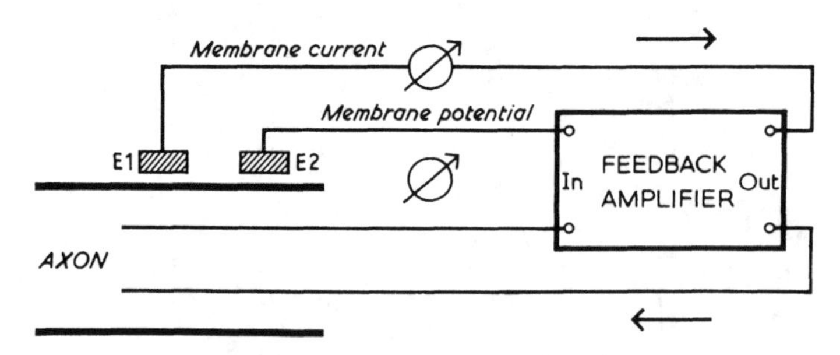

Fig. 6. Diagram of voltage clamp. The membrane potential is recorded between the internal electrode and external electrode E2. A feedback amplifier is connected to the two electrodes on either side of the membrane. Current from the amplifier maintains the potential at any desired level. Membrane current is measured from external electrode E1. Any change in the membrane potential is counteracted by current from the amplifier.

When varying voltage values are set by the experimenter, the membrane potential is displaced to new levels. The ionic current flow is then determined for each level and expressed as a function of time. The results show that when the membrane potential is changed, the membrane conductances are also changed. There is a transient inward current identified with the first phase of the action potential and attributed to current carried by Na^+ ions. If the external concentration of sodium is depleted, the sodium current is reduced, and if the sodium of the extracellular fluid is removed altogether, there will be no sodium current flowing. The inward current is followed by a persistent outward current due to the efflux of K^+ ions from the axoplasm. The increased potassium conductance continues until the external concentration of the ion reaches its equilibrium potential.

The voltage clamp experiments permit the accurate measurements of the changes in sodium and potassium conductance for each displacement of the membrane potential. The complete set of measurements over a physiological range of depolarization have been expressed in the form of equations that describe with remarkable accuracy the normal course of events that occur in the generation of the action potential. The critical factor which is of special interest is that the initial increase of sodium conductance is a transient effect that is rapidly inactivated or overtaken by the increased potassium conductance, thus tending to restore the membrane potential to its original level.

Repolarization

Under normal conditions restoration of the membrane potential begins almost as soon as the membrane is depolarized. First, the flow of current associated with the inward movement of sodium ions is reduced by inactivation of sodium conductance. Second, the outward current due to efflux of potassium ions greatly exceeds the inward current flow and raises the membrane potential to a level near that of the resting potential. Third, restoration of the original ionic concentrations and regain of potassium by the axon are brought about by activity of the metabolic pump.

INITIATION OF THE NERVE IMPULSE

Impulses may be initiated by electrical stimulation of the nerve; the stimulating current tends to reduce or abolish the potential difference across the membrane. The impulse starts off from the stimulated region, but is thereafter self-propagating and independent of the applied current. Once this has been accomplished, the membrane rapidly reverts to its original polarized state. Current flow through the membrane must be sufficient to lower the resting potential to a critical level. If the stimulating current is below threshold, excitability of the nerve fiber is raised, then rapidly declines without initiating an impulse. The decline of excitability following a subthreshold stimulus is known as accommodation and is due partly to inactivation of sodium conductance and partly by outward movement of potassium, which tends to raise the threshold. To study such excitability changes, a subthreshold conditioning stimulus is applied to the nerve and a second, or testing, stimulus applied afterward at varying time intervals. The intensity of the second stimulus required to initiate an impulse is a measure of the original excitability change in the nerve. Once the critical level of depolarization is reached, the ionic changes evoked give rise to the action potential. If the level is below threshold, there can be no action potential. For this reason the action potential is considered to be an all-or-nothing response.

Effects of stimulating currents

1. Cable properties of nerve. This term is used because nerve fibers are endowed with electrical properties analogous to those of a long-distance cable. The wires in a cable are like thin cylinders containing a central metallic core of high conductance, surrounded by thick insulating membranes. Current flow from one point of a cable to the next occurs over a considerable distance without significant attenuation. Nerve fibers, on the other hand, have high internal resistance, while current is lost because of imperfect insulation and leakage through the nerve membrane. The result is that a subthreshold stimulating current does not spread very far along a fiber before it becomes distorted or lost in the extracellular tissues.

2. Constant currents. Most methods of electrical stimulation produce their effects by lowering the potential difference of the polarized membrane to a level at which excitation occurs. Changes in excitability during the passage of a constant current are studied by allowing a battery to drive a current through two electrodes placed on the nerve. One electrode will be relatively positive in response to the other. At the anode, there will be an increased amount of positive charge resulting in hyperpolarization of the membrane; at the cathode, the membrane will become depolarized.

Strength-duration relationship

When a constant current is applied to a nerve, excitability increases at the cathode with the onset of current flow, but the current loses its effect during its continued passage owing to accommodation, and a greater strength of current is required to reach a threshold level. There is a minimum strength of current needed below which excitation cannot occur. This critical strength is called the rheobase. It is the strength of current necessary for lowering the membrane potential to a level at which it becomes capable of initiating a nerve impulse. If the duration of the current is shortened, the current strength must be increased in order to reach the threshold. The relationship between threshold strength and duration is shown in Figure 7. There is a minimum current strength below which excitation does not take place. One point in this curve is called the chronaxie–the shortest time required for a current twice the rheobase to reach excitation. In general, higher stimulus strengths are required with shorter durations.

3. Alternating currents. When a current strength rises to a maximum and falls again, the effect on the membrane potential depends upon the rate of change. An alternating current of very low frequency may be ineffective

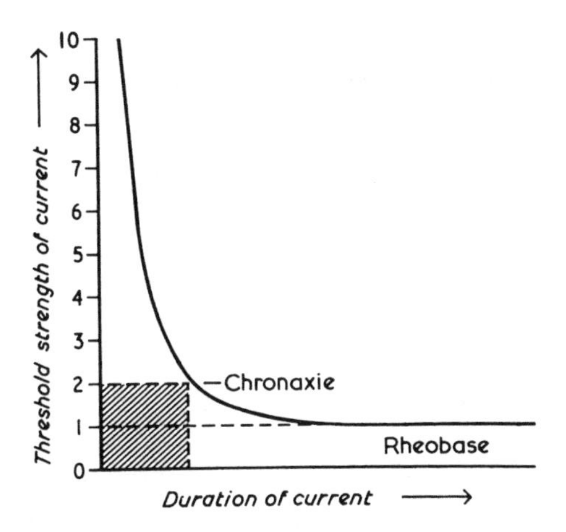

Fig. 7. Strength-duration curve, showing relation between minimum strength of current and the least amount of time during which it must flow in order to reach threshold. The rheobase is the strength needed to produce a nerve impulse. Chronaxie is the time needed for strength 2 x rheobase in order to excite the nerve.

and, likewise, one of very high frequency may be too brief to lower the membrane potential, since current flows in the opposite direction in each half cycle and thus cancels its effects out.

4. Square wave pulses. Currents of very short duration can be obtained from condenser discharges or electronic trigger circuits, which can switch the stimulus quickly on and off at any desired time interval. When a brief shock or square wave pulse is delivered to the nerve through a pair of electrodes, current flows through the membrane at the anode and out again at the cathode. The membrane potential is discharged by the outward current and at this point the potential difference may fall to zero and initiate a nerve impulse.

Meaning of threshold. As stated previously, the voltage drop required to convert a membrane potential into a spike potential is a critical level of depolarization called the threshold of excitation. Threshold is influenced by the following:

1. Type of current. A slowly-rising constant current or an alternating current of very low or very high frequency may not reach the required intensity to be effective.

2. Strength of current. A minumum strength of current is needed to excite a single nerve axon; a larger amount of current is needed during the relative refractory period or when the whole nerve is excited.

3. Distance between the electrodes. An impulse at any instant occupies a

certain length of axon. The optimal distance between a pair of stimulating electrodes is about 3 mm, as distances shorter than this may cause interference between the anode and the cathode.

4. Size of nerve fiber. As the diameter of a fiber increases, the resistance to current flow diminishes. Therefore, the threshold level of depolarization and excitation time are reduced in large fibers and increased in small fibers. Myelinated fibers have relatively low thresholds; unmyelinated fibers have high thresholds.

Propagation of the impulse

When a current of threshold strength is applied to a nerve fiber, an action potential is initiated in the vicinity of the stimulating electrode. This is the response of the fiber to lowering of the membrane potential and sudden influx of sodium ions. The sequence of changes that now give rise to the action potential results in a spread of current along the axon to adjacent parts of the membrane, while the part stimulated is rapidly restored to its resting level. Current can spread along the axon in either direction from the stimulated region.

Local circuit theory. Once an action potential has been initiated, the inside of the membrane becomes positive as the membrane approaches the equilibrium potential for sodium ions. This means that a potential difference is produced between the active region and the adjacent inactive membrane. Consequently, current flows along the inside of the axon and out again through the adjacent resting parts of the membrane and extracellular fluid to complete a circuit, as shown in Figure 8. As the current flows out, it discharges the membrane in the inactive region with the result that this region becomes active and, as soon as threshold is reached, another action potential is generated. A local circuit is once again produced, and the sequence is continued as a wave of depolarization along successive points of the axon. This is what is implied by the nerve impulse–a self-propagating process insofar as each section of the membrane becomes depolarized by the advancing action potential.

Recovery process. Impulses are propagated away from the point of stimulation or active region of the fiber. Recovery of the membrane potential to its resting level follows automatically as the outward movement of potassium ions exceeds the inward movement of sodium ions and as metabolic pumps complete the process of repolarization. During the passage of an impulse, the nerve fiber is refractory and therefore the action potential cannot travel backward to the original point of stimulation. Under experimental conditions, propagation occurs in both directions, away from a cen-

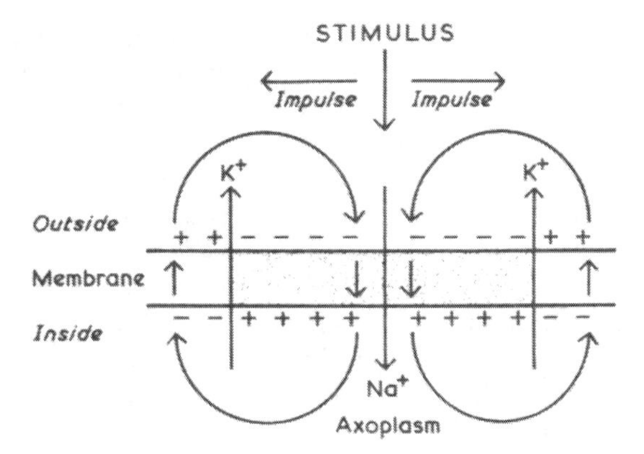

Fig. 8. Diagram of local circuit in unmyelinated nerve. Positive charges are lined up on the outside and negative charges on the inside of the resting membrane. The nerve is stimulated in the middle, causing an influx of Na^+ ions and outflow of K^+ ions. Current flows toward the region of electronegativity from the inactive membrane, as indicated by the curved arrows. The circuit is completed through the extracellular fluid. As successive regions of the membrane become depolarized, the impulse is propagated in both directions along the axon.

tral refractory region, but under natural conditions, in which the action potential is generated at one end of the axon, propagation must be unidirectional.

ELECTRICAL RECORDING OF ACTION POTENTIALS

The membrane changes induced by nerve stimulation can be recorded from a single axon, but the amount of extracellular current flowing in each local circuit is very small and high amplification is necessary. One technique, already described, entails the use of a giant axon from a squid and a fine microelectrode whose tip is small enough to penetrate the inside of the axon. Another more convenient method requires stimulation of a large number of axons simultaneously so that considerable extracellular current is generated; this can easily be recorded by placing a pair of silver wire electrodes in contact with the whole nerve.

Experimental method

If the sciatic nerve of the frog is dissected and kept in Ringer's solution, the nerve will survive for many hours and remain capable of propagating action potentials when stimulated. Usually, however, the isolated nerve is arranged in contact with an assembly of silver wire electrodes in a recording bath filled with mineral oil to prevent drying. One pair of electrodes is connected to a stimulator and a second pair is connected to the recording instruments–amplifiers, a loud-speaker, and an oscilloscope with a time-marker. When an electrical pulse is delivered through the stimulating electrodes, the current may be picked up directly by the recording electrodes which produce a vertical deflection called the *stimulus artifact.* This serves as a marker of the precise time at which the impulse response was initiated. In order to prevent distortion of the response, the stimulus artifact can be reduced in size by grounding an electrode which lies between the stimulating and recording pairs (Fig. 9). To observe the effects of physiological changes on the action potential, the mineral oil can be replaced by solutions containing appropriate drugs or different concentrations of ions.

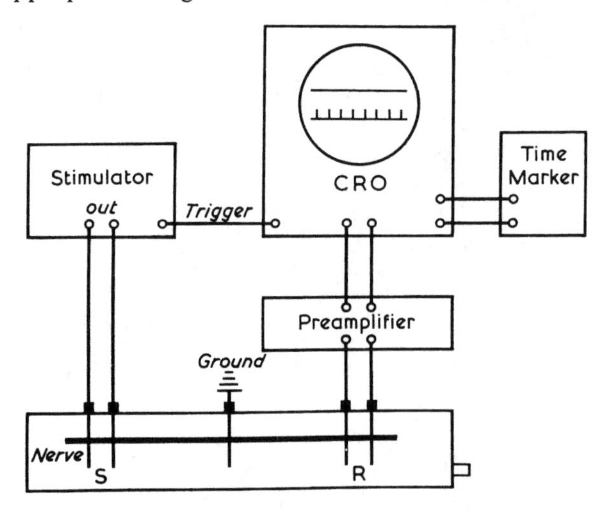

Fig. 9. Experimental arrangement for recording nerve action potentials. The isolated nerve rests on electrodes in a trough filled with mineral oil. Electrical pulses from the stimulator are applied to one end of the nerve, and activity is recorded from the other end; the nerve is grounded between stimulating (S) and recording (R) electrodes. The action potentials are passed through a preamplifier and displayed on one beam of an oscilloscope (CRO), which is triggered to the stimulator. A second beam is used for the time marker.

Diphasix action potential

The potential recorded between two external electrodes placed on a resting nerve is zero. No deflection occurs on the oscilloscope trace. If a threshold stimulus is applied to the nerve, a small deflection is seen following the stimulus artifact; this is the action potential. It represents a wave of surface negativity (internal positivity) traveling from the point of stimulation toward the recording electrodes. When the wave reaches the position of the first recording electrode, R_1 in Figure 10, a difference of potential is recorded between R_1 and the more distant electrode R_2. The first electrode becomes relatively negative, and current flows to it from the surrounding regions producing the initial deflection shown on the oscilloscope. As the wave passes along the nerve, the potential at R_1 declines and the deflection returns to the baseline. The wave now approaches the second recording electrode, reversing the direction of current flow as R_2 becomes relatively negative. The trace on the oscilloscope also appears in the opposite direc-

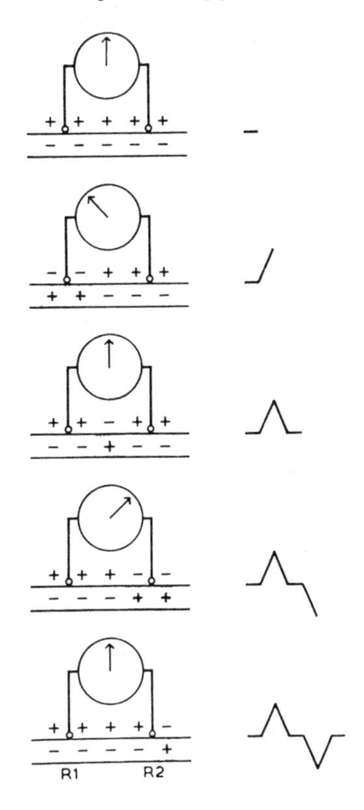

Fig. 10. Analysis of action potential wave. An action potential is shown at various points along a nerve. The direction of current flow is indicated on the meter and graphed on the right side of the figure. No deflection is seen until the action potential reaches the first recording electrode R1, which becomes negative with respect to R2. As the action potential passes between the two electrodes, the record returns to zero momentarily. When the action potential reaches the second recording electrode, R2 becomes negative with respect to R1, reversing the direction of current flow. Subsequently, the action potential travels away from R2 and the nerve repolarizes again.

tion, returning to the baseline as the action potential travels beyond the recording electrodes. The result is a diphasic action potential arising from the superposition of the two waveforms (Fig. 11).

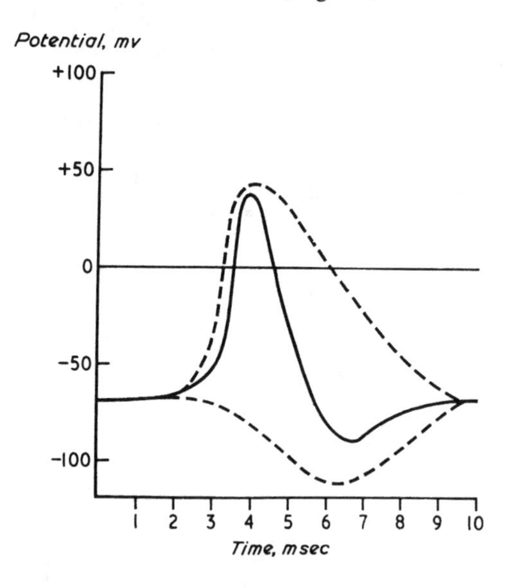

Fig. 11. Diphasic action potential. The main features are a rapid rising phase, a slower falling phase, and a prolonged period of hyperpolarization preceding recovery of the resting potential. Dotted lines show depolarizing and hyperpolarizing currents passing in succession at the recording electrodes.

Monophasic action potential

If the nerve is crushed between the recording electrodes, the action potential may be prevented from reaching the second electrode and only a simple monophasic potential is recorded as the wave reaches the first electrode and fails to propagate beyond it. The effect is the same as an injury potential or the steady potential recorded from the interior of the fiber. Monophasic recordings are useful for studying the changes resulting from nervous activity without the distortions due to reversal of current flow. The delay between the stimulus artifact and the beginning of the wave is a measure of the time taken for the impulse to travel from the point of stimulation to the active electrode.

Triphasic action potential

One disadvantage of recording from an isolated nerve is that the distribution of current flow during the passage of an impulse does not strictly represent the conditions of the natural environment. In the living body, the tissues surrounding a nerve form an extensive conducting medium known

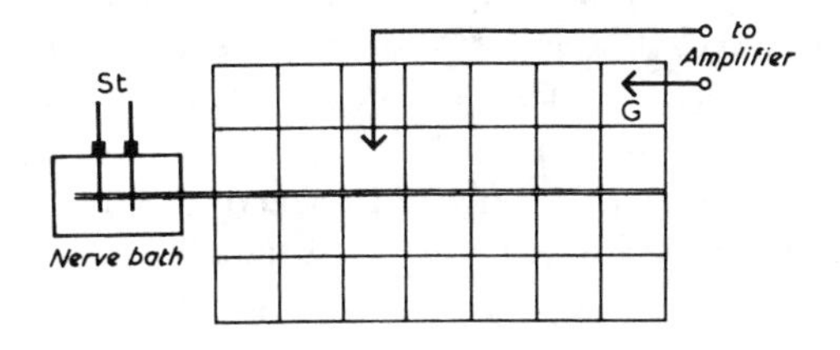

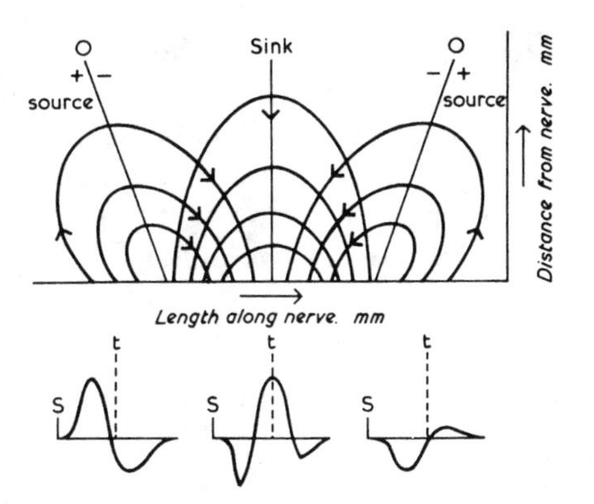

Fig. 12. Recording the electric field at different points in a volume conductor. Top: diagram of arrangement of stimulating electrodes (st) and nerve lying on filter paper soaked with Ringer's solution. The potential difference is recorded between the exploring electrode and another electrode (G) connected to the ground. Various points in the field are plotted by moving the exploring electrode nearer and further from the nerve. Middle: isopotential map and current flow. The amplitude and polarity of the action potentials are measured at a fixed time (1 msec) after the stimulus. Potentials of equal value are plotted to produce a series of curves; the zero potential lines are marked 0. Lines of current flow at right angles to the isopotential curves are indicated by arrows. Since current flows from relatively positive to relatively negative regions, the direction of flow is first outward, then inward, and finally outward as the impulse travels toward the right. Below: action potentials recorded at the same moment of time (t) at three different positions along the length of the nerve. When the nerve enters the conducting medium, the action potential is diphasic–first negative, then positive; the voltage is approximately zero at the time of the plot. Halfway along the length of the nerve, the action potential is triphasic–positive, negative, positive–indicating that the nerve acts as a source as the impulse approaches the exploring electrode, then as a sink as the impulse reaches it, and finally as a source as the impulse travels beyond it. At the end of the nerve, the action potential is diphasic–first positive, then negative.

as a "volume conductor." If recording electrodes are placed in this medium, current may be detected during the whole of the time that the impulse is traveling along the nerve. A map may be constructed showing changes in the form and polarity of the potential at different electrode positions in the field.

A simple experiment is illustrated in Figure 12. One end of the sciatic nerve lying in a nerve bath is stimulated, as described above; the rest of the nerve is placed across a sheet of filter paper soaked in Ringer's solution to represent the external conducting medium. A grid is drawn on the paper for plotting electrode positions and the potential difference between each position, and a common ground electrode is recorded. As an impulse is propagated along the nerve, current flows through the conducting medium, the direction of flow being first outward through the membrane, then inward, and finally outward. Since the electrical field is determined by the arrangement of the sources and sinks of current flow, the sequence is therefore positive-negative-positive. There are two reversals of membrane current so that the recorded potential is triphasic. The relative size and duration of the three phases of the triphasic potential will vary according to the recording electrode position. When current flows into the membrane, the nerve acts as a sink, indicated by the negative phase of the potential. As the impulse approaches this point or passes beyond it, the nerve acts as a source of current flow, accounting for the positive phases of the potential.

Compound action potential

The sciatic nerve of the frog, like any peripheral nerve, is made up of a large number of individual axons of various types and functions. Many of the axons have different thresholds of excitability for the production of action potentials. Consequently, by increasing the intensity of the stimulus from threshold to maximal levels, successive waves will appear in the recordings as the threshold is reached in the different fiber groups (Fig. 13). A maximal stimulus is one that is capable of exciting all the constituent fibers of the nerve simultaneously. A recording obtained in this way is called a compound action potential.

When the stimulating and recording electrodes are close together, it is difficult to separate the individual components, but if the recording electrodes are placed at a distance along the nerve, the faster impulses become separated from the slower ones and the compound action potential is spread out into a number of distinct elevations, each representing activity in a specific group of fibers (Fig. 14).

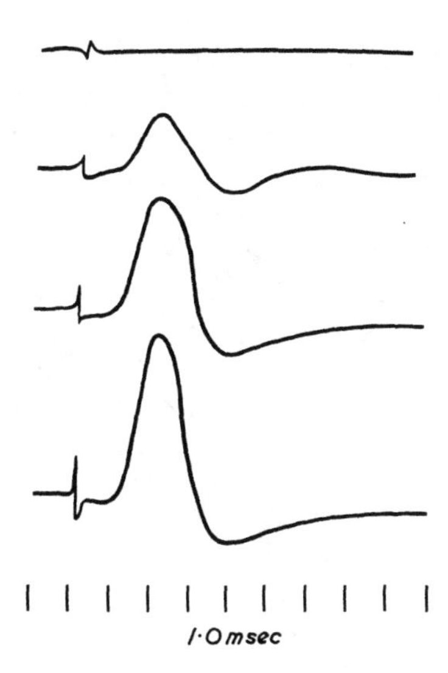

Fig. 13. Compound action potential of a frog's sciatic nerve. The strength of the electrical stimulus is increased from above downward. The first stimulus is below threshold strength. The maximal response in the last record follows a single stimulus 2.5 v. 0.01 msec.

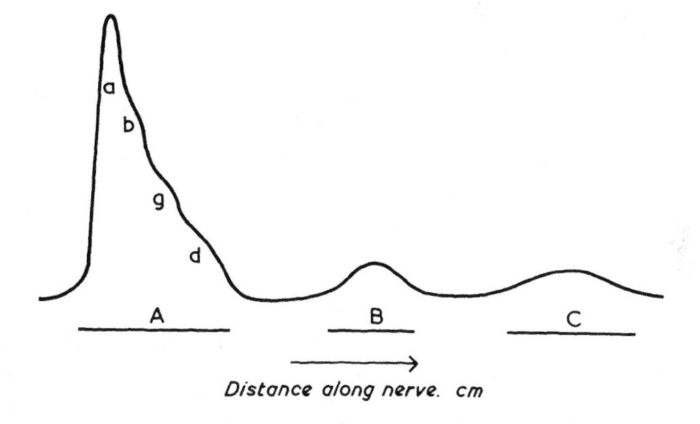

Fig. 14. Waveforms of complete action potential separated into components by leading off at different distances along the nerve trunk. The three main elevations are A, B, and C. The A fibers are represented by four distinct elevations, alpha, beta, gamma, and delta.

Demonstration of refractory period

Each time an action potential is propagated, it leaves the nerve in a refractory state for a few milliseconds during which time the nerve will not respond to a second stimulus. This fact can be demonstrated by delivering two stimuli of equal intensity but separated by variable time intervals. If the second stimulus is delivered within the absolute refractory period, only the first action potential is recorded. With longer intervals between the two stimuli, a second, smaller action potential increases progressively in size as the intervals become longer. Evidently, as the refractory state wears off, more and more fibers are excited again until the second action potential reaches the same size as the first (Fig. 15). By using the delay control on the stimulator and measuring the amplitude of the second action potential, the degree of recovery at each time interval can be determined.

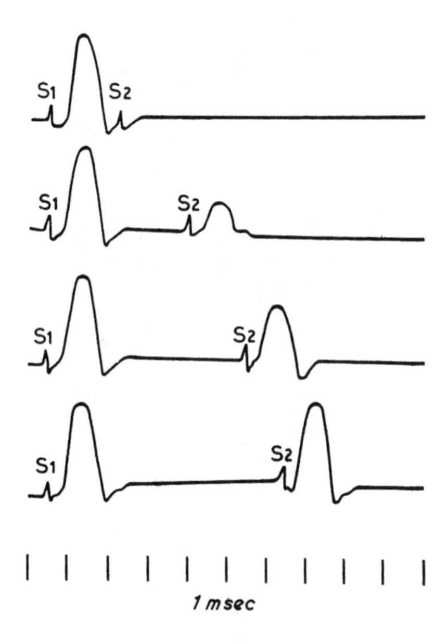

Fig. 15. Refractory period of nerve. In the top record, the second stimulus (S2) is applied during the absolute refractory period when it fails to give a response. As the interval between S1 and S2 is increased, the second response increases in amplitude until both reponses are the same size. The intervals between the stimuli are respectively 2.0 msec, 3.5 msec, 5.0 msec, and 6 msec.

NERVE CONDUCTION

Types of nerve fibers

Examination of the compound action potential reveals that all the fibers conveyed in a peripheral nerve trunk can be classified into three major groups, A, B, and C.

The A group has alpha, beta, gamma, and delta subdivisions related in a general way to their conduction properties. All are myelinated. The largest fibers are about 22 μm in diameter; they have the lowest thresholds and fastest conduction times. Functionally, they contribute to both the afferent and efferent pathways of the nervous system.

The B group forms a single elevation since there is little disparity between the size and conduction times of the individual fibers. All are myelinated. The B fibers are about 1 μm to 3 μm in diameter. They have correspondingly low conduction rates or about the same rate as those of the smallest fibers in the A group. B fibers are preganglionic in the autonomic nerves and are absent in many nerve trunks.

The C group comprises the later components of the compound action potential. The fibers are mostly unmyelinated and may be less than 1 μm in diameter. They have the slowest conduction times and the highest threshold to stimulation. Two classes of C fibers are sometimes described on the basis of differences in function, either motor or sensory: The sC fibers serve as postganglionic axons in the sympathetic system, and the d.r.C fibers are found in the sensory nerve trunks and dorsal roots.

Conduction velocity

This is the speed at which the wave of depolarization spreads along the axon or the time taken for an active region to depolarize an inactive region. The rate of current flow in a local circuit is determined by the passive electrical properties of the axon, including its resistance and diameter. Larger fibers have a lower internal resistance than small fibers. Hence, they tend to conduct impulses at faster speeds.

In unmyelinated fibers conduction is impeded by the high resistance of the axoplasm and the large capacity of the membrane. As a result, conduction velocity in the smallest axons may be as slow as 0.3 sec $^{-1}$. The myelinated fiber has a relatively larger diameter, lower resistance, and smaller membrane capacity. High conduction speeds (up to 120 m sec $^{-1}$ are therefore possible.

Myelin is an insulating material formed by Schwann cells situated along the length of an axon. As growth proceeds, the myelin is laid down in a spiral manner, wrapping itself round the axon to produce a multi-layered sheath; the Schwann cell, itself containing the nucleus, is pushed to one side. Between each Schwann call there is a gap called the node of Ranvier where the myelin is interrupted and the axon lies in contact with the external medium; the sheathed part of the axon is called the internode. Since myelin is a good insulator and prevents the spread of current, action potentials in myelinated nerves can be generated only at the nodes. When a fiber is depolarized, the local circuit flow is from one node through the axon to the next node and the action potential is said to "jump" from node to node. This is what is meant by saltatory conduction. Myelination makes possible the fast conduction of nerve impulses. The rate of conduction depends on the thickness of the sheath relative to the diameter of the axon and to some extent on the internodal distance.

Measurement of conduction velocity

A simple method of measuring the conduction velocity in the isolated nerve is to plot the time taken for an action potential to travel a known distance along the nerve and express the result in meters per second. The experimental details are as follows:

The sciatic nerve of a frog is dissected from underlying tissue and placed over the electrodes in a nerve bath filled with mineral oil or Ringer's solution. The nerve is anchored by threads to ensure good electrode contacts. Stimuli are delivered to one end of the nerve and recordings taken at two points along the nerve (X and Y in Fig. 16). The difference in latency

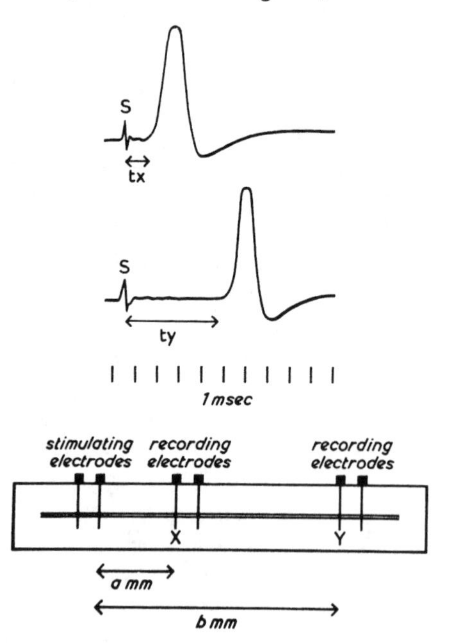

Fig. 16. Measurement of conduction velocity. Stimulating electrodes are applied to one end of the nerve and recording electrodes to points X and Y along the nerve. (a, distance between stimulus and point X; b, distance between stimulus and point Y.) The upper trace is a recording of action potential from point X, the lower trace a recording of action potential from point Y. Latencies tx and ty are measured from the time trace.

between the two action potentials is the time taken for the impulse to travel the distance between these two points. The conduction velocity (V) can now be calculated.

Distance to first recording point = a mm
Distance to second recording point = b mm
Latency of first action potential = t_x msec
Latency of second action potential = t_y msec

hence $V = b-a/t_y-t_x$.

Conduction velocity is increased by warmth and reduced by cold. It is also diminished during the relative refractory period and during the early period of regeneration following nerve section.

Conduction in human nerves

There are a number of sites on the surface of the human body called *motor points* where electrical stimulation excites motor nerve fibers to produce muscular contractions. Thus stimulation of the ulnar nerve in the forearm results in adduction of the thumb when the adductor pollicis muscle contracts. By means of electrodes applied to the thumb the motor response can be recorded on an oscilloscope, and its latency measured.

For recording sensory action potentials in the median nerve, the arrangement shown in Figure 140 (see Chapter XIX) can be used in normal subjects and on patients. Thus in carpal tunnel syndrome, the degree of nerve damage is assessed from the increase in latency found and from the reduction of the action potential. Severe compression may result in the absence of the sensory action potential on the affected side. In the majority of patients, slowing of conduction can be demonstrated in the segment of nerve distal to the wrist, although nerve conduction time in the forearm may be relatively normal (Fig. 17). This can be easily demonstrated by stimulating the median nerve at the elbow and recording the nerve action potential at the wrist. Conduction tests are valuable in assessing the progress of recovery following the clinical treatment of injuries to peripheral nerves.

Nerve block

Conduction in peripheral nerves can be blocked in several ways; local anesthetics, cold, heat, and pressure are the usual agents. Electrical methods interfere with generation of the nerve impulse itself:

1. A slowly-rising depolarizing current leads to inactivation of sodium conductance and to failure to produce reversal of the membrane potential.
2. Hyperpolarization of the membrane blocks conduction by preventing

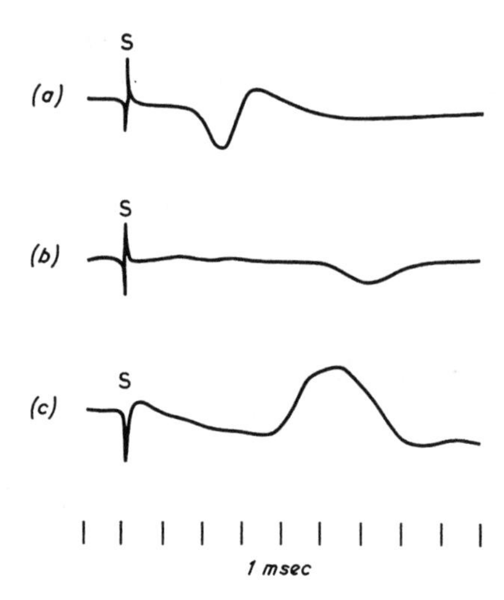

Fig. 17. Sensory and motor action potentials recorded from the median nerve at the wrist. (a, normal sensory action potential with latency about 2 msec; b, delayed and diminished sensory action potential—latency over 5 msec; c, normal motor action potential. Time base 10 msec. S, stimulus artifact.)

local circuit flow. This is sometimes called anodal blocking and can be demonstrated in the following way:

The compound action potential of the frog's sciatic nerve is recorded by the same experimental arrangement described above. A positive current is now fed to a portion of the nerve between stimulating and recording electrodes. The traces in Figure 18 show that as the strength of anodal current is increased, there is a progressive reduction in the amplitude of the action potential with prolongation of the conduction time until complete nerve block is achieved. The depression is maintained during current flow and may persist for a few seconds after the current is switched off.

PROPERTIES OF THE NERVE IMPULSE

1. All-or-none response. This refers to the propagated action potential in a single nerve axon. Once the stimulating current reaches threshold, the nerve impulse is conducted at full strength, its amplitude and shape being independent of the original stimulus. This indicates that the membrane potential is not further influenced by the stimulus that starts it off. It also suggests that the action potential, with its underlying cycles of sodium and potassium current flow, is a self-reinforcing event.

2. Refractory period. This is a silent interval during which the nerve loses

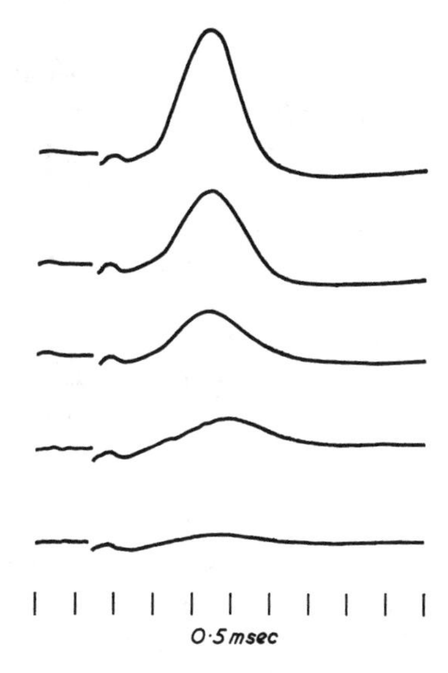

Fig. 18. Anodal block in isolated frog nerve. As the strength of anodal current is increased, the size of the action potential is diminished. The record at bottom shows the stimulus failing to excite any fibers.

its excitability for propagating a second impulse. The full recovery process may be divided into four phases:

(a) The absolute refractory period is the time during which no stimulus, however strong, can evoke a second impulse. It corresponds to the rising phase of the action potential and may be caused by inactivation of the sodium carrier mechanism. It lasts about 0.5 msec in mammalian A fibers, 1.0 msec in B fibers, and 2.0 msec in C fibers. Obviously, the membrane potential must be restored to some minimal value before it can be depolarized again.

(b) The relative refractory period lasts about 3.0 msec, during which time a stimulus stronger than the original may evoke a second impulse. This corresponds to the falling phase of the action potential when the outflow of K^+ ions exceeds the inflow of Na^+ ions. Hence, more stimulating current is required to produce a critical level of depolarization.

(c) The negative afterpotential represents a displacement of the membrane due to excess of K^+ ions outside the fiber as a consequence of increased potassium permeability. It appears as a small negative deflection following the decline of the action potential when the membrane is partly depolarized and its excitability slightly raised. This accounts for the subnormal period of the recovery process.

(d) The positive afterpotential represents a prolonged period of hyperpolarization before the membrane returns to its resting state. During this period K+ ions are driven back into the axon by increased activity of the metabolic pump in order to restore the intracellular potassium concentration. It appears as a positive deflection lasting up to 300 msec in myelinated fibers when membrane excitability is depressed. This accounts for the supernormal period of the recovery process.

3. Frequency. This refers to the number of impulses that can be sent along a nerve fiber in unit time. As a result of the refractory period, the frequency at which an axon can conduct impulses is limited. Allowing for the absolute phase alone, the maximum frequency cannot be more than 1000/sec in any fiber. Fortunately, the nervous system is not required to use frequencies as high as this. The large motor fibers supplying skeletal muscle rarely conduct at frequencies exceeding 120/sec, although some sensory fibers are capable of following stimulation frequencies up to 300/sec.

4. Two-way conduction. A nerve fiber can conduct impulses equally well in either direction. If, however, both ends are stimulated simultaneously the action potentials will collide in the center and be negated as each enters the refractory phase of the other. In the body, separate pathways are employed for conducting impulses to and from the central nervous system.

5. Heat production. The nerve fiber tends to swell and become opaque during the passage of an impulse. At the same time a measurable quantity of heat is produced. A.V. Hill observed two phases associated with the active state. In one a small amount of "initial" heat is liberated during the time of the action potential, presumably the result of the membrane permeability changes. In the other phase a larger amount of "delayed" heat accompanies the recovery process and the membrane returns to its resting state. During the recovery period energy is required to operate the metabolic pumps necessary for building up the ionic concentrations.

6. Transmitter release. When an impulse arrives at the axon terminal, it causes the release of a chemical substance or transmitting agent from a number of specific sites. The electron-microscope reveals the presence of vesicles at these sites. Since there is no direct continuity between one neuron and the next or between a nerve fiber and a muscle fiber, the electrical events associated with the nerve impulse cannot be conducted any further. Instead, the chemical substance released diffuses across the gap to the effector cell on which it acts. This mechanism by which impulses pass on the information they carry is known as synaptic transmission. A fuller description will be found in the next chapter.

SELECTED BIBLIOGRAPHY

Cole, K.S. An analysis of the membrane potential along a clamped squid axon. *Biophys. J.* 1:401-418, 1961.

Eccles, J.C. *The Physiology of Nerve Cells.* Baltimore: Johns Hopkins Press, 1957.
Frank, K., and Fuortes, M.G.F. Excitation and conduction. *Ann. Rev. Physiol.* 23:357-386, 1961.
Goldman, L. The effects of stretch on cable and spike parameters of single nerve fibres; some implications for the theory of impulse propagation. *J. Physiol.* 175:425-444, 1964.
Hodgkin, A.L. The ionic basis of electrical activity in nerve and muscle. *Biol. Rev.* 26:339-409, 1951.
Hodgkin, A.L. *The Conduction of the Nerve Impulse.* Springfield, Ill.: Charles C. Thomas, 1963.
Hodgkin, A.L., Huxley, A.F., and Katz, B. Measurement of current-voltage relations in the membrane of the giant axon of Loligo. *J. Physiol.* 116:424-448, 1952.
Katz, B. *Nerve Muscle and Synapse.* New York: McGraw-Hill, 1966.

CHAPTER III

Synaptic Transmission

A synapse is the junction between two neurons. The name was introduced by Sherrington to describe the contact between the terminal end of a nerve fiber and the membrane of another cell. Such contacts, which had previously been observed by Cajal and other histologists, frequently are referred to in the literature as "boutons terminaux," "end-feet," "varicosities," "bulbous enlargements," or simply "knobs." Under the microscope a nerve cell appears to be surrounded by clusters of synaptic knobs lying on the surface membrane of the soma and dendrites. If the axon of a nerve is cut, the synaptic knobs of that nerve degenerate and ultimately disappear without affecting the cell on which they terminate.

What happens to a synapse when an impulse reaches it was first demonstrated by Loewi in experiments on the frog's heart. Loewi showed that stimulation of the vagus nerve, causing slowing of the heart beat, released an inhibitory substance in the bathing fluid, which he identified as acetylcholine. He also showed that stimulation of the sympathetic nervous system, which accelerates the heart, caused the release of a different chemical transmitter, referred to as *sympathin* and later identified as noradrenaline. The importance of these discoveries was that they provided convincing evidence that the transmission of impulses across synapses was not a continuous electrical event but required the intervention of chemical mediators. The frog's heart happens to be a particularly sensitive tissue for study, as the transmitter substances are released in sufficient amounts to be assayed directly. In the central nervous system, only small quantities are released during synaptic activity so that the evidence for chemical transmission is more indirect and often inferred from analysis of the responsive neurons. In many cases, the identity of the transmitting agent is still unknown. The behavior of central synapses is further complicated by the fact that many presynaptic fibers are derived from different sources, some being excitatory and others inhibitory. A single nerve cell may have a large number of afferent fibers converging upon it, each capable of releasing transmitter substance and thus contributing to the total output of the cell. Clearly, the operating level of a neuron—i.e., the nature of the response at any instant—will depend upon the integrated activity of the various excitatory and inhibitory synaptic inputs.

The essential difference between an excitatory and inhibitory response of a neuron can be related to the ionic permeability changes produced in the postsynaptic membrane. An excitatory response occurs when the transmitter substance increases the permeability for sodium ions and the membrane potential is reduced to a critical level. The action is therefore a depolarizing one, and the potential change is called an excitatory postsynaptic potential (EPSP). An inhibitory effect occurs when the transmitter substance increases the permeability for potassium and chloride ions while the membrane remains impermeable for sodium ions; the membrane potential is displaced below the firing threshold. The action is therefore a hyperpolarizing one, and the potential change is called an inhibitory postsynaptic potential (IPSP). Thus synaptic excitation and synaptic inhibition are the results of ionic movements that displace the membrane potential in opposite directions.

The interaction between opposing current flows on the postsynaptic membrane accounts for the fluctuations of the membrane potential above or below the firing threshold. When there is sufficient concentration of excitatory transmitter, action potentials are generated by the cell; if the level of depolarization is kept below the threshold, the discharge is either reduced or the cell does not fire at all. Another way in which synaptic activity can modify the response of the cell is by reducing the amount of excitatory transmitter released from an afferent nerve terminal. This process is known as presynaptic inhibition. It is supposed that some axons make synaptic contact with the excitatory terminals of other presynaptic fibers instead of ending on the postsynaptic membrane of the cell. When these axons are stimulated, they produce a partial depolarization of the excitatory terminals, which in turn causes the amplitude of the presynaptic action potentials to decrease. As a result, the amount of excitatory transmitter released is also reduced.

These two inhibitory processes are extremely important to the way in which the nervous system operates. Since the majority of nerve cells receive synaptic inputs from a vast number of converging afferents, often from widely different sources, the output of the cells must be kept under constant control. Everywhere, excitation is opposed by inhibition. (The mechanisms by which inhibition is distributed to specific populations of neurons will be studied in subsequent chapters.) Despite the variety and complexity of these inhibitory mechanisms, it is apparent that synaptic transmission plays a fundamental role in their organization.

STRUCTURAL FEATURES

The synapse has two main elements, presynaptic and postsynaptic, separated by a cleft or gap. Each element is enclosed within its own membrane (Fig. 19).

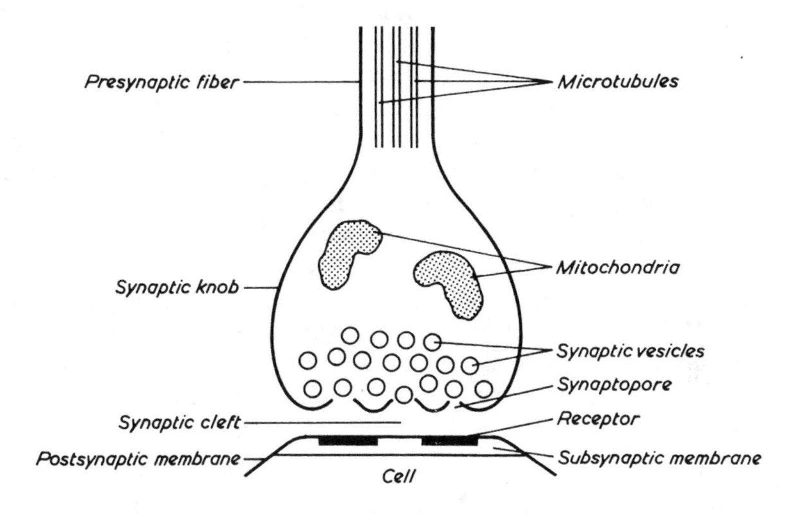

Fig. 19. Diagram of a synapse as revealed by the electron microscope. Note the synaptopores through which the vesicles can discharge their contents into the synaptic cleft (arrowed). The thickened portion of the subsynaptic membrane contains receptor sites for action of the transmitter substance.

1. The presynaptic element is formed by the swelling of an axon terminal. It has abundant mitochondria and numerous small vesicles containing the chemical substances responsible for transmission across the gap. Many of these vesicles are open to the surface allowing small amounts of transmitter to escape spontaneously. When the presynaptic membrane is depolarized by a nerve impulse, the number of releasing sites is increased and their contents are discharged into the synaptic cleft. Calcium is necessary for this action. The transmitter then diffuses across the gap.

2. The postsynaptic element is strictly the subsynaptic membrane or region of the cell lying under the synaptic knob. This is the site selected for the action of the chemical transmitter. The subsynaptic membrane is a receptor for transmitter molecules and is where the initial changes in membrane permeability occur. The rest of the postsynaptic membrane belongs to the soma and dendrites and is not affected directly by the transmitter.

EXCITATORY ACTIONS

For many years there were two schools of thought about the transmission of impulses from one neuron to the next. According to the electrical theory, the action potentials at presynaptic endings set up local potentials on the postsynaptic membrane. These potentials can summate and initiate an impulse down the axon when threshold is reached. The theory assumes that the two synaptic elements behave as a continuous electrical structure either because their membranes are fused and offer a low resistance pathway or because of the existence of protoplasmic bridges across the gap. Evidence for electrical transmission has been reported for giant fibers in crustacea and for Mauthner's cell of the goldfish brain. In the mammal, it is now accepted that the action potentials at presynaptic endings cause the liberation of a specific chemical substance, which reacts locally on the postsynaptic membrane to produce a change in ionic permeability. The membrane potential is thereby displaced from its resting state. The specific nature of the transmitter substance determines the direction of the displacement toward threshold or away from threshold and so leads to its excitatory or inhibitory effects.

Excitatory postsynaptic potential (EPSP)

At rest, the surface membrane of soma and dendrites, like that of the axon, is in a polarized state. The potential difference between the interior of the cell and an external indifferent electrode has been measured. For spinal motoneurons it is –70 mv; for pyramidal cells of the cerebal cortex it is –60 mv. The potential difference is reduced or abolished by the action of excitatory transmitter in the region of the subsynaptic membrane. As a result, this part of the cell becomes freely permeable to ions, including the inward flow of Na^+ ions. Current now flows from the synaptic knob through the synaptic cleft into the cell and out again through the adjacent postsynaptic membrane (Fig. 20). In this way the postsynaptic membrane is rapidly depolarized. The time course of these events shows that the inward current rises to a peak in about 0.5 msec, then slowly declines for many milliseconds afterward. During the period of current flow the EPSP is capable of summation, an important property of the synapse since the response may not be sufficient to attain the level of depolarization necessary to discharge the cell.

Subthreshold EPSPs are called *miniature potentials.* They do not propagate as action potentials do but spread passively along the membrane, diminishing in amplitude as the distance increases. In order to achieve spike

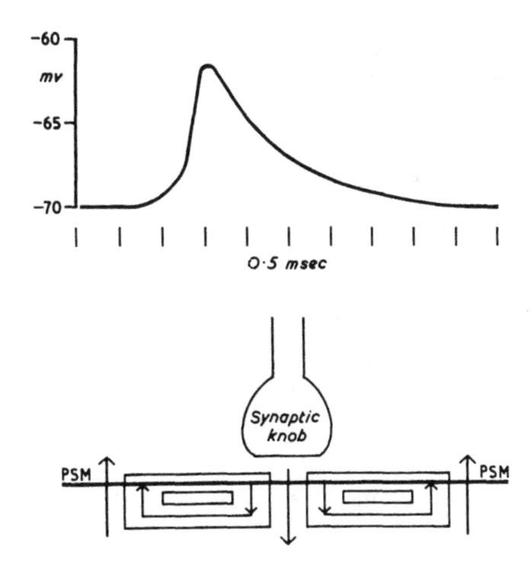

Fig. 20. Intracellular recording of excitatory postsynaptic potential. The resting membrane potential is reduced by a depolarizing current reaching a maximum level in about 0.5 msec and then slowly declining. The diagram below illustrates the synaptic region. Current generating the EPSP flows inward from the knob across the subsynaptic membrane and outward across the postsynaptic membrane (PSM).

production and discharge of the cell, a critical level of depolarization is essential. This level may be reached in one of two ways:

1. Temporal summation. This refers to the summation of successive potential changes in the postsynaptic membrane when one EPSP is superimposed on another. Thus if the level of depolarization produced by each impulse is subthreshold, the second potential change may add to the remainder of the first. Assuming the first EPSP lasts about 5 msec, temporal summation is effective during most of this time, and a spike may be generated. If the second impulse arrives too late, when the preceding EPSP has declined to a low level, no spike will be generated.

2. Spatial summation. This may be explained simply as the adding together of several subthreshold EPSPs produced at different sites on the postsynaptic membrane. By summation of the individual potentials, the critical level of depolarization may soon be reached for the cell to discharge. The attainment of threshold is therefore determined by the activation of a minimal number of synapses.

Generation of the spike potential

Our knowledge of the processes that cause a nerve cell to discharge has been derived mainly from intracellular recordings of mammalian spinal motoneurons. In order to produce a spike potential, the motoneuron can be excited in one of three ways—by synaptic activity, by antidromic stimulation, or by direct application of a stimulating current.

Synaptic (orthodromic) excitation

Intracellular potentials from motoneurons of the anesthetized cat were first recorded by Eccles and his associates following stimulation of the Group 1a afferent nerves from muscle spindles. The afferent nerves excited were from the synergic group of muscles whose collaterals make direct monosynaptic connections with the motoneuron under investigation. A subthreshold stimulus causes the generation of an excitatory postsynaptic potential. As the strength of the stimulus is increased, the size of the EPSP generated also increases and when this reaches a critical level of about 10 mv, a spike potential is generated—i.e., the motoneuron discharges an impulse. It can be seen from the record shown in Figure 21 that there is first a prepotential representing the EPSP, followed by the rising phase of the spike potential in which two further components can be recognized, an initial small spike and a later large spike. The small spike arises from the axon hillock, the initial segment or point of origin of the axon; this is the most excitable part of the postsynaptic membrane. The large spike arises

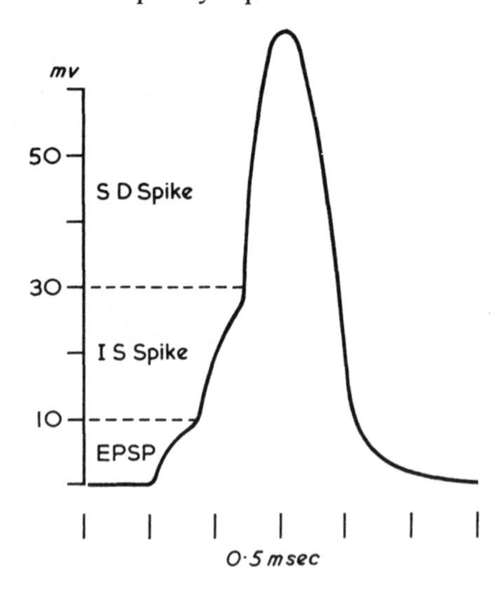

Fig. 21. Generation of spike potential. When the EPSP reaches a critical value, a spike potential is evoked first in the initial segment of the axon (IS spike). Further depolarization then produces the SD spike from the soma-dendritic membrane.

from the soma and dendrites of the motoneuron where the membrane has a much higher firing threshold (approximately 30 mv to 40 mv). The existence of these two components suggests that when an impulse is generated in a motoneuron there is a two-stage invasion of the postsynaptic membrane. The spike potential from the initial segment is generated first. Accordingly, the synapses terminating at this site are likely to be more effective than those situated on the soma and dendrites.

Antidromic stimulation

Intracellular potentials from a motoneuron can also be studied by stimulating the proximal end of a cut ventral root. The impulses are propagated antidromically along the axon into the cell. Analysis of the spike potential evoked in this way reveals that it has three separate components attributed to successive stages in the invasion of the different regions of the motoneuron. The first stage is generated in the axon itself and is called the M spike. It is an all-or-nothing response, resembling the action potential and caused by depolarization at a node of Ranvier. The second stage is generated in the initial segment and is called the IS spike. This potential may achieve an amplitude of 30 mv to 40 mv but it is set up when the membrane is depolarized to a threshold level of 6 mv to 10 mv. The third stage is generated by the soma and larger dendrites and is called the SD spike. This potential is produced by current flowing from the initial segment into the soma-dendritic membrane, which requires an additional depolarization of approximately 30 mv in order to attain a threshold level. The full-sized spike of the motoneuron is finally generated.

Direct stimulation

A spike potential can be induced artificially by passing a current through one barrel of a double microelectrode and recording through the other barrel (Fig. 22). Currents of various durations and intensities have been used to bring about a progressive depolarization of the neuronal membrane. The threshold level of depolarization for a directly applied current is approximately 10 mv, about the same level required for synaptic excitation. Likewise, the direct response has both IS and SD components, occurring slightly earlier than is the case with antidromic stimulation.

It will be apparent that the ionic mechanism generating a spike potential in a motoneuron resembles that of nerve fibers. Both are dependent on the attainment of threshold levels of depolarization. The origin of the spike

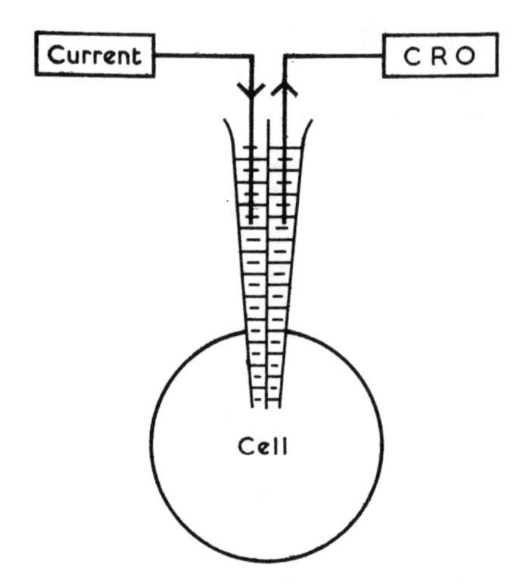

Fig. 22. Double-barrelled microelectrode for direct stimulation of cell. Current is applied from a battery to produce depolarization. The membrane potential is recorded from the second barrel. When the depolarizing current reaches a threshold level, a spike potential is generated and appears on the oscilloscope (CRO).

potential, however, is complicated by the fact that there are two zones on the postsynaptic membrane that show a remarkable difference in threshold for synaptically evoked impulses. The relatively high threshold of the soma-dendritic membrane contrasts sharply with the low threshold of the initial segment. As a consequence of this threshold difference, the synapses lying in contact with the initial segment have a greater chance of exciting the motoneuron than those situated more remotely on the dendrites. Synaptic transmission allows one nerve cell to influence another through a chemical mediator, which is either destroyed or reabsorbed by the presynaptic element. Removal of the transmitter from the receptor site obviously is of great importance since it ensures that the excitatory action is not unnecessarily prolonged. Under normal circumstances the total amount of depolarization produced in a motoneuron is governed by the number of inhibitory synapses that may be simultaneously active. This inhibitory mechanism will now be described.

INHIBITORY ACTIONS

Inhibition may be defined as any process that displaces the membrane potential away from its firing level. Originally, inhibition was defined as a depression of reflex excitability exerted on the motoneurons of antagonistic muscles, but all types of central inhibitory action may be explained satisfactorily on the basis of a common ionic mechanism. It has been shown that synaptic inhibitory action is brought about by the release of a chemical transmitter at the presynaptic terminals and that this induces a tendency toward hyperpolarization of the cell. The contribution made by inhibitory synapses on the final output of the cell may be seen as a reduction in the frequency of impulse discharge to the point where the cell ceases to fire altogether. If the inhibitory input to a neuronal circuit is very intense, there may be a long-lasting silent period, during which spontaneous activity is prevented and further transmission of impulses is blocked. Such effects are fundamental for keeping the activity of the nervous system under constant control.

Inhibitory postsynaptic potential (IPSP)

Inhibitory postsynaptic potentials are generated by ionic movements following the release of a chemical transmitter substance at a presynaptic terminal. The effect of the transmitter is to render the subsynaptic membrane permeable for the smaller anions and cations like chloride and potassium, but impermeable for the larger sodium ions. The increased ionic conductance is governed by the electrochemical gradient that drives K^+ ions out of the cell and Cl^- ions into the cell, the net current being outward at the subsynaptic membrane and inward through the adjacent postsynaptic membrane (Fig. 23). This flow of current induces a hyperpolarization of the

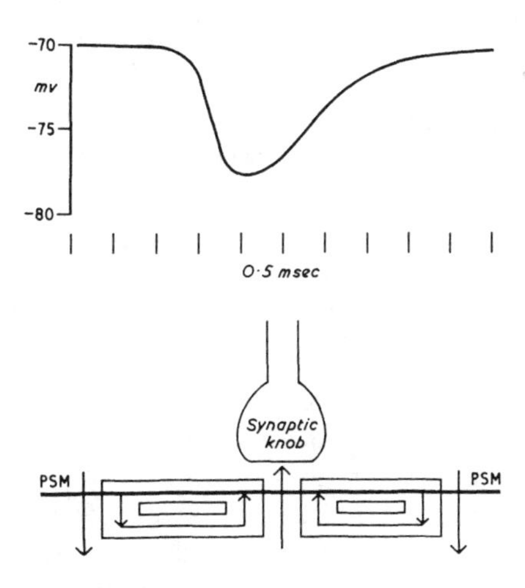

Fig. 23. Intracellular recording of inhibitory postsynaptic potential. The resting membrane potential is increased by a hyperpolarizing current. The potential reaches its peak about 1.5 msec after its onset, then rapidly declines. The diagram below illustrates the synaptic region. Current generating IPSP flows outward through the subsynaptic membrane toward the knob and inward across the postsynaptic membrane.

cell—i.e., the interior of the cell becomes more negative than it is at rest and therefore its responsiveness to excitatory impulses is diminished.

The time course of the current that produces the IPSP is determined from the intracellular recording depicted in Figure 24. It can be seen that the outward current reaches its peak in about 0.5 msec, after which there is a rapid decline over the next 2 msec. The amplitude depends upon the strength of stimulation. Like the EPSP. each synapse generates its own potential change, and the recording is simply a summation of the individual potentials. In motoneurons, whose resting potential is about −70 mv, the maximum amplitude recorded is usually about −80 mv—i.e., the IPSP is only about 10 mv higher. Nevertheless, the change makes it more difficult for the cell to fire an impulse, since the IPSP subtracts from any existing depolarization and keeps the membrane potential away from the firing level.

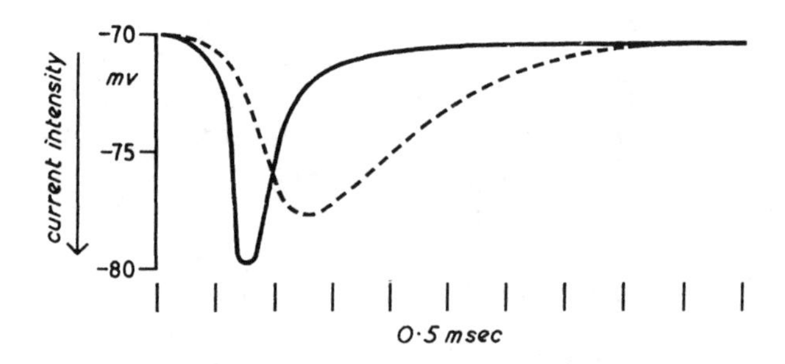

Fig. 24. Time course of inhibitory sybsynaptic current. Maximum intensity is reached in about 0.5 msec followed by a rapid decline with little residual current after 2 msec. The broken line represents the time course of the IPSP produced by the current.

Nature of ionic mechanism

In order to discover what particular ions cause an inhibitory current to flow across the subsynaptic membrane, two kinds of experiment have been performed. In one kind, various ions are injected into the cell by means of a double-barrelled microelectrode; in the other kind, the IPSP current is measured when the membrane potential is clamped at different levels.

1. Intracellular injections. It is possible to modify the size and polarity of the IPSP by passing an electrical current through a microelectrode filled with an appropriate ionic solution. For example, the IPSP is greatly influenced by the injection of Cl^- ions into a motoneuron; the increased intracellular concentration of chloride results in diffusion of Cl^- ions across the subsynaptic membrane to the outside. This has the effect of converting the

IPSP into a depolarizing current (Fig. 25). The polarity of the IPSP is reversed as the outside of the membrane becomes more negatively charged. Similar changes in the IPSP are brought about by lowering the intracellular potassium concentration. This has the effect of diminishing the outward flow of K^+ ions with progressive loss of positivity on the outside. As the membrane potential falls toward zero, the IPSP is reversed in sign from hyperpolarization to depolarization.

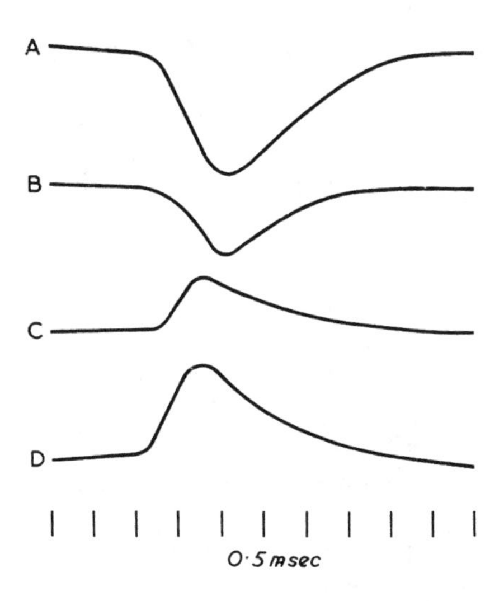

Fig. 25. Intracellular records showing reversal of IPSP by injecting Cl^- ions into the cell from a double-barrelled microelectrode. (A, normal IPSP with hyperpolarization in direction of equilibrium potential; B, Cl^- ions are injected into cell causing membrane potential to approach zero; C and D, reversal of polarity due to depolarization of the membrane and current flow inward.) Note the shorter time course of depolarizing potential as compared with the original hyperpolarizing potential.

2. Voltage clamp method. Current is passed from a battery through one barrel of the microelectrode in order to fix the membrane potential at a desired level; the other barrel is used for recording the potential changes during synaptic activity. At the normal resting potential of –70 mv, the IPSP is a hyperpolarizing response. If the membrane potential is artificially set at –80 mv, the IPSP is reduced to zero, because at this value the membrane potential equals the mean equilibrium potential for Cl^- and K^+ ions. It will be recalled that the equilibrium potential for K^+ is about –90 mv, and the equilibrium potential for Cl^- about –70 mv. If the membrane potential is made more negative than –80 mv, the IPSP is converted into a depolarizing response. The reversal point is therefore only 10 mv higher than the resting potential, which agrees well with the results obtained by other methods.

These experimental findings indicate that the IPSP is generated by permeability changes at the subsynaptic membrane, which permit the free passage of small-sized ions, particularly chloride and potassium ions, but exclude the passage of larger ions like sodium. As a consequence of these ionic movements, Cl^- and K^+ ions are driven along their electrochemical gradients toward their respective equilibrium potentials. The membrane potential recorded by an intracellular electrode shows that there is a net gain of positive ions on the outside. The IPSP is therefore a hyperpolarizing response, tending to drive the membrane potential toward the peak value of −80 mv (the mean of the Cl^- and K^+ equilibrium potentials). In motoneurons with a resting membrane potential of −70 mv, the driving force for the IPSP is only about 10 mv, but this is sufficient to displace the membrane potential away from the threshold level for firing an impulse.

Presynaptic inhibition

This refers to the influence an inhibitory nerve, usually an interneuron, exerts on the synaptic terminal of an excitatory nerve instead of making direct contact with the postsynaptic membrane. The inhibitory nerve endings produce a partial depolarization of the synaptic excitatory knobs, and thereby diminish the size of the presynaptic action potentials. As a result, there is less excitatory transmitter released, and the EPSP generated is reduced. The evidence is as follows:

1. Depolarization of excitatory terminals. This has been demonstrated by recording the antidromic compound action potential from a dorsal root in the lumbo-sacral cord of the cat; a standard test shock is used to stimulate the afferent terminals. The response is enhanced when a conditioning volley is delivered to a presynaptic inhibitory pathway, suggesting that the terminals are depolarized and made more excitable.

2. Reduction in amplitude of EPSP. This has been demonstrated from intracellular recordings of extensor motoneurons in the spinal cord of the anesthetized cat. The normal EPSP response is first set up by delivering a single test volley to the afferent nerves arising in the muscle. A preceding conditioning volley is then delivered to the afferent nerves supplying the antagonistic flexor muscles of the same limb. As shown in Figure 26, the amplitude of the EPSP is reduced following the conditioning volley on account of prolonged depolarization of presynaptic endings; the EPSP of the postsynaptic membrane remains unaffected.

3. Time course of depression. Also illustrated in the figure is the time course of the presynaptic inhibition, which shows that the EPSP depression begins with a C-T interval as brief as 5 msec, reaches its maximum at about

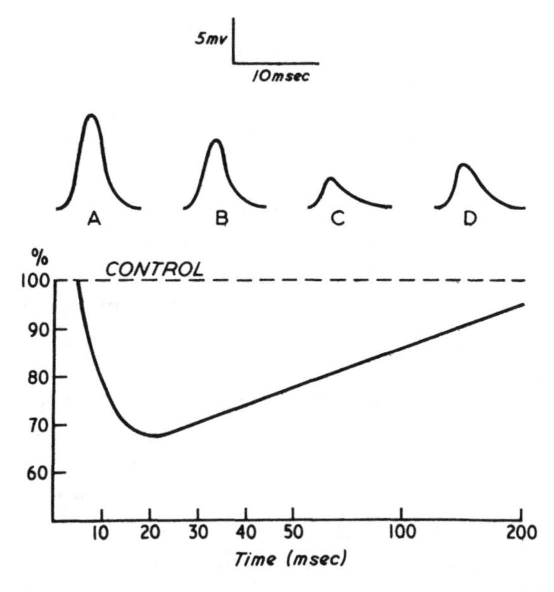

Fig. 26. Presynaptic inhibition. A is the intracellular record of EPSP evoked by stimulating the afferent nerve from an extensor muscle. B, C, and D represent the depression of EPSP produced by a preceding conditioning volley to afferents supplying the antagonistic muscle. The time course of the depression is plotted below. Ordinate is the percentage of the control EPSP and the abscissa represents intervals in msec after the conditioning volley. The plot shows maximum reduction in amplitude of EPSP with intervals of 15 msec to 20 msec and persistence of inhibition beyond 200 msec (Eccles, Eccles, & Magni, 1961.)

20 msec, and then slowly declines over a period of 200 msec or more. If, instead of a single burst, repetitive stimulation is employed, the depression is considerably increased and reaches its maximum value much sooner. Under such conditions a prolonged inhibition of reflex activity is produced.

4. Reduction of monosynaptic reflex. This can be demonstrated by recording the reflex motoneuron discharge from a cut ventral root. As the amplitude of the EPSP is reduced, the size of the monosynaptic reflex is also reduced.

It may be concluded from the above that presynaptic inhibition operates on the afferent excitatory nerve terminals and not directly on the postsynaptic membrane. On entering the spinal cord, many of the afferent fibers give off collaterals to interneurons, which in turn establish presynaptic contacts with the terminals of other afferent nerves. When active, these contacts produce a depolarization of the terminals, decreasing the amplitude of their action potentials and thus reducing the amount of transmitter released. The resultant depression of the EPSP accounts for the prolonged inhibitory effects exerted on the motoneurons. Presynaptic inhibition may play a significant role on spinal pathways concerned with the reflex control of posture and movement.

Disfacilitation and disinhibition

These terms describe two further mechanisms in the functioning of the central nervous system. They refer to the role of interneurons, an important network intervening between the incoming afferents and the motoneuron pools. The barrage of impulses flowing in this network provides a background of excitation or inhibition on which the final actions of the output neurons depend.

Disfacilitation occurs when the background of excitation is interrupted from any cause. The removal of tonic depolarizing impulses causes the membrane potential to be increased and its threshold raised, with a consequent reduction in neuronal firing comparable in some ways to the hyperpolarization produced by synaptic inhibition.

Disinhibition has the same effect as facilitation and occurs after removal of a tonic inhibitory background. The consequent change of membrane potential and greater amplitude of the EPSP result in an increase in the firing rate of the output neurons.

Numerous examples of both disfacilitation and disinhibition have been demonstrated in the brain as well as in the spinal cord. A diagrammatic representation of the basic circuit arrangment is shown in Figure 27. It will be apparent that apart from the straightforward action of afferent pathways on excitatory and inhibitory synapses, considerable modification can be imposed on the firing pattern of individual neurons by impulses operating through the internuncial system (see Chapter X).

GENERAL FEATURES OF SYNAPTIC TRANSMISSION

1. Spontaneous release. Presynaptic nerve terminals release small packages of transmitter substance to which the term "quanta" is applied. Such packages are discharged into the synaptic cleft at random intervals, as revealed by the generation of miniature potentials on the postsynaptic membrane. If the presynaptic terminal is allowed to degenerate, the miniature potentials disappear.

2. Impulse release. The arrival of a nerve impulse increases the rate of release and the number of active vesicles in the presynaptic terminal. Consequently, a larger number of packages are discharged into the synaptic

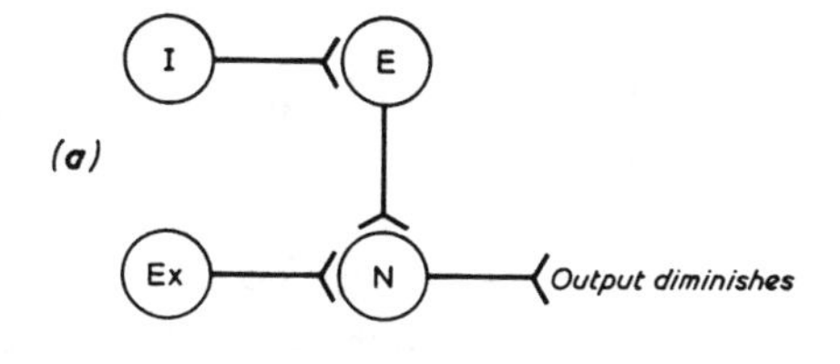

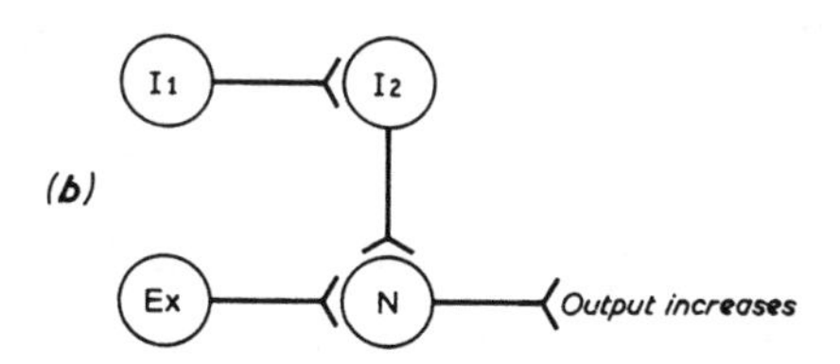

Fig. 27. Internuncial circuit arrangements. (a, Disfacilitation.) The inhibitory interneuron I inhibits the excitatory interneuron E. This reduces the excitability of the output neuron N and diminishes activation by the excitatory pathway Ex. The firing rate of N decreases. (b, Disinhibition.) The inhibitory interneuron I_1 inhibits the inhibitory interneuron I_2. This removes inhibition from the output neuron N and the pathway Ex can now activate it. The firing rate of N increases.

cleft more or less synchronously, producing a rapid build-up of transmitter. Calcium ions appear to be essential for this purpose.

3. Transmitter action. On reaching the postsynaptic element, the transmitter substance combines with a specific receptor and this results in a change of membrane permeability and flow of current. Pharmacological substances that act on the same receptor are known as competitive inhibitors.

4. Fate of transmitter. To ensure that its action is not unduly prolonged, the transmitter substance is removed from the synaptic cleft. This may be done in two ways:

(a) By active uptake. There is evidence that excess transmitter is transported back to the presynaptic terminal where it is stored again in the vesicles. In the sympathetic system, noradrenaline is probably removed in this way and thus becomes immediately available when required for continuous action.

(b) By destruction. Enzymes are present in high concentration near the receptor sites; their function is to destroy the transmitter by hydrolysis. At synapses where acetylcholine is the transmitter, the enzyme acetylcholine esterase splits the transmitter into choline and acetic acid. The choline is for the most part taken up by the presynaptic terminal for re-use. Drugs that interfere with the enzyme action prolong the effect of the transmitter or, by allowing it to accumulate, cause a more powerful response.

5. Transmitter-receptor reactions. When a transmitter substance reacts locally at a receptor site, it produces a change in membrane permeability that tends to displace the membrane potential from its resting state. If the displacement is in a depolarizing direction, the effect on the cell is excitatory, and if in a hyperpolarizing direction, the effect is inhibitory. In both cases there is an increase of ionic conductance, but what exactly determines the direction of current flow is open to several interpretations:

- Location of synapse. The contribution of a particular synapse to the activity of a cell may depend upon its site on the soma-dendritic membrane. In some instances, for example, that of Purkinje's cells of the cerebellum, excitatory synapses are most profuse on the dendrites, while inhibitory synapses terminate for the most part on the soma.
- Shape of vesicle. Differences in the form of synaptic vesicles have been described. Spherical vesicles are supposed to be excitatory, and flattened or oval vesicles inhibitory.
- Type of transmitter. The difference between an excitatory and an inhibitory action on the postsynaptic membrane is thought to depend upon the specific nature of the chemical transmitter. All transmitter substances released at central nervous synapses have not yet been identified. They are usually classed as cholinergic or adrenergic, according to the type of enzyme system that manufactures them. However, the same transmitter may have opposite actions at different synapses.
- Surface receptors. After its release from a presynaptic nerve terminal, the transmitter substance combines with a receptor on the subsynaptic membrane. There may be many different kinds of receptor on the same membrane, each having a specific action on the ionic conductance channels. Drugs that are known to modify synaptic transmission usually exert their effects directly on the receptors.

TRANSMITTER SUBSTANCES

A wide variety of chemical substances have been tested as possible candidates for central nervous transmission but only a few satisfy all the criteria. A true transmitter must develop naturally in the presynaptic element and be released by nerve impulses. The appropriate enzyme systems for its manufacture and subsequent removal must also be located in the region of the synapse. Furthermore, if the substance is ejected by iontophoresis, it should reproduce the ionic processes normally evoked by transmitter action. Substances that are believed to fulfill these criteria may be classified as acetylcholines, monoamines, or amino acids.

Acetylcholine

There is general agreement that acetylcholine is the transmitting agent at a number of sites in the central nervous system. It is released by motor axon collaterals to Renshaw cells in the spinal cord where a brief excitation evokes the well-known feedback inhibition of the motoneurons. In the brain, cholinergic synapses are found in the thalamus, basal ganglia, hypothalamus, and the afferent terminals to pyramidal cells in the cerebal cortex. The action of these synapses is for the most part excitatory, as shown by the depolarizing changes induced on the postsynaptic membrane. Recently, however, a number of reports have shown the iontophoretic application of acetylcholine to produce a marked inhibition of spontaneous firing in a proportion of cells located in the thalamus and brain stem reticular nuclei.

Monoamines

These substances are found in high concentration in extracts from nervous tissue, although their role in synaptic transmission has proved difficult to establish. They include noradrenaline, dopamine, and 5-hydroxytryptamine.

Noradrenaline is present in many parts of the central nervous system, especially in the hypothalamus and limbic regions, which contribute to the control of visceral functions. In the olfactory bulb and also in the dorsal grey area of the spinal cord, noradrenaline has a depressant action on spontaneously firing neurons; similar findings after intraarterial or intraventricular injections suggest that noradrenaline may function as a central inhibitory transmitter.

Dopamine is found mainly in the pigmented cells of the substantia nigra, nucleus interpeduncularis, and median eminence. It is manufactured in the body by conversion of levodopa through the action of the enzyme dopa decarboxylase. Dopamine is in turn the metabolic precursor of noradrenaline (Fig. 28). In Parkinson's disease, dopaminergic transmission is defective because of the loss of the pigmented neurons and depletion of dopamine. The administration of levodopa in the treatment of patients is based on the assumption that dopamine is an inhibitory transmitter in-

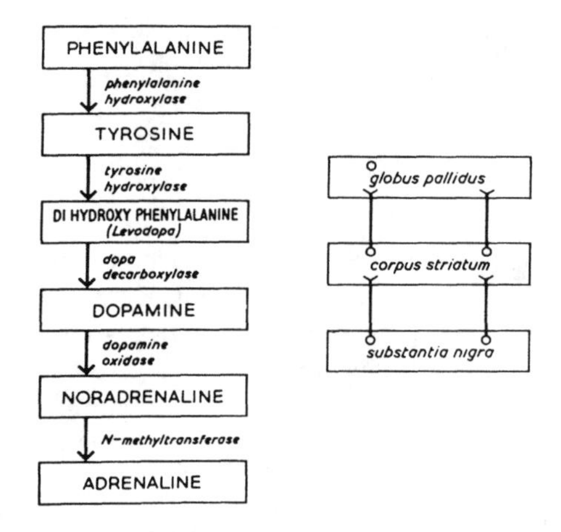

Fig. 28. Catecholamine synthesis. The enzymes responsible for catalyzing each step are indicated by the arrows. The diagram on the right shows dopaminergic pathways.

volved in the projections from the substantia nigra to the corpus striatum. While dopamine itself does not cross the blood-brain barrier, no such difficulty is encountered with its metabolic precursor.

The substance called 5-Hydroxytryptamine (5-HT), or serotonin, is found predominantly in the raphe nuclei of the brain stem and at the sites of their ascending and descending projections. If the raphe neurons are stimulated electrically, there is a marked increase of the monoamine content of perfusates collected during stimulation. Samples of the perfusate can be assayed on a strip of rat's stomach tissue. The distribution of 5-HT can also be visualized histologically under the microscope by a fluorescent technique that shows very high concentrations in the axon terminals. By these methods the projections from the raphe nuclei have been traced to the anterior hypothalamus, caudate nucleus, and sensorimotor cortex. It is believed that 5-HT is a true transmitter. It is synthesized from trytophane through metabolic pathways that have all been identified at the releasing sites. There is some evidence that tryptaminergic neurons are involved in mechanisms underlying the regulation of body temperature.

Amino acids

The third group of chemical substances considered as likely candidates for synaptic transmission includes a variety of amino acids like glycine, glutamic acid, and γ-aminobutyric acid (GABA); the latter is generally regarded to function as an inhibitory transmitter in the mammalian cerebral cortex. GABA is present in the brain in relatively high concentrations. It is released spontaneously from the terminals of inhibitory neurons and in greater quantities when these neurons are activated. Thus in microelectrode recordings from the brain of the anesthetized cat, it has been shown that the firing of single cortical cells can be suppressed by electrical stimulation of the pial surface. This effect is associated with an increased amount of GABA in the collecting fluid and suggests that it may be a possible transmitting agent in cortical inhibition. When GABA is applied iontophoretically, it faithfully reproduces the potential changes of synaptic inhibition that appear to be identical with the IPSP evoked by the natural transmitter.

In the spinal cord, however, GABA is thought to act differently at presynaptic and postsynaptic sites. A depolarizing action on primary afferent terminals has been reported by a number of investigators. This finding supports the view that GABA is the excitatory transmitter involved in presynaptic inhibition. On the other hand, its action on the postsynaptic membrane of motoneurons is one of hyperpolarization—i.e., an increase of the membrane potential responsible for postsynaptic inhibition.

In conclusion, it can be stated that there are a number of different chemical substances mediating synaptic transmission in the central nervous system. Most of the substances investigated can be shown to modify transmission in one way or another, but few fulfill the strict criteria for acceptance as a natural transmitter. It is of interest that the same substance can produce both excitatory and inhibitory effects, the result in all probablility depending upon the specific binding power of the receptors at different synaptic sites.

Integrated activity

Since a nerve cell is normally subjected to continuous bombardment from its synaptic input, the axonal discharge must represent the integrated activity of all the excitatory and inhibitory impulses that reach it. An excitatory impulse liberates a transmitter substance that causes an inward flow of current across the subsynaptic membrane; an inhibitory impulse has the reverse effect. Consequently, the membrane potential of the cell is in a state of fluctuation varying in value above or below the resting level of about −70 mv. Interaction between the opposing depolarizing and hyperpolarizing currents ultimately determines the direction of membrane displacement relative to its firing threshold. If a volley of impulses produces a continuous excitatory bombardment, the cell is maintained in a state of depolarization and fires spontaneously.

The frequency of the discharge, however, can be influeced at any time by the activity of inhibitory synapses that opposes the excitatory drive and, if sufficiently intense, can bring about cessation of firing. It is not essential for an equal amount of inhibitory transmitter to be released in order to control the output of the cell, because inhibition can be exerted to some extent presynaptically on the afferent terminals. Indeed, the amount of transmitter necessary to induce hyperpolarization is simply that required to keep the membrane potential below the critical level for initiating a spike. Another factor that is likely to affect the integration of depolarizing and hyperpolarizing currents is the situation of a particular synapse on the somadendritic membrane, in particular, its proximity to the initial segment or axon hillock where the threshold is lowest. Finally, by means of interaction occurring at any synaptic level in the nervous system, excitatory and inhibitory impulses compete with one another for domination of the pathways on which they converge.

SUMMARY OF SYNAPTIC PROPERTIES

Unidirectional transmission

A nerve impulse can be transmitted in only one direction, usually from the synaptic knob of one neuron to the postsynaptic membrane of the next. Antidromic stimulation of an axon may cause excitability changes in the soma and dendrites of a cell, but there is no conduction back across a synapse into the presynaptic fiber.

Synaptic delay

The time required for transmission across a single synapse is about 0.5 msec to 1.0 msec. A simple method of obtaining a value for synaptic delay is the recording of a motor nerve action potential following stimulation of a presynaptic fiber. If the motor nerve is also stimulated directly, the interval between the two action potentials gives an estimate of synaptic delay. Using an intracellular recording electrode, a more accurate method is the measurement of the time between the presynaptic stimulus and the onset of the postsynaptic potential. In the central nervous system, where impulses are frequently distributed over chains of interneurons, the conduction time is increased in proportion to the number of synapses in the chain.

Transmitter action

Presynaptic impulses release a chemical substance from storage vesicles in the synaptic knob. The excitatory and inhibitory reactions that follow are caused by the flow of ionic currents across the synaptic cleft, which tends to displace the membrane potential from its resting state. If the current causes a depolarization to a critical level, an impulse is generated in the postsynaptic membrane; a current in the reverse direction tends to hyperpolarize the membrane and inhibit the cell from discharging.

Location

The influence of a synapse on the activity of a cell is determined to some extent by its situation on the postsynaptic membrane. The region of greatest responsiveness is the initial segment of the axon where both excitatory and inhibitory synapses exert their largest effect. Elsewhere on the soma-dendritic membrane, the resting potential is more stabilized, while synapses located distally on the end of dendrites are the least effective. The long apical dendrites of cortical pyramidal cells are an important exception.

Integration

Under normal circumstances the output discharge of a single nerve cell reflects the total membrane change imposed by all the synaptic inputs impinging upon it. In the first place, interaction between excitatory and inhibitory synapses accounts for the background excitability of the cell. Second, to evoke an impulse discharge, the summation of all depolarizing and hyperpolarizing currents must reach a critical value to attain a threshold level. Third, when depolarization of the membrane is the dominant activity, for example, as a result of repetitive excitatory bombardment, impulse discharge is considerably enhanced. Fourth, summation of inhibitory effects, comprising both presynaptic and postsynaptic influences, can produce a long-lasting depression of the generator mechanism. Finally, to impose a degree of restraint on impulse discharges, inhibition can play its part simply by holding the membrane potential slightly away from its firing level.

SELECTED BIBLIOGRAPHY

Araki, T., and Terzuolo, C.A. Membrane currents in spinal motoneurons associated with the action potential and synaptic activity. *J. Neurophysiol.* 25:772-789, 1962.

Curtis, D.R., and Ryall, R.W. The synaptic excitation of Renshaw cells. *Exp. Brain Res.* 2:81-96, 1966.
Eccles, J.C. *The Physiology of Synapses.* New York: Springer-Verlag, 1964.
Eccles, J.C., Schmidt, R.F., and Willis. W.D. Presynaptic inhibition of the spinal monosynaptic reflex pathway. *J. Physiol.* 161:282-297, 1962.
Gray, E.G. Axo-somatic and axo-dendritic synapses of the cerebral cortex: An electron microscopic study. *J. Anat.* 93:420-433, 1959.
Hubbard, J.I., Llinás, R., and Quastel, D.M.J. *Electrophysiological Analysis of Synaptic Transmission.* London: Edward Arnold, 1969.
Kuno, M. Mechanism of facilitation and depression of the excitatory synaptic potential in spinal motoneurons. *J. Physiol.* 175:100-112, 1964.
Lloyd, D.P.C. A direct central inhibitory action of dromically conducted impulses. *J. Neurophysiol.* 4:184-190, 1941.
Paton, W.D.M. Central and synaptic transmission in the nervous system; pharmacological aspects. *Ann. Rev. Physiol.* 20:431-470, 1958.
Wilson, V.J., and Burgess, P.R. Disinhibition in the cat spinal cord. *J. Neurophysiol.* 25:392-404, 1962.

CHAPTER IV

Skeletal Muscle Contraction

The muscles of the body are classified as skeletal, cardiac, and visceral. All possess contractile properties—that is, the ability to shorten in length or to develop tension according to the purpose for which they are employed. Skeletal muscles operate the bones and joints, sustaining the weight of the body against gravity, providing the means of propulsion and also the equipment for performing feats of skill or strength. In many animals the speed and power of movements are essential for survival and, whether they are brought about reflexly or by voluntary action, they are always under the control of the central nervous system. For this reason skeletal muscles generally are spoken of as voluntary. Loss of power is evident whenever nervous control is impaired; the muscles waste and become paralyzed if denervation is complete.

Cardiac muscle is supplied by the autonomic nerves, which normally regulate its activity, but the heart is capable of independent contractions due to intrinsic rhythmic properties. Rhythmic action is especially well developed in visceral muscle—for example, in the smooth muscle investing the hollow organs, where it serves to maintain and regulate the flow of the contents of these organs. Visceral muscle continues to contract even when removed from the body and deprived of all its connections with the nervous system.

In 1850 Claude Bernard showed that a frog's skeletal muscle, paralyzed with curare, could still contract when stimulated directly; the drug prevented the normal response of the muscle to stimulation of its motor nerve. Arising from this observation, the problem of the transmission of the impulse from nerve to muscle was the subject of much debate until Dale and Feldberg demonstrated the presence of acetylcholine in the perfusing fluid following nerve stimulation, and Nachmansohn reported on its destruction by the enzyme cholinesterase. It is now established that transmission is accomplished by chemical mediation in which acetylcholine combines with receptors on the muscle surface in the region of the motor end-plate. The result of the combination is a rapid change in the permeability of the muscle membrane to sodium ions, the formation of an end-plate potential, and the initiation of a self-propagating muscle action potential.

The microscopic structure of skeletal muscle has been the subject of considerable investigation since Bowman first reported on the striated appearance of individual muscle fibers. The light microscope revealed alternating bands of dark and light material, designated A, I, and H bands, which together constituted a functional unit or sarcomere; a narrow band of refractile material separating two sarcomeres was called the Z line. To obtain a view of the striations better than that possible with direct microscopy, phase contrast was used. With this method individual bands of muscle could be made to appear dark. It could be seen that each muscle fiber was made up of a mass of myofibrils lying in parallel position to the long axis.

Huxley developed the interference microscope, using twin beams of light to bring out differences in the refractive index of the bands when the fiber was stretched or allowed to shorten. Under both of these conditions he found that the A bands stayed at a constant length and he supposed that they must slide past one another. This idea was strongly supported by subsequent electron-microscopic studies in which the individual myofibrils were seen to contain two sets of overlapping filaments extending across the A and I bands. The filaments were organized into separate arrays that did not change in length but appeared to slide past each other when the fiber contracted.

The mechanical properties of skeletal muscle have largely been determined from experiments on isolated nerve-muscle preparations. The muscle responds to a single shock with a brief "twitch," or a phase of contraction followed by a phase of relaxation; the muscle's response to repetitive stimulation is a maintained contraction, or "tetanus." The characteristic curves of muscle contraction are recorded by clamping the pelvic end of the muscle rigidly and attaching the tendon at its free end to a recording lever or transducer. Forces may be applied to the muscle by loading it with various weights and, as the muscle shortens allowing it to lift the weight and thus perform positive work. The amount of shortening is proportional to the initial length of the fibers, which is adjusted before stimulation by means of an electromagnetic stop; the time course of muscle shortening can also be accurately measured. Under these conditions the contraction of the muscle is called *isotonic*. In the body the skeletal muscles are arranged around a system of levers operating the joints; muscular contraction provides the propulsive force required for moving the limbs and for checking the movements when bringing the limbs to rest.

If both ends of the muscle are fixed to prevent shortening, the muscle responds to stimulation by developing tension. The rise of tension is usually recorded by means of a strain gauge that converts the pull of the muscle into an electrical signal. Tension developed by the muscle during a brief twitch is complicated by elastic components that tend to smooth out rapid changes. However, the amount of tension developed is much greater in a tetanus when the elastic elements have attained a fixed length. Muscle contraction recorded under these conditions is called *isometric*. Slow rises of tension can be maintained economically and are effective in sustaining the weight of the body against gravity. Isometric contractions are therefore utilized by muscles to give stability to the joints and to take up the mechanical strains imposed by posture.

NEUROMUSCULAR TRANSMISSION

There is no direct continuity between a motor nerve and its muscle. On arrival of an impulse at the nerve terminals the following sequence of events takes place–(1) release of acetylcholine, (2) depolarization of the end-plate, (3) propagation of a muscle action potential, and (4) muscle contraction.

Release of acetylcholine

Skeletal muscles are supplied by motor nerve fibers–the axons of anterior horn cells in the spinal cord or the motor nuclei of certain cranial nerves. Each axon breaks up into a number of terminal branches that lose their myelin sheaths and sink into shallow grooves on the surface of the muscle fibers. This is the end-plate region where nerve membrane and muscle membrane are in close contact and where transmitter action takes place.

Acetylcholine is released spontaneously in small packets from storage vesicles in the nerve terminals. This quantal release in the absence of nerve activity was first demonstrated by Fatt and Katz when they recorded miniature end-plate potentials too small in size to produce muscular contraction. If the nerve is poisoned to prevent the release of the transmitter, the miniature end-plate potentials are also abolished. The arrival of a nerve impulse causes a larger quantity of acetylcholine to be released and increases the number of active sites so that for a very brief moment a large amount of acetylcholine is discharged and the end-plate potential increases in size. This effect was shown to be dependent on the presence of calcium ions, while magnesium ions added to the muscle bath tend to depress the releasing process. The rate of release is affected also by the frequency of impulse arrival since the end-plate potential increases in size with repetitive stimulation.

Depolarization of the end-plate

After its release from the axon terminal acetylcholine combines with receptors on the muscle membrane, causing a marked increase of ionic permeability at this point. The resting membrane potential discharges almost to zero and an end-plate potential is produced. This has a simple monophasic waveform lasting only a few msec. before declining. It does not propagate as does an action potential but diminishes in size as the distance from the end-plate increases. The membrane depolarization, however, causes a flow of ionic current which short-circuits the surrounding muscle membrane to a level at which excitation occurs and a muscle action potential is set up.

The reaction that takes place at the end-plate can be blocked by treating the muscle with curare. This substance competes with acetylcholine for receptor sites and thus reduces the size of the end-plate potential. Curare does not interfere with the release of acetylcholine, but it is a very effective blocking agent, whose action results in muscular paralysis. The rapid decline of the end-plate potential is caused by inactivation of acetylcholine by the enzyme cholinesterase. The concentration of cholinesterase in the end-plate region is sufficient to account for the rapid hydrolysis of the transmitter into choline and acetate, which are inert. As soon as the end-plate membrane repolarizes, the muscle can respond again to another impulse. If, on the other hand, the enzyme itself is inactivated, the transmitter action will persist and lead to a greatly enhanced end-plate potential (Fig. 29). Drugs like eserine and prostigmine are potent anticholinesterases and in suitable doses give rise to a remarkable increase in muscular power.

Propagation of the muscle action potential

This is the muscle impulse or electrical signal that precedes muscular contraction. It is a self-propagating, all-or-none spike comparable in many ways to the nerve action potential. In the crustacean nerve-muscle system the motor nerve terminals are scattered along the surface of the muscle fiber, and the muscle responds to summation of the individual potential changes. In vertebrate muscle, the motor end-plate is centrally situated, and the impulse must be propagated toward both ends of the fiber and then

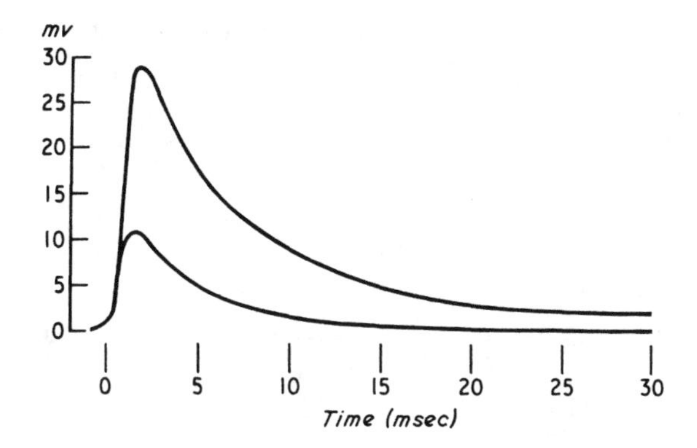

Fig. 29. End-plate potentials recorded from curarized muscle. Acetylcholine released by the nerve impulse depolarizes the end-plate. The potential rises sharply to about 10 mv and declines slowly. After the addition of prostigmine, the response builds up to about 30 mv and falls more gradually.

must pass into the interior in order to excite the contractile elements. The muscle action potential is initiated when the membrane potential at the neuromuscular junction is displaced to a critical level of depolarization. At rest the potential difference is about –90mv. This is rapidly reduced to –50 mv by the end-plate potential resulting in a local current flow. As in nerve fiber, the sudden influx of Na^+ ions and the efflux of K^+ ions are associated with the rise and fall of the spike. As the spike reaches its full size, the membrane potential is further reduced and finally reversed until the inside of the membrane becomes about 40 mv positive with respect to the outside–i.e., the peak of the spike may reach 130 mv in amplitude. Electrical recordings from the active muscle show that after initiation the spike is propagated by spread of local currents between the excited and resting parts of the membrane.

Excitation-contraction coupling

The electrical changes propagated over the surface of the muscle fiber also spread inward along the walls of the reticular tubules to reach the contractile elements. Calcium ions are released in the vicinity of these tubules prior to the onset of contraction. Active shortening of the fiber is brought about by interaction between the protein filaments, myosin and actin, which slide past each other within the myofibrils. During this process energy is released by the splitting of ATP, a measurable quantity of heat is generated, and a mechanical force is developed. Relaxation is brought about by removal of calcium back into the reticulum and the resynthesis of ATP to restore the resting state of the contractile elements.

THE MECHANISM OF CONTRACTION

Skeletal muscle is composed of bundles of striated fibers bound together by connective tissue. As a rule, the fibers are joined together at their extremities to form tendinous attachments. Each muscle fiber consists of a large number of myofibrils, which in turn contain the protein filaments referred to above. The mechanism by which contraction is brought about depends upon the relative movements of these two sets of filaments.

Structural organization of striated muscle

Observations from light microscopy

1. Low power. When examined under low power the muscle fibers appear to have transverse striations made up of alternating dark and light bands. These are due to differences of refractive index and are known as A and I bands respectively.
2. High power. With higher magnification the A band is seen to contain a lighter region in its center, less refractive than its surroundings, and called the H band. Likewise, the I band is divided by a region of different refractive index called the Z line. The position of these bands can also be determined by extraction experiments in which any one of the components can be dissolved by introducing an appropriate solution in order to reveal a conspicuous gap. A simplified diagram of the striation patterns is shown in Figure 30.
3. Phase contrast. A better optical system for viewing the muscle structure is obtained by using the phase-contrast microscope. This instrument gives an excellent image of the striations and allows the investigator to distinguish the individual myofibrils. In vertebrate muscle the region between two Z lines is usually regarded as the structural unit or sarcomere.

Observations from electron-microscopy

1. Protein filaments. Muscle fibers contain two kinds of protein filaments–thick filaments, largely composed of myosin, and thin filaments composed of actin. These filaments are arranged in alternating arrays across

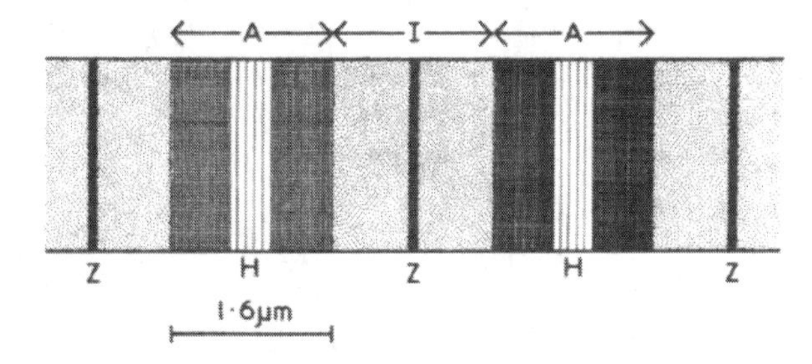

Fig. 30. Diagram of striation bands in skeletal muscle. A and I bands represent regions of high and low refractive index. The H zone is recognized as a region in the middle of the A band with a refractive index lower than that of the edges. The Z line is a narrow strip of high refractive index bisecting the I band.

the myofibrils. In longitudinal section, the thick filaments may be seen to extend from one end of the A band to the other; the thin filaments extend from the Z line into the A band as far as the boundary of the H zone. The arrangement is illustrated in Figure 31, from which it will be evident that the way in which overlapping occurs can account for the dark and light appearance of the striations:

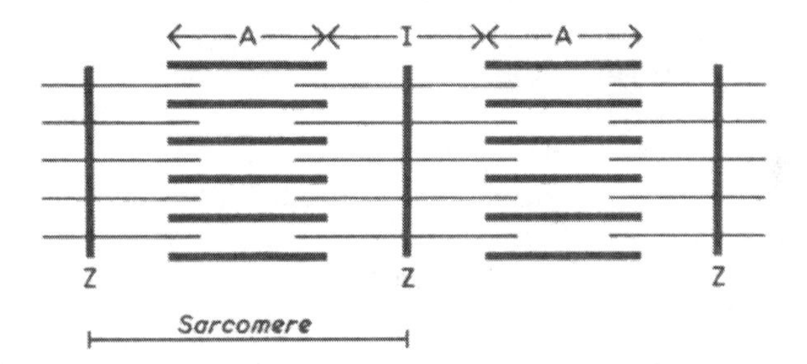

Fig. 31. Diagram of a sarcomere, showing alternating arrays of protein filaments. Thick horizontal lines represent myosin filaments; thin lines represent actin filaments. Note myosin is confined to the A bands, which do not alter in length when the muscle contracts. The actin filaments extend from the Z line to the H zone, yielding a region of double overlap at the center of the sarcomere.

The A band contains both dark and light regions. The darkness results from overlapping arrays of myosin and actin filaments; the lighter region (H band) has only thin actin filaments. The I band lies on either side of the Z line and contains only actin filaments.

In a transverse section through the A band where thick and thin filaments overlap, the arrangement is like that shown in Figure 32. Each of the myosin filaments is surrounded by six actin filaments, and each actin filament is related to three myosin filaments. A transverse section through the H band reveals a hexagonal array of myosin filaments, while a transverse section through the I band shows actin filaments only.

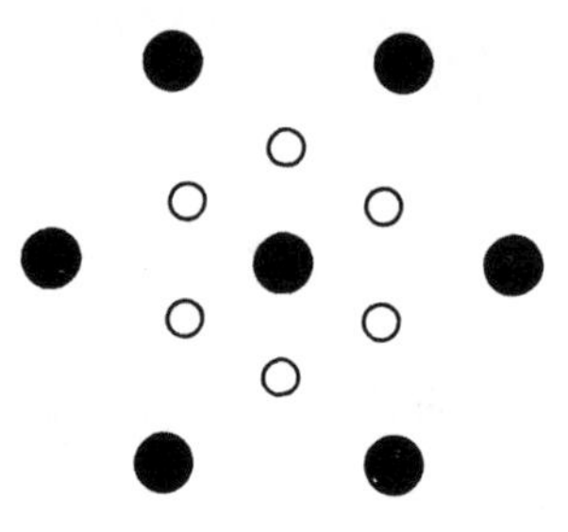

Fig. 32. Diagram of a cross-section of the A band, illustrating the relationship between thick and thin filaments. Closed circles represent myosin filaments; open circles, actin filaments. (After H.E. Huxley.)

2. Cross-bridges. These are thought to be projections of the myosin molecule that are capable of rotation and angular movements during contraction by virtue of their elastic components. They may serve as mechanical slots to keep the overlapping filaments apart during the sliding process or they may possibly act as force generators coupled to some stage in the cycle of ATP hydrolysis.

3. Reticular tubules. These are seen as a continuous network forming the background structure within each sarcomere. The tubules appear to surround the myofibrils and often have a characteristic triad formation consisting of a central tubule and terminal vesicles on each side. The network extends toward the surface of the muscle fiber and may even open into the membrane itself. For this reason it has been suggested that the tubules may form an electric pathway for transmission of the muscle action potential from the surface to the contractile elements of the interior.

The contractile process

Over the years a number of theories have been proposed to account for the contractile process and, although most of the basic principles have now been clarified, there are still many details to be established beyond all doubt. It is generally accepted that muscle contraction is the result of interaction between the myosin and actin filaments. The site of interaction appears to be in the cross-bridges between the filaments; the energy required for shortening and tension development is derived from the breakdown of

ATP. Recent evidence confirms the view that myosin acts as an enzyme for conversion of ATP into adenosine diphosphate and inorganic phosphate.

Sliding filament theory

Until direct observations were made with the electron-microscope, the contractile elements were thought to be continuous filaments that shortened by folding up inside the A band. It is now clear that striated muscle is organized into two overlapping sets of filaments and that the thick myosin filaments are confined to the A band. Huxley proposed that when a muscle is stretched, the two sets of filaments are pulled apart, increasing the width of the I band; when the muscle shortens, the two sets of filaments slide into one another, decreasing the width of the I band. In both of these circumstances the width of the A band remains unaltered (Fig. 33).

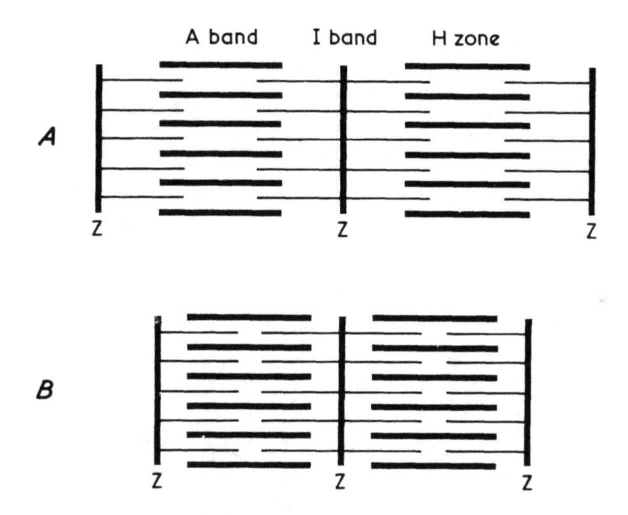

Fig. 33. Diagram illustrating the sliding filament theory. A, during stretch the two sets of dilaments are pulled apart by mechanical force; the width of the I band increases. B, during contraction the two sets of filaments slide into each other, decreasing the width of the I band. The sliding motion is the basis of the shortening that takes place in the distance between the Z lines of each sarcomere.

Molecular structure of myosin and actin

1. The Myosin filament has two components—light and heavy meromyosin. Light meromyosin consists of parallel strands of protein molecules. Heavy meromyosin projects from the sides of the filament to form cross-

bridges that link the myosin and actin molecules together. Each cross-bridge consists of a body and a head. The body may contain an elastic element that allows the cross-bridge to oscillate between the filaments; the head is hinged to the body in such a way that it can tilt or rotate away from its point of attachment ot the actin filament. Thus, the cross-bridges have flexibility–an important property in light of the proposal that shortening occurs by a series of ratchet-like movements or oscillations.

2. The actin filament has three components–actin, tropomyosin, and troponin. Actin molecules appear in the form of double strands coupled to molecules of ATP. These couplings are the active sites where hydrolysis of ATP takes place. Tropomyosin consists of protein strands which, together with troponin molecules, adhere closely to the active sites and normally prevent interaction between myosin and actin when the muscle is relaxed. Troponin has a high affinity for calcium ions.

Mechanism of the sliding process

The way in which the sliding process is brought about when muscle contracts is still somewhat speculative. It appears that the actin filaments are pulled inward by the cross-bridges, which utilize energy derived from the splitting of ATP. The precise mechanism of shortening and the nature of the chemical bonds between the filaments are still being investigated, but a knowledge of the molecular characteristics of the contractile elements has helped enormously toward a better understanding of the problem.

The sliding process can be considered to take place in two stages:

1. Actin filament reactions. Calcium ions released into the myofibril combine with troponin, which in turn causes aggregation of the tropomyosin strands. As a result, the active sites on the actin filament become exposed to the heavy meromyosin heads of the cross-bridges.

2. Myosin filament reactions. Interaction between the myosin head and the actin filament causes the head to rotate and in doing so pulls the actin filament with it. This action of the head is called the power stroke or force generator. In its new position, the head reacts with ATP, losing its bond with actin and thus becoming detached from it. The head now swings back to its original position, utilizing the energy from ATP, and the process is repeated as the head comes into contact with another active site (Fig. 34).

According to this concept of the sliding mechanism, the actin filaments are pulled into the A band by a force generated in the cross-bridges. It can be imagined that during activity, the myosin head oscillates back and forth pulling the actin filament a short distance in each cycle until the contraction is completed. Obviously, the force of contraction will be related to the number of cross-bridges in action at any one time.

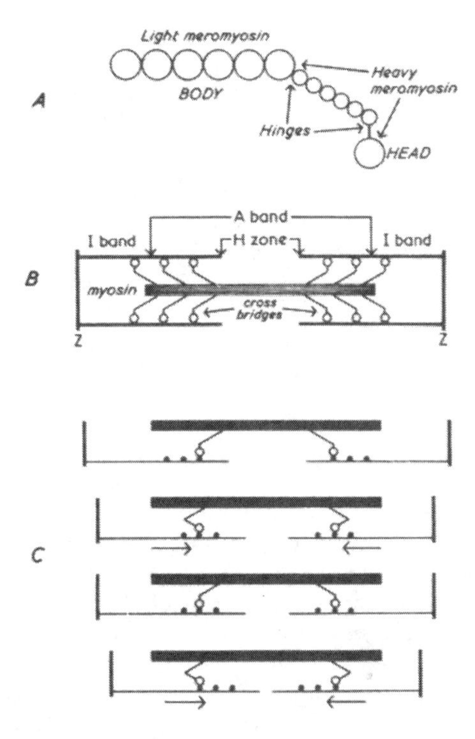

Fig. 34. Interaction between myosin and actin filaments illustrates a possible mechanism of contraction. A, myosin molecule shows structure of cross-bridge. The body is composed of light meromyosin, consisting of two peptide strands. The cross-bridge and head are composed of heavy meromyosin. The structure behaves as does an elastic element with double hinges. B, diagram shows cross-bridges of myosin filaments with heads in contact with adjacent actin filaments. C, proposed mechanism of shortening is illustrated. The arms of the cross-bridge extend toward the two ends of the actin filaments. The heads rotate on their hinges, thereby pulling the actin filaments toward the H zone of the A band. The heads then become detached and swivel back to the next active site on the actin filament when rotation begins again. (After A.F. Huxley.)

Sequence of events

The following may represent the sequence of events relating to the contraction-relaxation process:

1. The muscle action potential releases calcium ions from the reticular tubules.

2. Calcium ions combine with troponin, causing retraction of the tropomyosin molecules on the actin filaments. This allows the heads of the cross-bridges to bind with the active sites.

3. The heads of the heavy meromyosin molecules rotate, pulling the actin filament with them; this provides the force for the initial sliding process.
4. In its new position, each heavy meromyosin molecule combines with an ATP molecule, resulting in detachment of the head from an active site.
5. The energy released from hydrolysis of ATP restores the head back to its original position, where it can once again attach itself to another active site. The process is repeated until the actin filament is pulled toward the center of the A band, where the myosin and actin filaments overlap. Shortening of the muscle is completed when the Z line comes up to the ends of the myosin filaments.
6. Relaxation is brought about by the removal of calcium ions back into the reticular tubules and the return of tropomyosin to its resting state.

Summary

Skeletal muscle is normally excited by impulses in its motor nerve. Each nerve terminal lies in a groove on the surface of a muscle fiber. This is the motor end-plate. The terminal releases acetylcholine, a chemical transmitter that combines with receptor molecules at the end-plate, increasing the permeability of the membrane to all free ions. As the membrane is short-circuited, a local potential is set up. This initiates a new wave of excitation over the muscle membrane and into its interior. The end-plate potential is an essential intermediary between nerve impulse and muscle action potential. The electrical activity carried into the sarcoplasmic reticulum causes the release of calcium from the tubules, and this in turn starts the chain of chemical events that lead to muscle contraction.

Each muscle fiber consists of a mass of fibrils arranged into sarcomeres. The striations of a sarcomere are visible under the microscope as a system of refractive bands. The A band contains thick filaments of myosin overlapping thin filaments of actin, although the center has only myosin filaments. This central region is known as the H band. Actin filaments extend from the A band into the I band as far as the Z line. A sarcomere may be defined as the region between two adjacent Z lines.

When a muscle is excited or stretched, the two arrays of protein filaments slide over one another without altering their individual length. The energy for the sliding force is derived from the breakdown of ATP. It is possible that the heavy meromyosin component of the myosin molecule combines with active sites located on the actin filaments; the cross-bridges between the filaments act as force generators for the contractile process or for the development of tension. Relaxation occurs when this process is reversed by removal of calcium and resynthesis of ATP.

MECHANICAL PROPERTIES OF SKELETAL MUSCLE

Skeletal muscle has two main elements, contractile and elastic. The *contractile* element is responsible for shortening and the development of tension. When muscle shortens, the contraction is called "isotonic." If the muscle is fixed so that it cannot shorten, it develops tension between its attachments; this type of contraction is called "isometric." The *elastic* element is responsible for the compliance or stiffness of the muscle. It is found in series with the myofibrils and in the cross-bridges between the protein filaments; it is also present in the connective tissues and tendons. The series elastic element has an important effect on the mechanical properties of the whole muscle, for it smooths out any sudden contraction or rapid change of tension.

The physical characteristics of individual muscles in the body vary widely according to the functions they serve. For example, for the speed of movement which requires rapid contractions, the muscle fibers involved belong predominantly to the fast system and are called white muscle. On the other hand, stability for maintaining postures is provided by a tonic or slow system of fibers, frequently called red muscle because of the large amount of myoglobin in the sarcoplasm.

Recording methods

Many properties of skeletal muscle can be demonstrated on the frog's sartorius or gastrocnenius muscle when it is isolated from the body and kept in a moist condition.

The sartorious is identified by its narrow parallel fibers stretching from the pelvic girdle to the tibia. The muscle is placed on a bed of electrodes for stimulating and recording, the pelvic end is clamped and the free tibial end is attached to a lever or strain gauge, which can take up various loads. For isotonic recording the position of the lever is adjusted to register the resting length of the muscle before applying a load. The length of the muscle can be altered by using an electromagnetic stop or by tightening the micrometer screw on the strain gauge. For isometric recordings, the muscle is prevented from shortening by the stop or by using a stiff spring to increase the tension of the strain gauge (Fig. 35).

The gastrocnemius muscle lies on the back of the leg. A thread is tied round the Achilles tendon, which is then cut away from the calcaneum. The muscle is dissected free to the knee region. Next, the sciatic nerve is identified and followed upward to the vertebral column until it is clear of all adherent tissue. The preparation is placed in a muscle bath and a strong pin

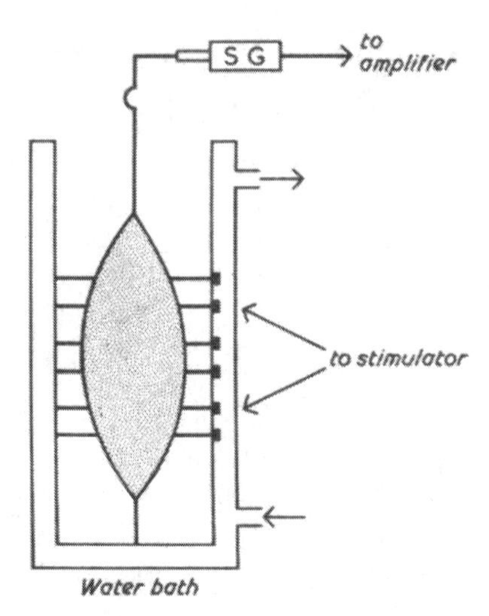

Fig. 35. Diagram of apparatus for recording isometric contractions. The muscle lies on an array of stimulating electrodes surrounded by a water bath for controlling temperature and bathing fluid. The pelvic end is clamped and the free end connected by a linen thread to a strain gauge (SG), which converts changes of tension into changes of voltage. The initial length of the muscle is adjusted by a micrometer screw attached to the strain gauge; various loads are added by stretched springs.

is inserted through the knee joint; the thread from the tendon can now be tied to the recording lever or to the hook on the strain gauge. The sciatic nerve is placed on a pair of stimulating electrodes and immersed in Ringer's solution or mineral oil to prevent drying (Fig. 36). The muscle may be stimulated either by a single shock or by a series of shocks.

Characteristic curves of muscle

The mechanical properties of skeletal muscle can be expressed in terms of four principal curves—active state, stress-strain, tension-length, and force-velocity.

Active state curve

In this condition the isometric tension is plotted against time. The response of the whole muscle to a single electrical stimulus applied to its motor nerve is called a "twitch." A rise of tension can be detected about 12 msec after the stimulus, this interval being called the latent period. Activity reaches full intensity after about 20 msec, the duration of the contraction

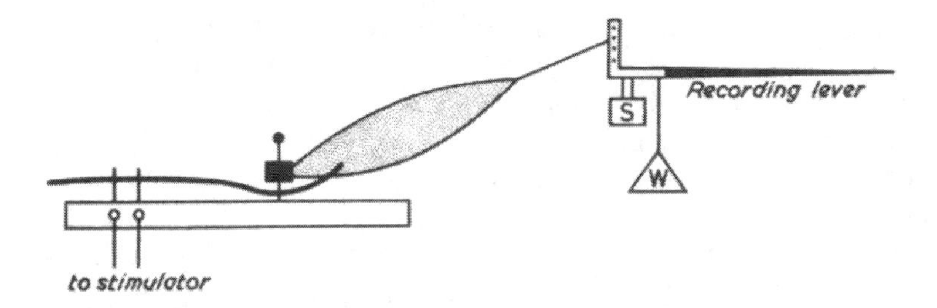

Fig. 36. Diagram of apparatus for recording isotonic contractions. Gastrocnemius-sciatic preparation. The nerve lies on a pair of stimulating electrodes in a Ringer bath. The muscle is fixed by a pin through the knee joint. The tendon is attached by a thread to a light lever carrying the writing point. The load taken by the muscle is adjusted by the stop (S), which also determines the initial length of the muscle. Forces are applied by weights (W) in the scale pan.

time when peak tension is recorded. The falling phase of the active state curve corresponds to the relaxation time of the muscle. In an isotonic twitch, when the muscle is allowed to shorten and pull on a lever, the time course of the mechanical response is much longer, owing to inertia of the recording system. As can be seen from Figure 37, the action potential is over before the contraction begins.

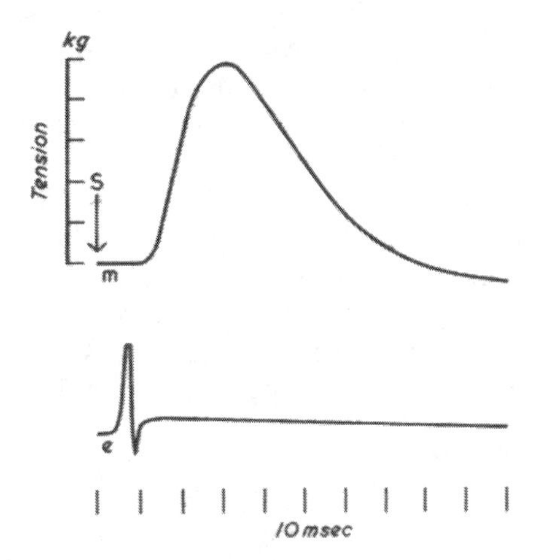

Fig. 37. Simultaneous recordings of mechanical (m) and electrical (e) responses of gastrocnemius muscle elicited by a single shock (S) to the sciatic nerve. The active state curve or twitch has a brief contraction phase followed by a longer relaxation phase. The muscle action potential precedes the onset of the mechanical effect.

Stimulus strength. A minimal degree of stimulus strength is required to produce a visible response; this level of strength is called the threshold stimulus. When the strength is increased progressively above threshold, the amplitude of the muscle curve is increased as more and more muscle fibers are excited. A maximal stimulus is one in which all the muscle fibers are excited simultaneously and the curve represents a maximal contraction. Increasing the strength of stimulus beyond the maximal level cannot increase the size of the twitch. However, the active state curve can be readily influenced by changes in temperature or by adding drugs to the muscle bath. For example, all the time relationships are increased by a rise in temperature; the falling phase is delayed by a decrease in temperature and by drugs such as adrenaline.

Summation. This means the adding together of individual muscle contractions. When two stimuli are delivered to the motor nerve, the muscle responds by giving two twitches. If the second stimulus is delivered before the first twitch is completed, summation of the two responses occurs during either the contraction or relaxation phases. This effect is illustrated in Figure 38, which shows the interval between two maximal stimuli being progressively reduced. It will be seen that the total response is much greater than that of a single twitch, except at very short intervals when the second stimulus falls during the latent period–i.e., when the muscle is in a refractory state.

Tetanus. This term describes the mechanical fusion of individual muscle contractions that occurs when the motor nerve is stimulated repetitively at various frequencies. A partial tetanus occurs with stimulus frequencies between 5/sec and 25/sec, and a complete tetanus occurs with higher stimulus frequencies (Fig. 39). Mechanical fusion is the result of the viscous property of sarcoplasm, which maintains the active state between the contractions. A plateau level of activity is reached at a critical fusion frequency when the individual contractions cannot be distinguished and the muscle gives a steady pull on its tendon. In contrast, the muscle action potentials accompanying such contractions always remain discrete, corresponding to the arrival of each nerve impulse. Under isometric conditions, the tension developed by a full tetanus may be about four times that of a single twitch.

Stress-strain curve

The contribution made by the series elastic elements to the tension exerted by the whole muscle may be expressed by a family of curves. A load is attached to the muscle, which is then stimulated isometrically, thus stretching the elastic component. On release of the electromagnetic stop, the elastic component shortens. A plot is constructed by measuring the amount of shortening produced with different loads. By repeating the experiment and

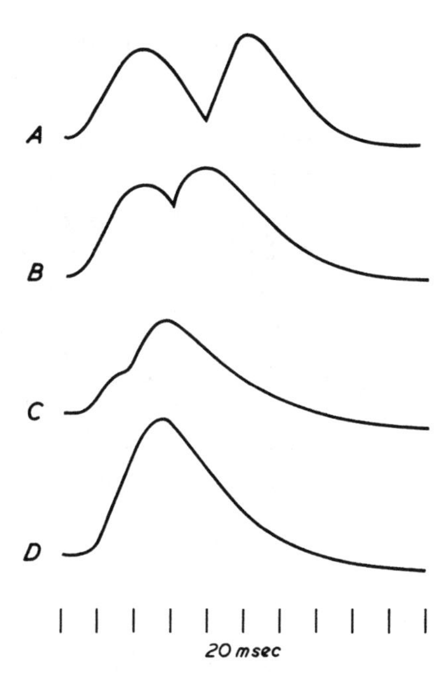

Fig. 38. Summation of muscular contractions. Paired maximal stimuli are delivered in succession to the motor nerve. From above downward, intervals between stimuli are 80 ms, 60 ms, 40 ms, 20 ms. A, two separate contractions; B, the second contraction is superimposed on the falling phase of the first; C, the second contraction appears during the rising phase of the first; D, complete summation of the two contractions. Time is recorded in 20 msec intervals.

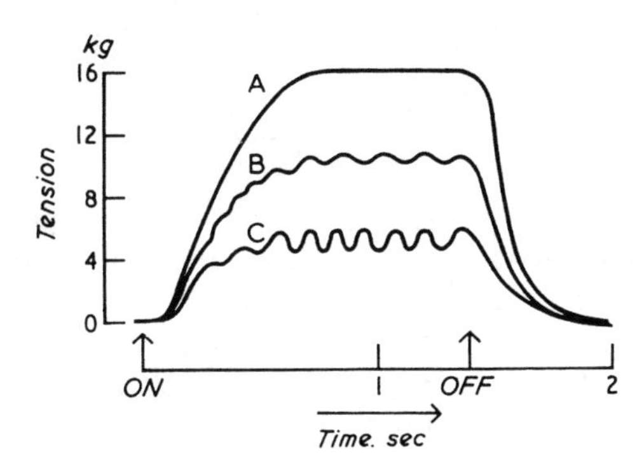

Fig. 39. Effect of repetitive stimulation of sciatic nerve on tension developed in gastrocnemius muscle. Rise of tension increases with increased frequency of stimulation. A, full tetanus; B, incomplete tetanus; C, partial tetanus in which the individual contractions can be seen. The period of stimulation is indicated by arrows.

releasing the stop at various times during the muscle twitch, a family of curves is obtained, as shown in Figure 40. It will be evident that all the curves from the same muscle are approximately the same, indicating that the series elastic component has constant properties during both contraction and relaxation phases. The series elastic component is not influenced by the stimulus as is the contractile component, but passively follows the changes in length that occur during muscle activity.

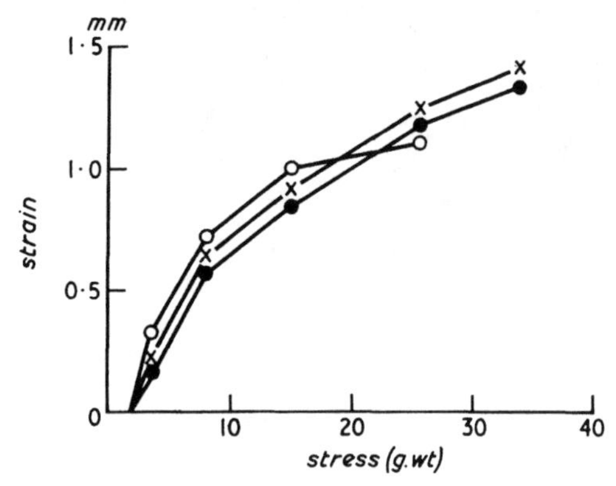

Fig. 40. Stress-strain curves of the series elastic component at various times after a single stimulus. The muscle is prevented from shortening by the stop and therefore develops isometric tension. The stop is withdrawn at the times indicated in the figure, and the muscle is allowed to shorten. The curves show a rapid movement, caused by the elastic component, followed by a slower phase, caused by the contractile component. Ordinate, muscle length (mm); abscissa, load (g.wt.). Times of release: X, 200 ms; •, 280 ms; 0, 480 ms. (After D.R. Wilkie, 1956.)

Tension-length curve

This curve is plotted by measuring the tension developed by a tetanized muscle set at different lengths before stimulation. The curve shows that the greatest tension is developed when the muscle is set at its resting length or the length it normally has in the body (Fig. 41). The relationship between tension developed and initial length is roughly linear—i.e., for each successive increase of length the muscle develops more and more tension. If, however, the muscle is stretched too much the tension falls off, and stretching the muscle beyond a third of its resting length may cause physical damage.

The tension-length curve of a sarcomere has been obtained by clamping the central region of a single muscle fiber and holding it at various lengths during stimulation. Huxley showed that the greatest tension was developed

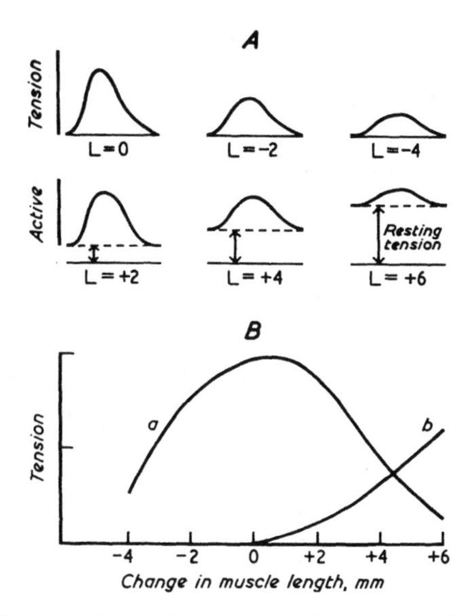

Fig. 41. Tension-length curves. Isometric tension developed by muscle following supramaximal stimulation.
A, upper traces are recordings, respectively, at resting length (L = O) and when the muscle is shortened by 2 mm and 4 mm before stimulation; lower traces show active and resting tensions developed when the muscle is stretched.
B, plot of total tension (a) exerted by stimulated muscle at various initial lengths. Note that the greatest tension is developed when the muscle is near the length at which it operates in the body. The curve on the right (b) represents resting tension recorded when the muscle is stretched. The difference between the two curves represents the active tension.

when the thick and thin filaments overlapped; tension decreased as the overlap diminished and fell to zero when there was no overlap. These findings are consistent with the sliding filament theory, thus supporting the view that maximal force exerted by the fiber is proportional to the number of cross-bridges between the filaments.

The tension exerted by cardiac muscle can be explained in much the same way, the force of contraction increasing with the initial length of the fibers, which in turn depends upon the venous return and intra-cardiac volume. Thus the development of tension seems to be a function of muscle length, a relationship that must be regarded as a fundamental property of muscle contraction.

Force-velocity curve

This curve relates the work done by a muscle to the speed of contraction. It shows that the amount of work done in raising a load (P/Po) increases as the velocity is reduced. This is true up to a certain limit when the work falls off (Fig. 42). Conversely, when there is no load on the muscle the velocity reaches its highest value. The relationship between speed of contraction and loading is more or less the same for most types of skeletal muscle, and in the human being it is a common experience that lighter objects can be moved more rapidly than heavier ones. The shape of the curve, however, varies to some extent with temperature since energy is liberated when the muscle shortens. With rapid contractions, a high proportion of the energy released is used to overcome the internal viscosity of the muscle, while much less energy is dissipated with slower contractions. Furthermore, as chemical reactions take time, less chemical energy is available for rapid rates of contraction and fatigue sets in relatively early.

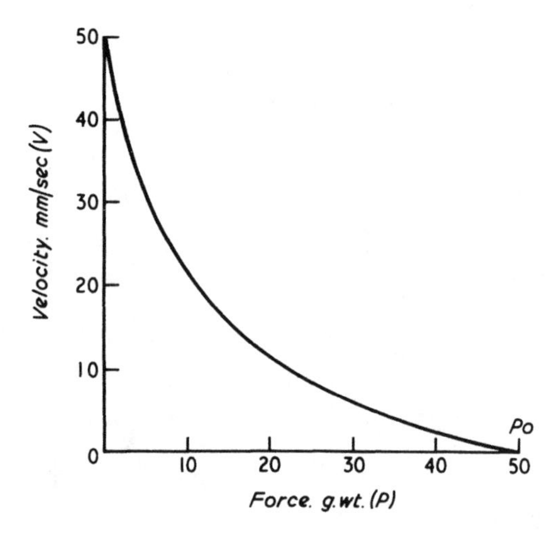

Fig. 42. Force-velocity curve. Speed of shortening is plotted against load. Velocity (V) decreases as load is increased. V = O when $P = P_0$, the maximum force that the muscle can hold. If forces greater than P_0 are applied, the muscle lengthens.

Each of these curves represents the response of the isolated muscle preparation under a given set of conditions. They do not take into account how the muscle will react to the various factors that can influence muscle contraction in the intact body. Taken together, however, the curves appear to express the range and limitations of the contractile mechanism as well as indicate the optimal conditions for developing mechanical power.

FUNCTIONAL PROPERTIES OF SKELETAL MUSCLE

The motor unit

This is the functional unit of the locomotor system. It may be defined as consisting of a single motor cell, its axon and terminal branches, and all the individual muscle fibers supplied by them. The actual number of muscle fibers in a particular motor unit depends upon the type of muscle and may be calculated from the innervation ratio or the total number of muscle fibers divided by the total number of motor axons. In muscles involved in delicate movements like those of the eye, the innervation ration is only 1:4. But in the large postural muscles, which do not require a fine degree of control, the innervation ratio is 1:150. Some motor units are more slowly contracting than others because of differences in the motor axon size, which in turn is related to conduction velocity.

How muscle contractions are graded

When a single axon is excited, all its related muscle fibers contract in an all-or-nothing manner. Hence the weakest possible contraction that can be elicited is the twitch of one motor unit. When more force is required, e.g., to produce a movement, more and more motor units are brought into action and each unit discharges more frequently. With stronger grades of contraction, the motor unit twitches summate to give a partial tetanus and ultimately a full tetanus. Finally, the most powerful contractions are brought about by greater synchronization of the individual motor units, resulting in more units reaching the peak of their activity at the same moment of time. It will be apparent from the above that muscular activity can be finely graded from the most delicate to the most powerful contractions and that the behavior of the motor units really depends upon the patterns of impulse discharge in the motoneurons. These may be analyzed as follows:

1. Number. The larger the number of motor units in action, the greater the force of contraction. The nervous system makes this possible by recruitment of cells in the motoneuron pools, maximal recruitment occurring when all the motoneurons supplying a muscle are discharging. The total pull exerted by the muscle is simply the sum of the tensions generated by each motor unit.

2. Rate. Each motoneuron can vary its rate of discharge from very low levels to very high levels. At low rates, some degree of summation takes place but the contractions remain sub-tetanic. At higher rates, the muscle response enters the tetanic range. The more rapid the rate of discharge, the greater is the tension exerted by the muscle.

3. Phase. At any moment of time some motoneurons may be discharging while others in the pool are resting. If the majority of motoneurons are discharging together, the pull on the muscle will be so much greater; the motoneurons are said to be "in phase" or synchronous. The advantage of synchronous activity is seen in such movements as lifting a heavy weight or striking a hard blow with the fist. Obviously, such forces are too strong for normal muscular efforts and they cannot be maintained for long periods without fatigue. Relatively smooth performances are achieved by low discharge rates of motoneurons firing asynchronously or out of phase, so that when one group of motor units is contracting another group is relaxing.

Mechanisms of control

The character of impulse discharge in the motor nerves is regulated by local mechanisms acting on the motoneuron pools. Activity can be modified in several ways:

1. Control by the muscle itself. A large number of muscle spindles lie in parallel position to the main muscle fibers. The spindles are stimulated when the muscle is stretched. Afferents from the spindles enter the spinal cord through the dorsal roots and make synaptic contact with the motoneurons supplying the muscle. Thus any stretch applied to the muscle results in a reflex contraction that tends to resist the stretch and to maintain the muscle at a constant length. This mechanism is called the stretch reflex. It is a self-regulating closed-loop, relying upon sensory feedback from the muscle and it accounts for the silent period of the electromyogram when the muscle is made to contract.

2. Control by the gamma efferents. Each muscle spindle contains a number of intrafusal muscle fibers innervated by gamma motor efferents. The sensory portion of the spindle lies between the contractile poles of the intrafusal fibers. When the latter contract, shortening of the poles gives rise to a spindle discharge, which in turn operates the stretch reflex. The muscle will thus maintain its contraction until the spindles cease to discharge. Impulses from higher centers in the brain modify the discharge rate of the gamma efferents and thereby help to grade the contraction of the muscle according to the task required of it.

3. Control by Renshaw cells. Branches from motor nerve axons run back into the anterior horn of the spinal cord to excite internuncial Renshaw cells, which in turn inhibit the motoneurons. This negative feedback circuit stabilizes the output discharge of the motoneurons supplying a muscle and its synergists.

Muscular effort and fatigue

Muscular effort is a good example of physiological integration whereby several systems of the body contribute toward a common end. During moderate exercise, such as the running of short distances, the muscles take up more oxygen, use more energy, and produce more heat. Consequently, there are greater demands on the circulation to increase the blood flow to the active muscles and to remove heat and waste products more rapidly. The respiratory system also plays its part to keep pace with the extra oxygen requirements. As the speed of running increases, the capacity of the individual is probably determined by the state of the muscles themselves, the limiting factors being the speed of contraction and the force of contraction. In sprint running, speed is more important than force since there may be little time for bodily adjustments, while the metabolic products of the exercise can be dealt with afterward during the recovery period. On the other hand, in short-term exercise, e.g. weight-lifting, performance is determined mainly by the force of contraction of the active muscles.

Whatever the nature of the muscular effort, prolonged and powerful contractions strain the reserves of the muscles and lead to fatigue. In this condition the muscles become too weak to make the necessary movements or develop tremors, which impair the movements. When fatigue approaches the limits of endurance, the muscles fail to give a twitch when motor impulses reach them. There is no failure of the nervous mechanism and the muscle action potentials are generated as before. It is therefore likely that the site of fatigue is in the contractile elements, which become weaker owing to depletion of their chemical reserves. This can be demonstrated by interrupting the blood flow through the contracting muscles, by means of an inflatable cuff placed around the arm, when fatigue sets in after a few minutes and no recovery is observed until the cuff is removed and the blood flow is restored.

ELECTRICAL EXCITABILITY OF SKELETAL MUSCLE

The introduction of electro-diagnostic techniques has proved to be of immense value in the assessment of muscular abnormalities in patients. The techniques most frequently used involve measurements of excitability thresholds and oscillographic studies of muscle action potentials.

Electrical stimulation

The simplest electrical method of testing the excitability of a muscle is to measure the minimum strength of stimulus required to evoke a response and to compare this strength with that required to excite the muscle on the opposite side.

1. Responses to Galvanic stimulation. A muscle stimulated with a constant current responds by a brief twitch when the current is switched on and a second twitch when the current is switched off. This is known as a "make or break" stimulus, the former starting from the cathode and the latter from the anode. Galvanic stimulation is effective only when the strength of the stimulus reaches threshold. The smallest voltage required to produce a contraction is termed the *rheobase*. If the duration of the current is reduced, a larger voltage must be used (cf. Chapter II). The time required for a stimulus twice the strength of the rheobase to reach threshold for a given muscle is termed the *chronaxie* of that muscle. Chronaxie is therefor a measure of excitability and is often greatly increased when the muscle fibers are denervated or affected by disease.

2. Responses to Faradic stimulation. In this form the stimulating current is of short duration and must therefore be of sufficient intensity to cause contraction. Since muscle fibers have a much higher threshold than nerve fibers, a brief Faradic stimulus may be ineffective after denervation, whereas the longer duration of a Galvanic stimulus can produce a response. This provides a simple test for denervated muscle and is known as the reaction of degeneration—loss of response to Faradic currents and persistence of response to Galvanic stimulation. Plotting strength-duration curves can be usedful in following the progress of recovery after reinnervation when the normal pattern of the curve is gradually restored.

Electromyography

The electrical activity of skeletal muscle may be recorded by means of surface electrodes or with needle electrodes inserted into the muscle under investigation. The activity may be induced by voluntary effort or evoked by stimulating the motor nerve.

1. Surface electrodes. These are usually silver discs filled with electrode jelly and strapped to the skin overlying the muscle. Figure 43 shows the normal motor unit discharge on bending or extending the elbow. One pair of electrodes lies over the biceps and another pair over the triceps. In each record, the upper trace shows the response of the contracting muscle, and the lower trace the response of the antagonistic muscle. Activity is increased when the movements at the elbow are carried out against resistance.

2. Needle electrodes. In clinical electromyography a concentric or coaxial electrode is used. This is a bevelled needle of stainless steel in which a platinum wire runs down the inside and is insulated from the shaft except

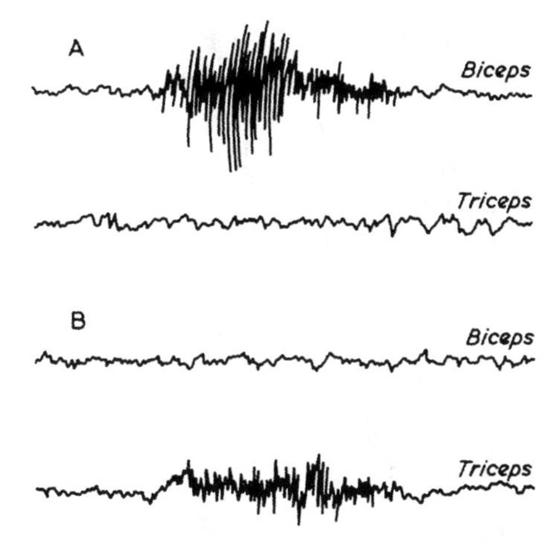

Fig. 43. Electromyograms recorded with surface electrodes from human subject.
A, upper pair of traces shows response of biceps when the elbow is bent against resistance; triceps electrically silent.
B, lower pair of traces shows response of triceps when the elbow is extended against resistance; biceps electrically silent. Paper speed 10 mm/sec.

at the tip. Potential differences between the inner core and the outer sheath are amplified, monitored on a speaker, and displayed on an oscilloscope. By this method it is possible to record the activity of only a few motor units or even a single motor unit in contrast to the large number of potentials picked up by surface electrodes. The action potential of the motor unit has certain characteristics which help to detect deviations from normal muscle patterns. These are related to the waveform, amplitude and duration of the potentials. Furthermore, the sound produced during voluntary contraction is well sustained.

Normal skeletal muscle is electrically silent when completely relaxed. This is evident from the top trace in Figure 44. The lower trace shows a powerful motor unit discharge when the subject is making a maximal voluntary effort. The units fire at a rate of 50 to 60 per second and are usually diphasic or triphasic. In abnormal conditions, the size and number of the muscle action potentials are often reduced and the discharge is poorly maintained during voluntary contraction. When the motoneurons themselves are affected by disease, surviving units are sometimes of greater amplitude than normal and thus are easily recognized.

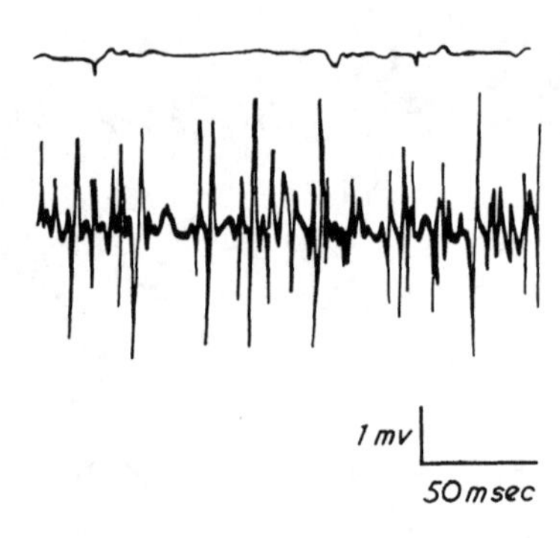

Fig. 44. Motor unit potentials recorded through a concentric needle electrode. The upper trace shows the absence of potentials in relaxed muscle. In the lower trace, motor unit activity on maximal voluntary effort is recorded.

SELECTED BIBLIOGRAPHY

Alexander, R.S., and Johnson, P.D., Jr. Muscle stretch and theories of contraction. *Amer. J. Physiol.* 208:412-416, 1965.

Brown, G.L., Dale, H.H., and Feldberg, W. Reactions of the normal mammalian muscle to acetylcholine and to eserine. *J. Physiol.* 87:394-424, 1936.

Buller, A.J., and Lewis, D.M. The rate of tension development in isometric tetanic contraction of mammalian fast and slow skeletal muscle. *J. Physiol.* 176:337-354, 1965.

Carlson, F.D., and Wilkie, D.R. *Muscle Physiology.* Englewood Cliffs, N.J.: Prentice-Hall, 1974.

Castillo, J. del, and Katz, B. Changes in endplate activity produced by presynaptic polarization. *J. Physiol.* 124:586-604, 1954.

Gage, P.W., and Quastel, D.M.J. Competition between sodium and calcium ions in transmitter release at a mammalian neuromuscular junction. *J. Physiol.* 185:95-123, 1966.

Hubbard, J.I., and Willis, W.D. The effects of depolarization of motor nerve terminals upon the release of transmitter by nerve impulses. *J. Physiol.* 194:381-407, 1968.

Huxley, A.F. Muscular contraction. *J. Physiol.* 243:1-43, 1974.

Katz, B., and Miledi, R. A study of spontaneous miniature potentials in spinal motoneurons. *J. Physiol.* 168:389-422, 1963.

PART II

The Sense Organs

CHAPTER V

Principles of Sensory Activity

An army commander without an intelligence service is virtually crippled in carrying out his duties, however versatile his troops. The same is true of the brain. A knowledge of what is going on inside and outside the body is absolutely imperative if the brain is to perform anything useful. Such knowledge is provided by the receptors or sense organs, of which there are a vast number. Always on the alert, they may be called into immediate action by a specific event. Information is transmitted by the receptors by means of nerve impulses which are in the form of patterns or signals. (How these are processed to cause subsequent effects will be the subject of later chapters.) The purpose of this chapter is to describe the various types of receptor and to discuss the basic mechanisms by which their discharges are conveyed to the central nervous system. Essentially, a receptor is a modified nerve ending capable of detecting a change in the environment and of converting that change into a generator potential and impulse activity.

Work on receptor functions goes back to the nineteenth century when Johannes Müller proposed that sensory activity was due to the excitation of receptors by a specific form of energy change. Sherrington suggested a classification of the sense organs based on the kind of information they transmit, while Head introduced clinical terms to describe disturbances of sensory function in man. The use of an amplifier by Adrian in 1926 made it possible to record action potentials in single nerve fibers. In conjunction with Zotterman, Matthews, Bronk, and other pioneers, many of the properties of sensory receptors were disclosed, including the all-or-nothing concept and the phenomenon of adaptation. It was shown that the nature of the nerve impulse was the same no matter what kind of receptor discharged it. The next step forward came with the discovery of the receptor potential by Katz in 1950. Recording from the frog muscle spindle, he observed a slow wave preceding the action potential when the muscle was stretched and concluded that this potential represented a link between the mechanical stimulus and the propagated impulse discharge. The conversion of a mechanical stimulus into an electrical potential suggested that the receptor performed as a transducer, and it was soon apparent from the work of Grundfest and others that transducer action depended upon depolarization of the nerve ending by mechanisms already described. It was the local flow of current caused by the receptor potential that in turn generated the action potential in the nerve fiber.

Most of the general features of sensory activity have now been fully documented. The action potentials in each nerve fiber carry information from the sense organs in the form of a code or series of pulses. A peripheral nerve contains many groups of fibers in which the patterns of activity are both spatially and temporally organized. Consequently, the total input signal is enormously variable and the patterns can be modified by stimulation in a number of ways:

1. The speed of conduction is variable according to the size of the fiber and the degree of myelination.

2. The frequency of impulse discharge is related to the intensity of the stimulus. A strong stimulus increases frequency and also excites higher threshold fibers that may not respond to a weaker stimulus. As a result, more and more receptors are recruited.

3. The duration of the signal also depends upon stimulus intensity; it can be brief with a weak stimulus and persistent with a strong stimulus.

4. During prolonged stimulation of a receptor the frequency of discharge may decline to a lower rate. This is known as sensory adaptation. Some receptors are slowly adapting, others rapidly adapting, so that the complete picture of the information conveyed may be somewhat complex.

The function of the central nervous system is to process the information it receives. The system can be used advantageously in two possible ways—either by utilizing simple reflex action or by alerting the mechanisms of consciousness. The vast majority of sensory impulses, especially those arising from the internal structures of the body, are normally concerned with regulatory functions that do not enter into consciousness. With intense stimulation, however, and under pathological conditions, such impulses may evoke a sensation of pain. On the other hand, impulses arising from the body surface and those elaborated by the special sense organs are the principal channels for the various forms of sensation that are part of our feelings and impressions of the outer world.

CLASSIFICATION OF THE SENSE ORGANS

Sherrington introduced a classification of the sense organs based on their development from the three principal layers of the embryo.

1. Exteroceptors. These are developed from the primitive ectoderm. They

may be divided into the cutaneous or general senses and the special senses for sight, hearing, taste, and smell.

2. Proprioceptors. These are developed from the primitive mesoderm. They are found in the muscles, tendons, and joints and also include the receptors of the labyrinth. They are known clincially as the deep senses and give information about movements and posture.

3. Visceroceptors. These are developed from the primitive entoderm. They are located in the internal organs including the heart, lungs, and abdominal viscera, and in the areas around the blood vessels and glands. Their function is to regulate the internal environment but they may also be responsible for visceral pain.

Types of receptor

There are many different types of sensory receptor each of which is adapted for detecting a specific energy change. The commonest type is the one that responds best to mechanical changes; some types are sensitive to chemical or thermal changes, while others are more highly selective and respond only to a particular form of stimulation. Despite the wide range and complexity of the receptors found in the body, their basic structure is essentially the same, namely, a central core of modified nerve terminals enclosed in a nucleated sheath or fibrous capsule. The simplest kind of receptor is a network of free nerve endings with no other structures associated with them. At the extreme opposite, the special senses have accessory structures in which the receptor forms only a small part of the complete organ.

Cutaneous Senses

The skin is richly endowed with various forms of sensory receptors whose function is to transmit information about the external environment and to detect any changes that might be harmful to the individual. They may be classified as mechanoreceptors, thermoreceptors, nociceptors, and pain terminals.

Mechanoreceptors

All these receptors are excited by non-injurious mechanical disturbances of the skin.

Lamellated receptors. The best known are Meissner's corpuscle and the Pacinian corpuscle (Fig. 45). They are similar in design. Each corpuscle consists of a concentric sheath of thin lamellar cells arranged around the axon terminal; the sensory nerve fiber loses its myelin sheath before entering the capsule. Meissner's corpuscle is a rapidly adapting receptor particularly abundant in the fingertips and other areas of the skin where the sensation of touch is highly developed; it is excited by deformation of the epidermis when the skin is lightly stroked. The Pacinian corpuscle is also rapidly adapting but lies deep in the subcutaneous tissue; the many layers of its capsule are sensitive to compression from outside. The corpuscle responds to touch, pressure, and vibration, depending upon how the compression is applied.

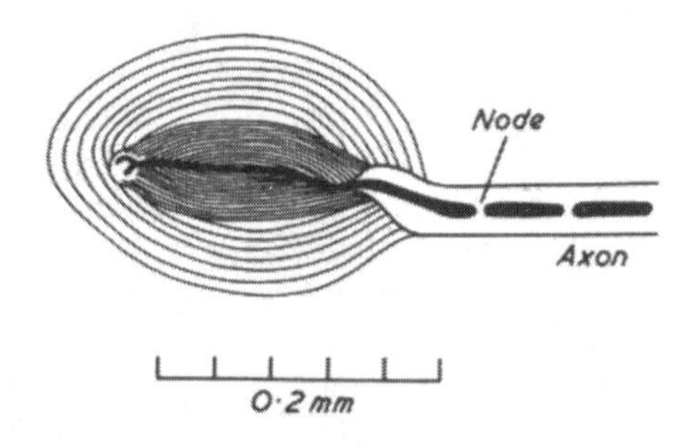

Fig. 45. Diagram of Pacinian corpuscle. The axon loses its myelin sheath before entering the central core of the corpuscle where it is thin and straight. The first node of Ranvier lies outside the capsule. The laminated connective tissue surrounding the core is sensitive to mechanical compression.

2. Hair follicle. This is a tubular depression of the skin lined by epithelial cells from which the root of the hair grows. The fine nerve endings surrounding the base of the follicle are sensitive to mechanical distortion caused by displacement of the hair. Two types are generally described, one being mainly a velocity detector and the other responding to steady-state conditions. The vibrissae of animals are highly specialized tactile receptors for detecting rapid changes of position.

3. C mechanoreceptor. Responses produced by light contact with the skin are believed to arise from mechanoreceptors innervated by unmyelinated C fibers. Although they fatigue easily, they respond best to long contact or sudden withdrawal of contact.

4. Merkel's tactile cells. These are specialized cells containing secretory granules and cytoplasmic protrusions forming a neurite complex in the epidermis. When associated with a hair follicle, the Merkel's tactile cell may

modulate the activity of the sensory nerve fiber, producing slow adaptation to displacement of the hair. The function of the secretory granules is uncertain.

5. Ruffini's corpuscle. At one time Ruffini's corpuscle was identified with warm receptors, but it is now classified as a mechanoreceptor responding to continuous stretching of the skin. It is found in the dermis and deeper tissues and is slowly adapting.

6. Krause corpuscle. This is another variety of mechanoreceptor at one time identified with cold receptors but structurally similar to Meissner's corpuscle. It responds to rapid deformations of the skin such as those set up by low frequency vibrations.

Thermoreceptors

Cold and warm spots have been described in human skin. They may be detected as distinct points from which the corresponding sensations are aroused, although histological identification of specific receptors has proved very difficult. Cold spots are found in the superficial layers of the skin. They may be stimulated by local cooling of the skin surface, by intravenous injection of cold solutions, or by chemical action of substances like menthol. They are not excited by moderate mechanical stimuli. The sensory nerve endings belong to the A delta myelinated group of fibers with conduction velocities less than 20 m/sec. Warm spots are not localized with the same degree of precision. They may be stimulated by local heating or by chemical actions. The sensory nerve endings belong mainly to the unmyelinated C group of fibers with conduction velocities less than 1 m/sec.

Nociceptors

It is doubtful whether there are specific receptors that respond to noxious stimuli only. Some mechanical and thermal receptors may have higher thresholds than others so that they respond when the stimulus becomes excessive. It is no doubt important for survival that appropriate warning be given of events that threaten damage to the skin. Two types of nociceptor have been described:

1. Mechanical nociceptors are excited by sharp objects but are rapidly adapting and fatigue easily.

2. Thermal nociceptors are excited by extremes of cold or heat. They are slowly adapting and give persistent discharges which can also be painful.

Pain terminals

Free nerve endings, consisting of a fine network of unmyelinated axons, are found in the superficial layers of the skin, interweaving between the cells and serving overlapping fields. Free nerve endings are also found in the cornea, which is known to respond to touch, warmth, and cold in addition to pain. For this reason it is thought that the pain terminals are not specialized to react to a single form of energy, but to virtually any form of stimulation when this is sufficiently intense. It was noted above that extremes of heat and cold can produce pain and so can a variety of mechanical and chemical stimuli. The wide range of sensory nerve fibers carrying impulses from the skin is no doubt associated with different qualities of pain. Thus, a sharp stabbing pain results from stimulation of the A delta group of fibers, whereas a dull ache or burning pain and tickling sensation are produced by stimulation of C fibers. All are slowly adapting.

Proprioceptors

Deep sensation arises from receptors in muscles, tendons, ligaments, and joints. Their discharges are important for the reflex control of muscular contraction, the maintenance of posture and balance, and for the coordination of voluntary movement. All these activities occur for the most part on a subconscious level. In addition, proprioceptors transmit to consciousness information on the position of the limbs and joints, on passive movements, and on vibration, deep pressure, and pain.

Muscle spindles

With few exceptions all skeletal muscles contain a variable number of muscle spindles situated in parallel position with the long axis of the muscle fibers. Many spindles, found in the smaller muscles of the body, serve discrete movements, although such spindles are equally conspicuous in the large postural muscles. The essential feature of the muscle spindle is that it has both muscular and nervous components. Each spindle consists of a fusiform capsule surrounding a bundle of about 2 to 12 intrafusal muscle fibers supplied by gamma efferent nerves. Leaving the capsule are the afferent fibers of the primary and secondary nerve endings.

The intrafusal muscle fibers are of two types designated nuclear bag and nuclear chain (Fig. 46).

1. Nuclear bag fiber. This is a very small skeletal muscle fiber with a

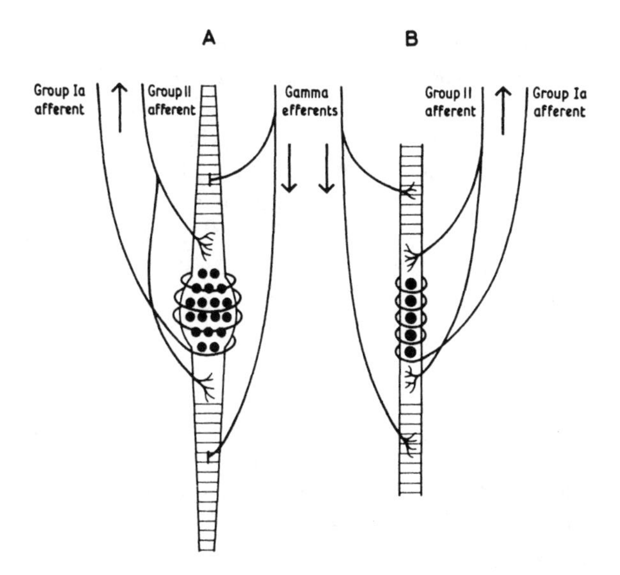

Fig. 46. Intrafusal muscle fibers of muscle spindle. Simplified diagram shows the two types of intrafusal fiber and their innervation. In A, nuclear bag, note that the central region consists of a cluster of nuclei with contractile material extending to the poles. Gamma efferent motor fibers terminate in discrete end-plates. Sensory axons forming spiral and flower spray endings emerge from the spindle; they belong, respectively, to Group Ia and Group II afferents. In the nuclear chain, B, the central region has a single chain of nuclei with contractile material on either side. Motor fibers terminate as trail endings. Sensory innervation is mainly by flower sprays.

central nucleated region called the equator and two contractile poles. The central region consists of a cluster of nuclei arranged in the form of a bag and a non-contractile portion on either side of it from which the sensory nerve endings arise. The contractile portion of the fiber extends the whole length of the spindle and may even project beyond it.

2. Nuclear chain fiber. This is shorter and thinner than the nuclear bag fiber and it does not project beyond the poles of the spindle. The nuclei in the equatorial region are arranged in the form of a chain, leaving the diameter of the fiber unchanged throughout its length.

The nerves to the spindle are both sensory and motor. The *sensory* nerves are large myelinated afferent fibers arising from two types of nerve endings:

- The annulospiral or primary ending is composed of a large number of bulbs connected together by fine cylindrical tubes. It spirals around the equatorial region in the form of incomplete rings and supplies both types of intrafusal fiber. The axons emerging from the spindle belong to group Ia in the classification of muscle afferents introduced by Lloyd.
- The flower spray or secondary ending takes the form of spirals or fine

sprays on each side of the equator; it is found predominantly on the chain fibers. The axons emerging from the spindle belong to group II muscle afferents, which are smaller in diameter than group I.

The *motor* nerves supplying the intrafusal fibers are derived from the gamma efferents and appear to have two kinds of endings–end-plates and trails. A single motor axon can terminate in either kind of ending. End-plates are usually found on nuclear bag fibers and trail endings on nuclear chains, although this arrangement does not apply to all spindles.

Golgi tendon organs

First described by Golgi, these receptors usually are found between a muscle and its tendon, in the intermuscular connective tissue and capsules of joints, and less often in the tendon itself. Golgi tendon organs have a relatively simple structure, consisting of fine, unmyelinated sprays enclosed in a delicate membrane. The afferent nerve fibers all belong to group Ib of Lloyd's classification. Since tendon organs lie in series with the main mass of muscle, they are sensitive to any sudden change in muscle length.

Joint receptors

The connective tissues forming ligaments and joint capsules are supplied with numerous receptors occurring as free nerve endings or as encapsulated structures. One type of ending, named after Ruffini, consists of a medium-sized nerve axon which breaks up into several spray terminals. Another type looks like a modified Pacinian corpuscle.

Visceral receptors

The viscera and their attachments are sensitive to a variety of stimuli, including pain-producing agents. A number of specialized receptors have been described, but many appear to have no definite structure and so cannot be identified histologically.

Chemical receptors

1. Gastric receptors sensitive to changes in pH are distributed to all parts of the stomach. They are supplied by the terminals of unmyelinated vagal afferents.

2. Cardiovascular chemoreceptors are present in the carotid and aortic bodies, which possess a rich innervation surrounding a network of sinusoidal blood vessels. They are sensitive to chemical changes in the blood, such as anoxia, excess Co_2, and acidemia.

Mechanoreceptors

1. Pacinian corpuscles are found in large numbers in the mesentery and along the mesenteric nerves and vessels. They are readily excited by slight mechanical stimuli due to movements of the intestine. The fibers supplying the corpuscles belong to the A beta group of splanchnic nerve afferents.
2. Stretch receptors in the stomach are excited by mild distension or by contraction of the stomach wall. They are known as tension-signaling devices in series with the contractile elements. Receptors responding to mechanical stimulation are also present in the wall of the urinary bladder; they are slowly adapting.
3. Stretch receptors in the lungs are formed by the terminal ramifications of vagal afferent fibers. They respond to inflation during the inspiratory phase without adaptation; the greater the volume of air breathed in, the higher the frequency of impulse discharge. Deflation receptors are stimulated in forced expiration.
4. Baroreceptors are found in the carotid sinus, aortic arch, atria, and roots of the great vessels. Carotid sinus baroreceptors are terminals of the sinus nerve, a branch of the glossopharyngeal nerve. All other baroreceptors are supplied by the vagus nerve. The function of baroreceptors is the detection of changes in the systemic blood pressure and the venous return to the heart.

Nociceptors

There are no clear-cut boundaries to distinguish nociceptors from other visceral receptors. The viscera are well supplied with free nerve endings arranged in a plexus of fine, unmyelinated axons ramifying in the muscular walls and mucous membranes. Most are sensitive to over-distension or to the presence of irritants, but the search for a specific pain receptor has not been fruitful.

PROPERTIES OF RECEPTORS

Adequate stimulus

This can be defined as the stimulus that most easily evokes a response from a receptor. It implies that some types of receptor will respond to only one form of stimulation—ie., they are specific, while other types may react to one kind, but they can also be excited by other kinds of stimulation. Thus, mechanoreceptors are highly sensitive to deformations and pressure, but are also sensitive to thermal or chemical stimuli. The idea of receptor selectivity, however, is subject to certain qualifications and applies only to normal intensities of stimulation. Almost any kind of stimulus can be effective if it is sufficiently intense or likely to cause tissue damage; such stimuli tend to activate high-threshold nociceptors and give rise to pain. The adequate stimulus for visceral receptors is exceptional. Stimuli that would normally elicit pain if applied to the skin, as occurs with cutting or burning, do not produce the same sensation from viscera. The stimulus adequate for visceral pain may be a chemical irritant, such as strong acids, or a mechanical stimulus causing over-distension of a hollow organ or vigorous contraction of the muscular wall.

Modality

The kind of sensation aroused in human consciousness is called its modality. According to Müller's doctrine of "specific nerve energy," each modality is represented by a specific receptor and a specific area in the brain that becomes excited. For instance, if a touch corpuscle is stimulated, the sensation of touch is aroused because impulses reach the area of the brain controlling tactility. The doctrine does not take into account other factors that can influence sensation—by altering its quality, enhancing it, or blocking it altogether. Furthermore, the nature of the stimulus does have some bearing on the kind of sensation aroused—itch, tickle, and vibration being examples of sensory modalities that do not appear to have specific receptors.

- The modalities of sensation from the skin are touch, pressure, vibration, cold, warmth, and pain. Within each of these are distinguished various qualities, such as "sharp" and "blunt" for the touch sensation, "burning" for warmth, "tickle" and "itch" for pain.
- The modalities from muscles, tendons, and joints are termed the deep sensations and include the sense of position and the appreciation of passive movement.
- The modalities of sensation from the viscera are complicated by anatomi-

cal considerations, since the viscera lie in cavities innervated by somatic nerves. Examples of true visceral sensibility are hunger and thirst, yet there are no specific receptors for these modalities. Pain of visceral origin is frequently accompanied by somatic pain because of involvement of the overlying structures.

Generator potential

The function of a receptor is to transform the energy derived from a stimulus into a discharge of nerve impulses. The process involves an intermediary stage or generator potential. The stimulus first causes a change in the permeability of the receptor membrane, resulting in a local depolarizing potential. The flow of current spreads only a short distance and soon dwindles without propagation just like the end-plate potential of muscle fibers. The receptor therefore acts like a transducer, transforming an external energy source into an electrical event. A weak stimulus generates a small flow of current, and a strong stimulus a larger flow of current. Hence the greater the intensity of stimulation, the greater the amplitude of the generator potential. When the amplitude reaches a critical size, the generator potential initiates an action potential in the sensory nerve fiber.

The generator mechanism has been demonstrated in many different kinds of receptor, the one most easily studied being the Pacinian corpuscle because of its size and the fact that the concentric layers of its capsule can be dissected away to expose the nerve endings. If the corpuscle is treated with a local anesthetic to block the initiation of the action potential, only the generator potential is recorded. It can be shown that the amplitude and rate of rise of the generator potential are related to the strength of the applied stimulus. It can be further shown that a second potential is generated when the stimulus is discontinued–an effect known as the "off" response.

Action potentials are initiated at the first node of Ranvier, which lies inside the capsule of the Pacinian corpuscle. As long as the generator potential persists, the action potentials will continue to be formed since the node of Ranvier will repolarize and discharge again. The greater the amplitude and rise rate of the generator potential, the more quickly will these processes occur. This means that the frequency of impulse discharge is directly proportional to the intensity of the stimulus, a relationship that applies to most sensory receptors.

Other receptors in which the generator potential can be readily recorded are the frog muscle spindle and the stretch receptor of the crayfish. During stretching the sensory endings become depolarized to a maximal value, then decline to a lower level that persists throughout the stretch. These two phases in the generator mechanism coincide accurately with the rate at which action potentials are generated in the sensory nerves.

Receptor discharge

Action potentials in single fiber preparations of the frog sciatic nerve were first recorded by Adrian soon after the introduction of the vacuum tube amplifier; in this pioneer experiment the gastrocnemius muscle was stimulated by stretch. Action potentials from cutaneous fibers were obtained by touching areas of skin or by the movement of hairs. All the sense organs responded in much the same way and a good deal of new information was discovered about the elementary properties of sensory discharges. Adrian concluded that the potentials were similar to those in motor nerves and obeyed the all-or-nothing law. He demonstrated how sense organs registered stimulus intensity by variations in the frequency of impulse discharge, the upper frequency range being limited by the time required for recovery between each impulse. Another important factor relating to single fiber discharges was the variability of inter-spike intervals. Receptors did not always yield a steady discharge in which one impulse followed another at a fixed period of time. When the stimulus was a weak one, a longer recovery time was necessary for the fiber to reach the threshold level of that stimulus, and the inter-spike interval was correspondingly long. With strong stimuli the inter-spike intervals were shorter.

The information transmitted by a single fiber to the central nervous system is of value for accurate localization of the stimulus. The area supplied by an axon and its terminal branches is called its sensory field, but within any given field there is usually a good deal of overlap between different sensory fibers. Accordingly, the total contribution to the sensory input is proportional to the number of fibers activated. This in turn is determined by the intensity of the stimulus. A nerve trunk contains a large number of fibers, each representing a direct line of communication from its sense organ and capable of discharging independently with respect to frequency and temporal pattern. In summary, the effects of increasing the intensity of the stimulus are larger generator potential, increased frequency of discharge, shorter recovery time, reduced inter-spike intervals, and greater number of receptors activated.

Sensory adaptation

The frequency of receptor discharge depends not only upon the intensity of the stimulus but also upon its duration. In other words, when a stimulus is applied, the receptor responds at first by discharging at a relatively higher frequency. Then, if the stimulus is prolonged, the discharge rate falls to a lower level or the receptor may not respond at all. This decline in impulse frequency during continuous stimulation is defined as sensory adaptation. The phenomenon was clearly demonstrated by Adrian on a frog muscle spindle preparation and is illustrated in the graph of Figure 47. The curve shows the initial response to stretching the muscle and the decline of frequency over a period of 20 sec. when the load was kept constant. The frequency fell to half its initial value in about 10 sec.

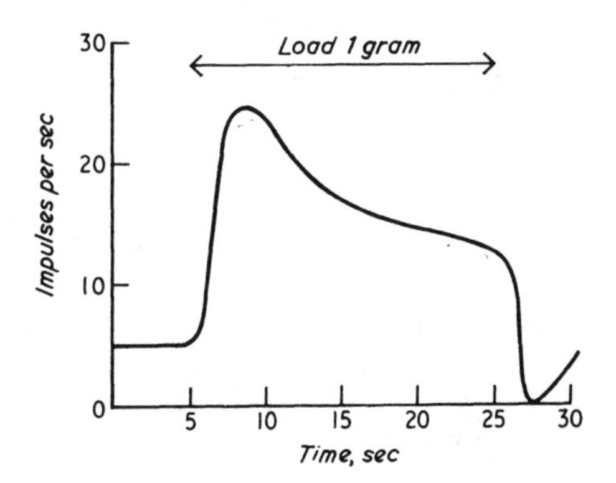

Fig. 47. Sensory adaptation. Curve shows impulse frequency produced by stretching a muscle with a constant load. Frequency rises rapidly to a maximum then falls to a lower level, which is maintained until load is removed. (After E.D. Adrian, 1928.)

Some receptors adapt to a far greater extent than others. The muscle spindle is an example of a slowly adapting receptor. It is able to maintain a level of discharge for long periods of time, thus keeping the nervous system continuously informed about the state of the muscles. For this reason it is appropriately called a *tonic* receptor. Other types of slowly adapting receptors include the Golgi tendon organ, thermoreceptors in the skin, and the chemoreceptors of the cardiovascular system. The importance of adaptation to receptor function is best illustrated by those endings that generate pain, since if the discharge declined too quickly, pain would lose much of its protective value.

The Pacinian corpuscle is a good example of a rapidly adapting receptor. When the capsule is stimulated by compression, the layers of connective and elastic tissue act as a filter to cut out a long-lasting stimulus. If the layers are removed by dissection, leaving only the central core, the decapsulated remnant behaves like a slowly adapting receptor. The Pacinian corpuscle signals the onset and cessation of compression, responding only to the mechanical change or the dynamic phase of the stimulus. Rapidly adapting *phasic* receptors cannot be effective during continuous stimulation so that any further information transmitted by the stimulus is lost to the nervous system. Thus we cease to be aware of the clothes touching our bodies almost as soon as we have put them on.

The mechanism of adaptation may be partly explained as depression of the generator potential which, in the case of phasic receptors, is kept below firing level by adjustment of the visco-elastic structures within the capsule. Clearly, the mechanism is not caused by fatigue, as the latter is a slower process due to exhaustion of metabolic resources or to oxygen deficiency.

PATTERNS OF DISCHARGE

Spontaneous activity

This implies that a receptor may discharge in the absence of any external stimulus. In the muscle spindle preparation a number of prepotentials have been observed in the absence of any tension. Normally they are incapable of initiating a nerve impulse, but when several such potentials are generated simultaneously they may cause the discharge of a propagated spike. Retinal ganglion cells maintain a spontaneous discharge in the absence of light and, likewise, spontaneous activity has been recorded from auditory nerve fibers in the absence of sound. The source of the spontaneous activity is believed to be ionic movements causing fluctuations in the level of receptor excitability.

Phasic activity

The application of a stimulus may give rise to three different types of responses–a discharge may be evoked from a receptor that was silent prior to stimulation; the discharge may reflect a sudden change in the character of the applied stimulus; or a receptor may discharge when the stimulus is switched off.

1. Silent receptors. At threshold intensity, stimulation evokes a brief burst of spikes. The response is instantaneous and ceases when the stimulus is discontinued. If the same stimulus is repeated, a second burst of spikes is recorded. Because of the all-or-nothing character of the spikes (uniform amplitude and duration), the only information they carry is their number and the time intervals between them. However, as previously mentioned, the presence of impulse activity in one nerve fiber and its absence in other fibers will be recorded in the central nervous system as an event that has occurred at a particular localized point.

If the intensity of the stimulus exceeds threshold, the frequency of impulse discharge is increased. This relationship is illustrated in Figure 48 which shows the response to mechanical vibration recorded from the afferent fiber of a Pacinian corpuscle. It can be seen that the discharge patterns follow the increasing stimulus rates induced by vibration.

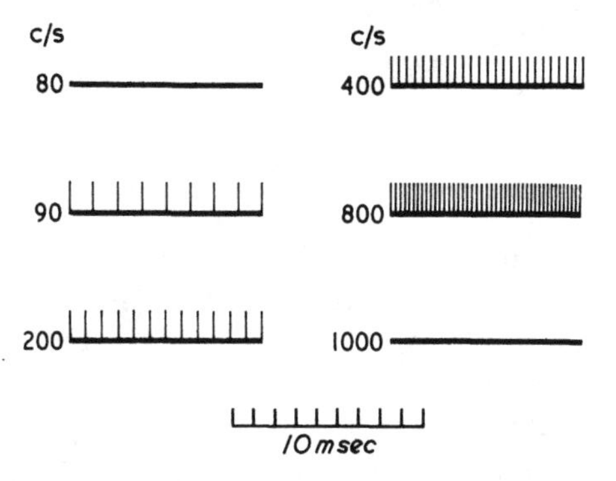

Fig. 48. Response of Pacinian corpuscle to vibration. The discharge rate is increased with increasing frequency of vibration. Frequencies below 90 c/s and above 800 c/s are ineffective. (After C.C. Hunt, 1961.)

2. Change in character of the stimulus. A receptor that is already firing at a steady rate may change its firing frequency in response to a sudden change in the stimulus. For example, when a muscle is stretched from one length to another, its spindles discharge more rapidly during the applied stretch than they had previously; the frequency then declines to match the new steady state (Fig. 49).

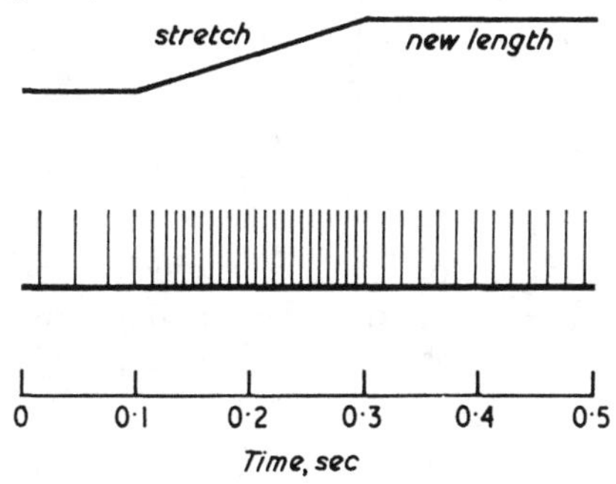

Fig. 49. Change in the character of the stimulus. Action potentials were recorded from muscle spindle. The initial discharge is increased as the muscle is stretched. Note change in firing frequency when the muscle is held at the new length.

3. On-off effect. Some receptors respond initially to an applied stimulus and also discharge again when the stimulus is discontinued; other receptors

respond only to cessation of the stimulus. This "on-off" behavior has been noted for the Pacinian corpuscle but is also found in other receptors. In the retina, photoreceptors excited by illumination give "on" discharges and those excited when light is withdrawn give "off" discharges. In the ampulla of the vestibular organ, a flow of fluid in one direction excites a discharge from the hair cells and a flow in the opposite direction inhibits the discharge.

Tonic activity

This implies that a receptor may discharge continuously to signal a steady state. Tonic activity is not confined to sensory mechanisms and is perhaps equally important in motor systems such as the action of the vagus on the heart or the vasomotor control of the blood vessels. However, many sensory receptors are capable of sustaining a steady discharge during constant environmental conditions. The muscle spindle is a good example of tonic activity, as it provides a stream of impulses for maintaining postural tone in skeletal muscles. It is generally agreed that the end-organ concerned is the nuclear chain and that its discharge continues even during the contraction of the muscle in proportion to the influence of its gamma motor innervation.

SENSORY CODING

The various impulse patterns discharged from a receptor provide the first relay of a signal transmitted to the central nervous system. As each receptor can be regarded as one source or channel of information, sensory coding can be defined as the study of the discharge characteristics in a number of parallel channels. Initially, the neural code evaluates the nature of the stimulus; it is concerned with the modality and specificity of the stimulus, the time and speed of onset, its duration and intensity, and finally, any change imposed by the environment or by commands from the central nervous system itself. The mechanisms used for sensory coding may be considered under two headings—temporal and spatial.

Factors related to time (temporal patterns)

1. Time of onset. This is determined by the time taken for the generator potential to reach threshold and is therefore related to the size and velocity of the stimulus.

2. Frequency. This is a measure of the number of impulses per second. Most receptors display graded increases of frequency when the stimulus intensity is increased until maximal amplitude of the generator potential is attained. A distinction must be made between the frequency measured per time unit and the mean frequency averaged over a period of time. The period may make quite a difference since the discharge may be affected by adaptation or by a changing background. The range of frequencies is limited by the recovery time of the generator mechanism and by the refractory period of the nerve fiber.

3. Inter-spike interval. This is a measure of the time intervals between individual impulses and is usually displayed in the form of histograms. The time intervals can vary within a single burst, being shorter or longer at the commencement of the burst, suggesting that information is encoded for more than one parameter of the stimulus.

4. Trains of impulses. These are produced by repetitive discharges from the receptor and follow each other in bursts at variable intervals. Information may be carried by the sequence of these intervals occurring in some readable order. In other words, certain temporal patterns are distinguishable by repetition of a particular time interval occurring between the trains. Such temporal patterns may be important for signaling the quality of a sensory event.

5. Duration. This is often directly related to the strength of the stimulus and implies that a greater number of impulses will be transmitted during any given period of time.

Factors related to the receptor (spatial patterns)

1. Differential sensitivity. Some receptors are specific—i.e., they show a high sensitivity for one type of stimulus and are relatively insensitive to all other types. This property is important for accurate localization of the stimulus and also for discrimination between two or more closely related stimuli. Other receptors are non-specific and will respond to almost any form of stimulation that is excessive or damaging.

2. Number of receptors. Increasing the intensity of the stimulus generally recruits more receptors and therefore increases the size of the receptive field. It follows that information is carried to the central nervous system simultaneously along several channels. Since impulses recorded from a nerve trunk will consist of many overlapping waveforms, special computer techniques are required to separate the channels in order to trace the individual patterns. It is now recognized that multiple-channel coding schemes may be employed by the nervous system for conveying information about the area of activity and the intensity of the stimulus.

Sensory processing

As information from our sense organs can be transmitted only in the form of impulse patterns, the nervous system must be capable of analyzing the signals so that they can be used advantageously. All the different nerve tracks are spatially organized throughout the sensory pathway, but further processing occurs at different levels, new impulse patterns arise, and modifications may be imposed as a result of descending influences from the brain. Consequently, the signal that reaches the sensory cortex is likely to be quite different from that entering the spinal cord. Information carried in a peripheral nerve represents the first stage of sensory processing. At this stage the signal is faithful to the changes induced in the sense organs and most closely reflects the character of the applied stimulus. The total input pattern of a peripheral nerve is derived from a combination of active channels, each of which may be firing at a different frequency and with different inter-spike intervals.

A method that is now used to separate the individual channels is to display them in the form of cross-correlation histograms. This means simply displaying the relative times of occurrence of impulses in each of the channels. Differences in the peaks of the histograms indicate certain features of the receptor discharge, including threshold, refractoriness, conduction velocity, and adaptation.

In summary, it can be stated that the various receptors of the body have access to independent carrier channels that signal events about the environment. This information is conveyed as coded patterns of nerve impulses in the peripheral nerves to the central nervous system. Each channel uses temporal features for carrying the code; these include the time of onset and cessation, the duration of a burst of impulses, the number of impulses within a burst, the rate of impulse repetition, and variations in the sequence of train intervals. When several channels operate together there is a greater opportunity for extracting information from a more extensive area of the environment.

PERIPHERAL ORGANIZATION

Receptive fields

The area supplied by a single sensory axon and its terminal branches is known as its receptive field. The nerve endings are concentrated more in the center and less toward the periphery of the field. Therefore, a weak stimulus evokes a greater discharge from the center and a smaller discharge with longer latency from the periphery. When the stimulus becomes more intense, the number of fibers excited is progressively increased, causing the response to spread to adjacent receptive fields. Mapping experiments performed by recording from single nerve fibers and lightly stroking the skin reveal that there is considerable overlap between one field and the next. Hence if one fiber is sectioned, little or no sensory loss can be detected. On the other hand, there is no doubt that receptive field organization of the sensory nerves is very important for localization of the stimulus and makes fine discrimination possible.

Weber-Fechner Law

The ability to discriminate between small changes in stimulus intensity was first investigated by Weber and later expressed mathematically by Fechner. The law named after them attempts to define a relationship between the magnitude of stimulus increments and the magnitude of the sensation evoked. Weber suggested that the ability to discriminate between different stimulus intensities depended upon a constant ratio between them. He discovered that the smallest difference between two weights that could be detected (just noticeable difference) was a constant fraction of the weights themselves. Thus it was possible to distinguish a weight of 30 g from 31 g or 60 g from 62 g, the fraction 1/30 remaining constant for a given modality. This can be expressed as $\Delta I/I = C$ in which I is the intensity of the stimulus, ΔI the smallest noticeable difference, and C a constant. Fechner developed the law by assuming that each noticeable difference corresponded to a unit of sensation, from which he derived a measurement based on a logarithmic scale. Simply expressed, the relationship between sensation and stimulus became $S = K \log I + C$, in which the magnitude of sensation S is proportional to the logarithm of the stimulus intensity.

The Weber-Fechner Law is an attempt to provide a measurement of sensation whereby we discern small differences or make judgments such as one weight feeling heavier than another weight or one light appearing brighter than another light. It describes the relationship on a logarithmic scale which is applicable to most types of sense organ. The graph in Figure 50 shows the general form of this relationship. It can be seen that the plot approximates a straight line over a physiological range, but does not hold good for very weak or very strong stimulus intensities. This means that the scale has a limited value. Thus, other mathematical relations have been proposed for quantitative studies of discrimination.

Peripheral afferent pathways

Impulses from the sense organs are conveyed by their axons to cell stations in the spinal and cranial ganglia. These are collections of large bipolar

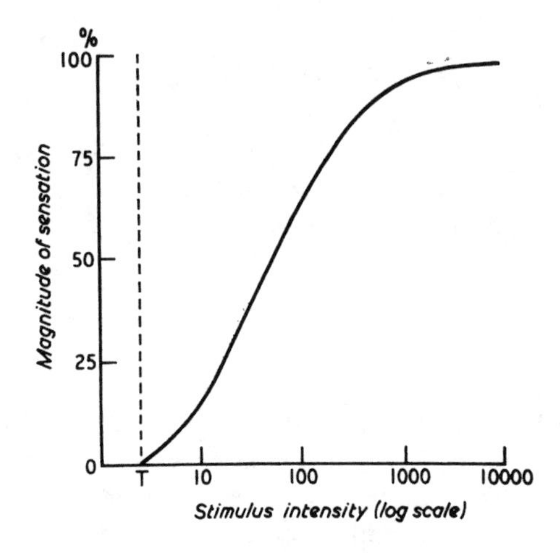

Fig. 50. General form of relationship between stimulus intensity and power of discrimination. The ordinate represents the magnitude of sensation as a percentage of maximum; the abscissa, the intensity of applied stimulus. The relationship is logarithmic in that each increment of stimulus intensity does not produce similar increases in discriminative sensation. Note that the linear relationship is over physiological range, but the deviation of response is close to threshold (T) and at very strong stimulus intensities.

cells found on the posterior roots of the spinal nerves and on the sensory roots of the trigeminal, facial, auditory, glossopharyngeal, and vagus nerves. Each ganglion cell gives off a single fiber which divides into two processes—a peripheral limb distributed to the sense organ and a central limb proceeding in the posterior root before dividing into branches running into the spinal cord or brain stem.

1. Types of afferent fiber. The spinal and cranial posterior nerve roots are the principal portals of entry for all classes of sensory nerve fibers. A few afferent fibers may gain entry into the cord via the anterior roots, although the latter are predominantly efferent in function.

- Impulses from the skin and superficial structures travel in the cutaneous nerves. The size of the different axons falls into several groups, ranging from large myelinated A fibers to the small unmyelinated C fibers.
- Impulses from the muscles, tendons, and joints travel in the nerve trunks supplying these structures but, of course, are sensory in function. They are classified in groups I to IV. Group I fibers are the largest in diameter (20 μm to 12 μm), arising mainly from the primary endings of muscle spindles (Group Ia) and from Golgi tendon organs (Group Ib). Group II fibers (12 μm to 4 μm) arise from the secondary endings of the muscle spindles. Group III consists of small myelinated fibers, 4 μm to 1 μm in diameter, and Group IV contains mainly unmyelinated C fibers, both categories arising from free nerve endings.

• Impulses from the viscera travel in the autonomic nerves. They may take devious routes through more than one plexus and through several ganglia before reaching their cell stations. They do not form synapses in the ganglion. Visceral afferent fibers are classified into three groups, A beta, A delta, and C, the latter being unmyelinated.

2. Conduction velocities. Another classification of the afferent fibers is based on the distribution of their conduction velocities. The cutaneous afferent fibers have a bimodal distribution, one peak at 70 m/sec, or 6 times the diameter, and the other at 20 m/sec, or 4 times the diameter. Spindle afferent fibers also have a bimodal distribution with peaks at 100 m/sec and 50 /sec, respectively. The method is not without criticism since the diameter of any afferent fiber is not uniform throughout its length, and conduction velocities will tend to be slower where the fiber branches or breaks up into its sensory terminals.

3. Dermatomes. Afferent fibers entering the spinal cord have a segmental distribution. The area of skin supplied by a single posterior nerve root is called a dermatome. Mapping of the dermatomes in the animal was performed by Sherrington by cutting three roots above and three below the root to be examined. The area of remaining sensibility was then tested. This procedure was necessary because the dermatomes of adjacent roots overlap greatly. Mapping experiments may be useful in studying the laminations of spinal tracts and in following the projections of the body surface to their destinations in the brain. Charts of the dermatomes are important for clinical examination. Thus, the presence or absence of anesthesia in the skin may help to localize the level of a spinal lesion.

4. Visceratomes. The segmental distribution of visceral afferent fibers is not as precise as the somatic input, and overlapping is far more extensive. Afferents from the heart and lungs pass into the spinal cord through the posterior nerve roots of the upper five thoracic segments. Afferents from the abdominal organs enter the spinal cord between the fifth and ninth thoracic segments, and those from the pelvic organs, which pass through the sympathetic chain, enter the spinal cord between the tenth thoracic and second lumber segments. A knowledge of the visceratomes has a practical application in surgical intervention for the relief of pain.

CENTRAL CONTROL OF RECEPTORS

A brief mention will be made here of the fact that certain receptors are supplied with efferent fibers from the central nervous system. Fibers that are excitatory in function tend to lower the threshold of the receptor and thus increase its sensitivity to stimulation; fibers that are inhibitory in function tend to depress the excitability of the receptor and thus reduce the afferent discharge. A good example of a receptor under continuous central nervous control is the muscle spindle. Stimulation of the small motor nerves supplying the intrafusal muscle fibers causes an increased afferent discharge from the spindle. Hence, central nervous mechanisms that control the activity of these gamma efferent nerves must also exert a profound influence on the functions of the receptor.

The stretch receptor of the crayfish is often quoted as an example of a receptor supplied with inhibitory fibers. The discharge elicited by subjecting the receptor to stretch ceases completely during stimulation of the inhibitory axon. The inhibitory process resulting in hyperpolarization of the receptor membrane is caused by selective increase of ionic permeability to K^+ and Cl^- ions. Efferent fibers to the vestibular receptors have also been described. Little is known about their function except that they may play a part in modifying adaptation. Thus ballet dancers and skaters attain a very high degree of adaptation to spinning and other high-speed movements. Efferent fibers to the organ of Corti are believed to complete a feedback loop capable of altering the threshold of sensitivity of the cochlear hair cells.

SELECTED BIBLIOGRAPHY

Adrian, E.D. *The Physical Background of Perception.* Oxford: Clarendon Press, 1947.

Bishop, G. H. Neural mechanisms of cutaneous sense. *Physiol. Rev.* 26:77-102, 1946.

Boyd, E.A., and Roberts, T.D.M. Proprioceptive discharges from stretch receptors in the knee joint of the cat. *J. Physiol.* 122:38-58, 1953.

Catton, W.T. Mechanoreceptor function. *Physiol. Rev.* 50:297-318, 1970.

Granit, R. *Receptors and Sensory Perception.* New Haven, Conn.: Yale University Press, 1955.

Hunt, C.C., and Takeuchi, A. Responses of the nerve terminal of the Pacinian corpuscle. *J. Physiol.* 1-21, 1962.

Loewenstein, W.R. Biological transducers. *Scientif. Amer.* 203 (2): 99-108, 1960.

Mountcastle, V.B. The problem of sensing and the neural coding of sensory events. In *The Neurosciences,* Quarton, G.C., Melnechuk, T., and Schmitt, F.O. (Eds.). New York: Rockefeller University Press, 1967.

Segundo, J.P. Communication and coding by nerve cells. In *The Neurosciences,* Quarton, G.C., Melnechuck, T., and Adelman, G. (Eds.), 569-586. New York: Rockefeller University Press, 1970.

Stein, R.B. and Matthews, P.B.C. Differences in variability of discharge frequency between primary and secondary muscle spindle afferent endings of the cat. *Nature* 208:1217-1218, 1965.

Terzuolo, C., and Washizu, Y. Relation between stimulus strength, generator potential and impulse frequency in stretch receptor of crustacea. *J. Neurophysiol.* 25:56-66, 1962.

CHAPTER VI

Organ of Balance

The neural mechanism for balance is served by hair cells that have specialized structural properties yet retain many features common to other types of mechanoreceptors. They are stimulated by displacement caused either by gravitational forces or by fluid motion and are therefore direction-sensitive detectors. The hair cells respond to sudden changes in the position of the head or to movements of the body in space. By exciting the sensory nerve endings that supply them, they give rise to complex muscular reactions designed to restore natural orientation and thus maintain balance. These are called the labyrinthine reflexes. The information provided by the labyrinth or organ of balance is used in collaboration with that from other receptor systems, including the input from the eyes and from the muscles, tendons, and joints. Accordingly, the organ of balance has a very important role to play in the regulation of posture.

FUNCTIONAL ANATOMY

The labyrinth is contained in a series of cavities within the petrous part of the temporal bone. It consists of three membranous parts called the utricle, saccule, and semicircular canals (Fig. 51).

The *utricle* is a membranous sac containing endolymph. It communicates with the semicircular canals by five openings, one being common to the superior and posterior canals; a fine duct leads from the utricle into the saccule. The walls of the utricle are lined with a layer of cubical cells, which at one point form a thickened mass called the macula. This is the receptor area, consisting of supporting cells and flask-shaped hair cells enclosed in a gelatinous membrane. Small crystals of calcium carbonate, called otoliths, are embedded in the membrane and lie in contact with the tips of the hairs protruding into the cavity (Fig. 52). The otoliths act as a load on the hair cells, causing the hairs to bend in response to any change of position in the gravitational field. The hair cells are supplied by sensory axons of the ves-

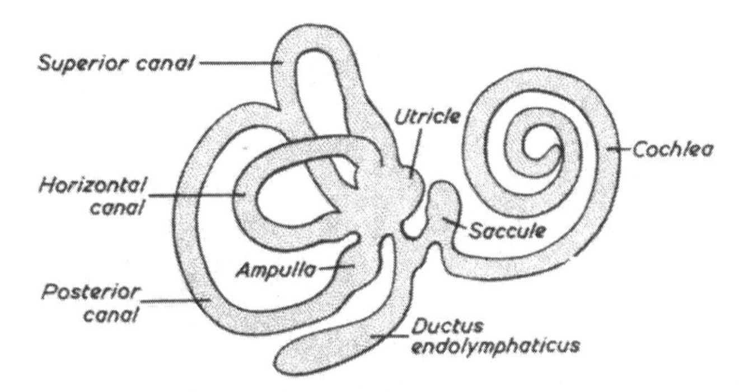

Fig. 51. Diagram of labyrinth. The three semicircular canals open into the utricle, one opening being common to the superior and posterior canals. Each canal has an ampulla at one end in which the receptor organ is situated. The saccule and its pouches are shown communicating with the cochlea.

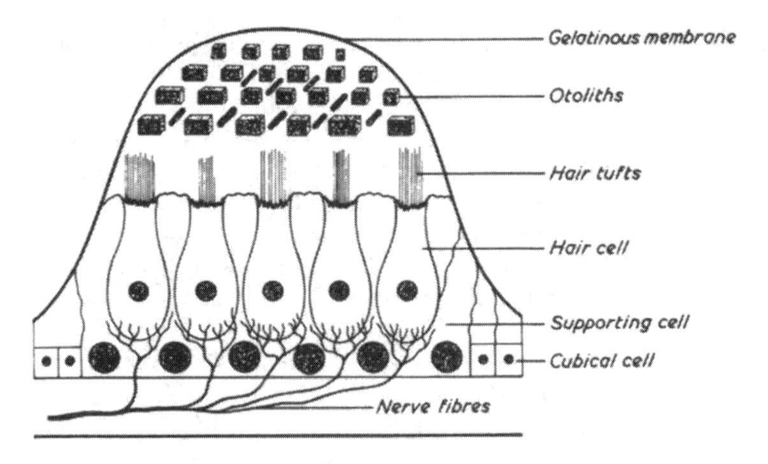

Fig. 52. Diagram of the macula. The sensory epithelium in the utricle is covered by a gelatinous membrane containing crystals of calcium carbonate (otoliths). This loading acts by gravity on the hair cells supplied at their base by terminal fibers of the vestibular nerve.

tibular nerve. Different patterns of excitation occur when the hairs are bent to one side or pulled upon by the otoliths. When the head is in the upright position, the nerve impulse discharge is minimal; on the other hand, when the head is bent backward or the body is upside down, the otoliths pull on the hair filaments and the discharge is greatly increased. In this way the brain is informed about orientation in space.

The *saccule* is a small membranous sac communicating with the utricle. There are two other openings in its cavity, one forming a blind pouch and the other leading into the cochlea. The receptor organ is similar to that of the utricle except that the axons are derived from filaments of the auditory nerve. The otoliths respond to slow mechanical vibration.

The *semicircular canals* are three in number on each side, and are known respectively as the superior, posterior, and horizontal canals. They are arranged at right angles to each other so that they represent all three planes in space. Considering the position of the canals on the two sides, it will be seen that the right and left horizontal canals are external, while the anterior vertical or superior canal of one side lies in the same plane as the posterior vertical canal of the opposite side (Fig. 53). Each bony canal contains a membranous duct filled with endolymph and communicating with the utricle, one opening being shared by the vertical canals. Near its junction with the utricle, each canal has a dilatation or ampulla in which the receptor organ is situated. Here a thickening of supporting cells and sensory hair cells forms a transverse projecting ridge called the crista on top of which is a gelatinous membrane or cupula, pivoted like a swing door. Movements of the endolymph displace the cupula toward the ampulla or away from it and therefore act as a stimulus to the sensory nerve fibers innervating the hair cells. Each fiber represents the peripheral limb of a bipolar cell in the vestibular ganglion. The central limb enters the trunk of the vestibular nerve, which accompanies the auditory nerve into the pons.

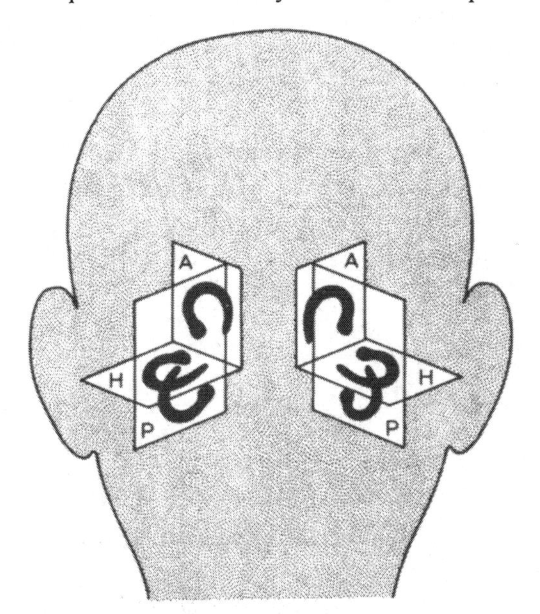

Fig. 53. Diagram showing the planes of the semicircular canals in the head. The horizontal canals (H) of the two side lie in the same plane. The anterior vertical canal (A) of one side lies in a plane parallel to the posterior vertical canal (P) of the opposite side.

EFFECTS OF LABYRINTHINE STIMULATION

Action of the utricle

The utricle is responsible for positional sense. The mechanism of stimulation is the pull of gravity on the otoliths when the head is tilted or moved in space. The hair cells appear to be arranged like a fan with marginal cells pointing in opposite directions; the arrangement allows for the precise detection of movements away from the upright position, and the result is that different patterns of impulses are discharged to the nervous system for carrying out any reflex adjustments that might be necessary. It is possible to distinguish two types of response to gravity—tilting of the head and linear acceleration (or deceleration).

1. Tilting. Stimulation is minimal when the head is in the upright position. By tilting the head to one side, the frequency of impulse discharge on that side increases with the angle of tilt; the discharge declines slowly if the head remains in the new position. The macula receptors adapt slowly because the discharge depends upon the angle of tilt and not upon the time that the head remains tilted. In animals, the action of the utricle can be studied by recording potentials from single fibers of the vestibular nerve or by inserting an electrode into the vestibular nuclei in the brain stem. Adrian found in the cat that when the head was level the receptor discharged at about 6/sec, but the frequency increased to 95/sec, with a lateral tilt of 20° (Fig. 54).

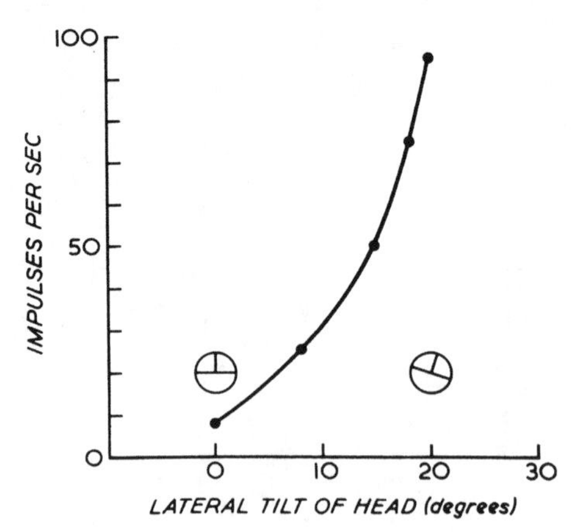

Fig. 54. Response of utricle to lateral tilting of head of a decerebrate cat. Impulses were recorded from the left vestibular nucleus, while the head was tilted to the left. As tilt increases, impulse frequency rises correspondingly. Angle of tilt is shown in the inset. (After E.D. Adrian, 1943.)

2. Linear acceleration. Reflex responses to linear acceleration may be observed when an animal is suddenly moved in space. For example, if a blindfolded cat is suddenly lowered with its head down, its fore limbs become extended to facilitate safe landing. Downward movement is more effective than upward movement. The utricle responds to a change in the rate of movement–i.e., at the commencement of the movement (acceleration) and also when the movement slows down (deceleration). Similar responses may be experienced by pilots in fast aircraft, especially when pulling out from a dive.

Role in posture

Tonic labyrinthine reflexes contribute with other reflex mechanisms to the control of posture. When standing on a horizontal platform with the eyes closed, most subjects tend to sway since the center of gravity falls in front of the ankle joints. If the platform is tilted, the sway is increased and the body will tend to fall forward. Balance is maintained because tilting stimulates the otolith organ causing reflex modification of tone in the antigravity muscles. In order to demonstrate labyrinthine reflexes, the following experiments may be performed in the cat:

When the animal is placed on its back, the otoliths pull on their hair cells, producing increased extensor tone in all four limbs. Or when the animal is prone with the head down, extensor tone is reduced to a minimum. Third, the reflexes are present after cutting the posterior roots of the spinal cord to eliminate proprioceptive influences from the muscles themselves. And finally, the reflexes are abolished after section of the vestibular nerves.

Action of the semicircular canals

The semicircular canals are responsible for rotational sense; they give information about rotational movements of the head occurring in the plane of each canal. The sensory end-organ is the cupula, and the stimulus is simply the flow of endolymph, which displaces the cupula in a certain direction. The signals transmitted by the vestibular nerve have a widespread influence on the central nervous system, passing into the cerebellar and reticulospinal pathways for reflex motor adjustments and to the nuclei serving the ocular muscles for corrective movements of the eyes. Signals also pass upward to the cerebral cortex to produce sensations of rotational direction, which can sometimes be confusing or result in symptoms of motion sickness.

The planes of rotation

The three semicircular canals are arranged in planes at right angles to one another. The horizontal canal is stimulated by rotation around the vertical axis of the body—ie., turning to the right or to the left; the superior canal is stimulated by rotation in the median plane as in tumbling forward or backward; the posterior canal is stimulated by rotation in the transverse plane as in performing a cartwheel. Endolymph flow occurs chiefly in the pair of canals whose axis most nearly approximates the axis of rotation, while the endolymph in the other canals is little affected. According to the direction of rotation, the flow of endolymph may be toward the ampulla (leading) or away from the ampulla (trailing). When the human head is in the upright position, the horizontal canals are inclined downward, forming an angle of about 30° with the horizontal plane.

Displacement of the cupula

The action of endolymph flow on the cupula has been visualized in experiments on fish by introducing a drop of oil into a canal. Displacement of the cupula may be observed as the droplet moves with rotation. Normally, the cupula occupies a mid-position on the crista where pressure on the hair cells is minimal. If the cupula deviates to one side, it will pull on the hair cells and stimulate the sensory nerve endings; if it deviates to the opposite side, the nerve endings will be depressed. In the case of the horizontal canals, the hair cells are stimulated when the cupula is bent toward the ampulla and inhibited when the cupula is bent away from the ampulla. The reverse holds good for the vertical canals. The combination of excitation from one semicircular canal and inhibition from the other forms the basis for the interpretation of the direction of the movement and of the resultant muscular reactions.

Effects of rotation on the horizontal canals

The way in which the sensory receptors provide information about rotational movements may best be understood by considering what happens when the head is rotated in the horizontal plane. The direction of the movements is shown by the arrows in Figure 55.

1. Rotation begins. Owing to inertia of the fluid, movement of the endolymph tends to lag behind the movement of the canals, thus generating a force between the cupula and the endolymph. Consequently, if the head is rotated to the right, the cupula in each horizontal canal will deviate to the

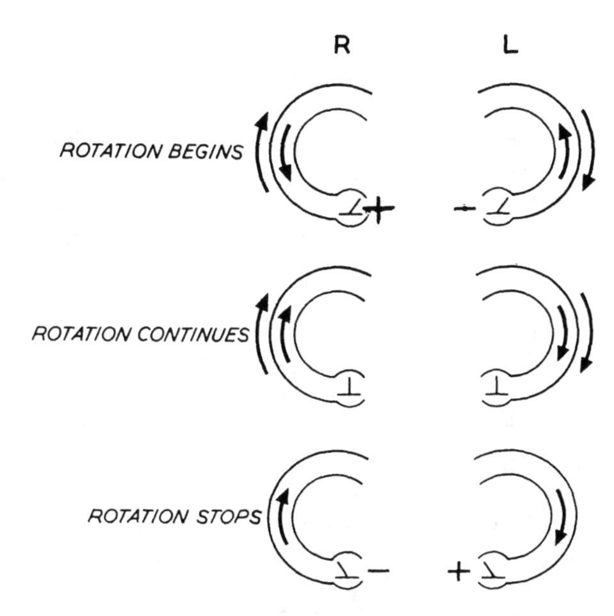

Fig. 55. Effects of rotation on horizontal semicircular canals. When the head is rotated in a clockwise direction, the cupula is deflected toward the ampulla on the right side and away from the ampulla on the left side owing to inertia of endolymph. As rotation continues, endolymph flow matches movement of the head, and the cupula returns to neutral position. When rotation is stopped, endolymph momentum causes deflection of each cupula in reverse direction.

left. This will be toward the ampulla on the right side and away from the ampulla on the left side. Accordingly, the sensory hair cells will be stimulated on the right side and depressed on the left.

2. Rotation continues. Providing the rate of rotation is kept steady, the endolymph flow soon catches up and moves with the canals. The cupula returns to its original neutral position and stimulation subsides.

3. Rotation stops. Owing to its own momentum, the endolymph continues to flow and a force is generated in the reverse direction. The cupula in each canal will now deviate to the right. This will be away from the ampulla on the right side and toward the ampulla on the left side. Finally, after about 25 sec to 30 sec, each cupula returns to its resting position and the sense of movement ceases.

Similar actions take place in the vertical canals when the head is rotated in other planes. The cupula is displaced from its neutral position by any change in the movement of endolymph—i.e., when the head begins to rotate or when it stops rotating. The effective stimulus is called angular acceleration or angular deceleration. Because the canals are arranged as functional pairs, the sensations evoked by deceleration will be in a direction opposite to the original movement. The semicircular canals thus provide accurate information about angular movements of the head, such as those occurring under natural conditions during walking and running or when the head is turned quickly to one side. On the other hand, erroneous information may be provided if the stimulus exceeds the normal physiological range of the sense organ. This may result in severe disorientation which poses a serious problem to aviators when landing an aircraft or performing various maneuvers in flight.

EFFECTS OF ROTATION ON EYE MOVEMENTS

The labyrinth exercises a powerful control over the eye muscles. The purpose of this control is to maintain a stable image when the head turns from one direction to another. With each turn, signals from the labyrinth cause the eyes to move in an equal and opposite direction to that of the head, thus allowing the eyes to fixate on an object at least long enough to give a clear image. Without such compensatory movements of the eyes, the retinal images would rapidly change and vision be impaired during the slightest movements of the head. Automatic control is achieved by vestibulo-ocular reflexes transmitted through the vestibular nuclei and medial longitudinal bundle to the brain stem nuclei, innervating the extra-ocular muscles. Other pathways arising from the utricle proceed to the flocculonodular lobe of the cerebellum and are then carried in cerebellar-ocular projections.

Measurement of eye movements

If two silver disc electrodes are attached to the skin near the outer canthus of each eye, a resting current flowing through the tissues can be recorded in an electro-oculogram. The resting current appears to be orientated in the visual axis. When the eye moves, it acts as a rotating dipole so that the current field around it moves also. Traces are obtained while the subject is being rotated with his eyes closed and, after rotation has ceased, when he opens his eyes and attempts to fix his gaze on an object held in front of him. The movements of the eyes induced by rotation are called *nystagmus* and occur as a result of stimulation of the semicircular canals. The direction of the movements is determined by the particular canals stimulated–horizontal, in which the eyes move from side to side; vertical, in which the movement is upward and downward; or rotatory, in which several planes are combined. Nystagmus has two components:

1. A slow movement results from impulses reaching the eye muscles from the vestibular nucleus. As the head rotates, the eyes turn slowly in the opposite direction in order to maintain their fixation.

2. A quick movement is produced by descending impulses from the cerebral cortex acting on the vestibular nucleus. Having turned as far as possible, the eyes swing quickly back to fix on a new object. Thus the quick movement is in the same direction as the rotation.

Bárány chair experiment

The subject is seated in a chair with the head tilted forward about 30° to stimulate the horizontal canals. The subject is then rotated for 20 sec at 30 revolutions per minute. The chair is stopped abruptly and observations are made on the different types of vestibular reactions.

Post-rotational nystagmus

The diagrams in Figure 56 explain the direction of the slow component when the subject is rotated to the right. At the beginning of rotation, the

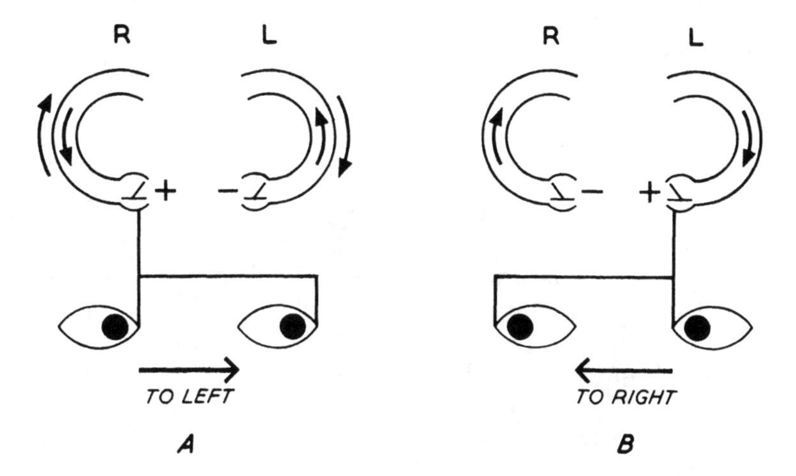

Fig. 56. Diagram showing direction of eye movements when subject is rotated to right (R). Horizontal canals have been stimulated. A, at the beginning of rotation, cupula is deflected toward right ampulla, stimulating vestibular nerve on right side (+). Impulses pass to right internal rectus and left external rectus. Eyes therefore move to the left. B, when rotation is stopped, cupula is deflected toward the left ampulla, causing eyes to move to the right. This is the slow component of post-rotational nystagmus.

cupula is displaced from its neutral position owing to inertia of the endolymph. This will be toward the ampulla on the right side and away from the ampulla on the left side. Consequently, the nerve endings on the right side will be stimulated. The relationship between the horizontal canals and the extra-ocular muscles is also shown in the Figure: Each canal is connected to the internal rectus muscle on its own side and to the external rectus of the opposite side. The increased discharge on the right side causes the eyes to move to the left.

As rotation continues, the endolymph moves with the head and the cupulae return to their resting position because there is no longer any deflecting force. The eyes regain their central position.

When the rotation is stopped, the endolymph continues to flow owing to its momentum and the cupula in each canal is displaced, this time in the opposite direction. The left canal is now stimulated causing the eyes to move to the right.

Clinical nystagmus

The slow component of nystagmus is caused by impulses reaching the eye muscles from the vestibular nerve. In clinical terminology, nystagmus always refers to the quick component. Thus a left horizontal nystagmus is one in which the quick or correcting movement is to the left, and the slow or vestibular movement to the right.

TESTS FOR LABYRINTHINE FUNCTION

The reactions induced by these tests occur in normal subjects with intact labyrinths. Failure to respond to stimulation is therefore diagnostic of disease in the vestibular apparatus itself, the vestibular nerve, or its central nervous connections. Abnormal labyrinthine stimulation may arise from a number of causes and appear as attacks of dizziness, nausea, impairment of vision, motor disturbances, and perceptual illusions. The latter are particularly relevant to problems of high-speed flight and to the selection of aircrews, since it is necessary to exclude those who are susceptible to spatial disorientation.

Rotating chair

1. Past pointing. The subject is seated in a chair and shown a card upon which is a black cross. The subject touches the center of the cross with the

forefinger, closes his eyes, and tries again. The chair is then rotated. As soon as rotation ceases, the subject opens his eyes and touches the cross as before, then closes his eyes and repeats the attempt. The points at which the forefinger touches the card are compared before and after rotation. Alternatively, the subject is required to draw a vertical line, when it is found that his arm deviates out to the side stimulated.

2. Coriolis acceleration. The subject is seated in a chair with his head tilted forward and eyes closed. The chair is then rotated. As soon as the subject feels the sensation of rotation disappear, he nods his head sharply backward and forward until a new sensation is recognized. This effect tends to produce disorientation from stimulation of the vertical canals as may occur, for example, if a pilot suddenly turns his head when performing a steep dive.

3. Walking. As soon as the chair movement is stopped, the subject attempts to walk along a straight line in front of him. He may have difficulty in maintaining his balance, even when the eyes are open, and tends to sway to one side or he may fall owing to over-compensation.

4. Autonomic disturbances. Some subjects are more prone than others to the discomforts of labyrinthine stimulation. Impulses may spread from the vestibular nuclei to other centers in the brain stem involved in the control of the autonomic system. As a consequence, they may suffer from all the symptoms of motion sickness, including nausea, vomiting, tachycardia, and sweating.

Caloric tests

Since one ear is examined at a time, the caloric tests are helpful in the diagnosis of vestibular disease. In patients with a unilateral lesion, the usual reactions to caloric stimulation are observed only on the intact side. Thus the vestibular nerve is frequently damaged in acoustic nerve tumor; the patient fails to respond to stimulation on the affected side, but giddiness and nystagmus may be elicited from the normal side. The principle of the caloric tests depends upon the production of convection currents by using a syringe to inject the ear with cold or hot water. The change of temperature sets in motion the endolymph of the semicircular canal and so displaces the cupula from its neutral position.

1. Cold water test. The subject sits with the head tilted backward 60°, bringing the horizontal canal into a vertical position. The cold water irrigating the ear causes a downward movement of the endolymph, away from the ampulla (in the direction of the arrow, Fig. 57A). This is equivalent to stimulation of the opposite side. Accordingly, if the right ear is irrigated, nystagmus will appear after a short latent period with the slow component

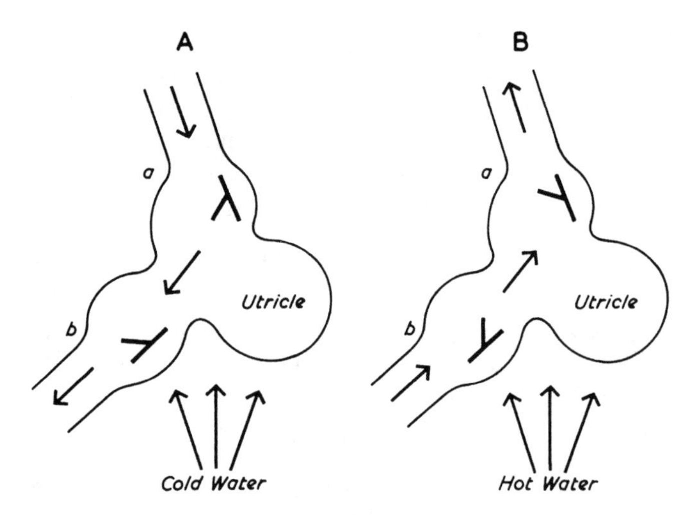

Fig. 57. Caloric test. Water is introduced through a tube in the external meatus of the ear. The horizontal and anterior vertical canals are stimulated, depending upon the position of the head with respect to gravity. A, cold water increases the density of the endolymph, causing convection currents downward in the direction of the arrows. If the right ear is stimulated, the slow component of nystagmus is to the right.
B, hot water causes the endolymph to rise. The cupula is deflected in the opposite direction. If the right ear is stimulated, the slow component of nystagmus is to the left. (a, anterior vertical canal; b, horizontal canal.)

to the right. The subject may complain of giddiness and if allowed to stand tends to fall to the right.

2. Hot water test. The subject sits with the head tilted forward 30°. The hot water irrigating the ear causes an upward movement of the endolymph, deflecting the cupula toward the ampulla (in the direction of the arrow, Fig. 57B). If the right ear is irrigated, nystagmus appears with the slow component to the left, and the subject tends to fall in the same direction. The effects of hot water are therefore the opposite of those of cold water. The temperature change of the endolymph first reaches the horizontal canal which is nearest to the external auditory meatus, but the movements extend into the anterior vertical canal so that a combined horizontal and rotatory nystagmus may be observed.

Labyrinthectomy

The disturbances following unilateral labyrinthectomy are caused by the unopposed action of the intact side and involve the utricle as well as the semicircular canals. In animal experiments, nystagmus disappears after a few days, but muscular tone is diminished on the affected side, causing the trunk to be curved and the animal to have difficulty in maintaining balance when its head is turned in any direction. Birds are unable to orient themselves in flight and tend to hop or walk around in circles. In the human being, damage to the labyrinth or the vestibular nerve may occur from injury, infection, or tumor. The immediate effects are mainly subjective, arising from an imbalance of vestibular influences on the central nervous system. The patient soon learns to compensate for these effects when walking and resists sudden movements of the head in order to avoid distress from attacks of vertigo.

The disturbances following bilateral labyrinthectomy are much more severe, especially in animals that rely on their vestibular reflexes for maintaining posture and balance. Such animals are unable to land on their feet when falling through the air and have difficulty in orientation when placed under water. In the human being whose labyrinthine reflexes have been permanently impaired, there is no nystagmus or vertigo, and compensation takes place as the patient learns to use other sensory mechanisms for the coordination of his muscles. Balance is always possible and the gait is steady with the aid of vision. In the dark, however, the patient is greatly handicapped and physical activity is limited.

In summary, the labyrinthine mechanism depends upon bilateral impulse discharges generated by the hair cells. The utricle gives information about the precise position of the head, obviously important in the human being for maintaining erect posture. The utricle also provides the information necessary for maintaining balance when the body is suddenly moved in space, as in walking or jumping. The utricle is therefore sensitive to both tilting and linear acceleration. The semicircular canals register the movements of the body in space. They detect angular acceleration or deceleration in the plane of a rotatory movement, whereby the subject can make appropriate adjustments to prevent falling off balance. Impulses from the labyrinths combine with those from other sense organs to maintain normal posture and provide an awareness of body orientation with respect to the immediate environment.

VESTIBULAR PROBLEMS IN AVIATION

Strong vestibular stimulation occurring during flight may give rise to spatial disorientation and motion sickness. Incidents have been described in which signals from the pilots' labyrinths produced false information about the correct orientation of the aircraft and, as a result, illusions in the pilots' perceptions. The judgment of a pilot in handling an aircraft may be so adversely affected as to jeopardize his safety; no doubt such misinterpretations account for a proportion of flying accidents. In addition, the recurrence of disorientation incidents in the same individual is a recognized cause of flying stress, often leading to symptoms of anxiety and loss of confidence.

Perceptual disorientation

The greatest precision in judgment is required during the approach and landing of an aircraft. The pilot must judge height and distance in relation to air-speed and the angle of the flight. In most landings, he makes use of visual cues that have to be learned by repeated experience, but in poor visibility or in unfamiliar terrain, errors of judgment are a distinct possibility.

Spatial disorientation

There is usually no problem in maintaining control of an aircraft when flight instruments or other visual cues are available. Disorientation may arise, however, even with an experienced pilot, when he is exposed to excessive accelerations or turning forces. Disorientation is the result of erroneous cues from the labyrinths, which may tempt the pilot not to believe his instruments and cause him to make inappropriate corrections.

1. Errors from the utricle. A sudden change of flying speed will produce a force that will be superimposed on the downward force of gravity. The resultant vector force is at an angle to the other two, giving a false impression of tilting. This may account for the illusion of nose-down or nose-up attitudes with respect to the horizon.

2. Errors from the semicircular canals. Sensations of false rotation may arise immediately after a steep turn or spin. Angular rotations can be especially disturbing when movements of the head are superimposed on the rotation of the aircraft. The pilot may feel that the aircraft is spinning in the opposite direction and may be tempted to pull out of the spin when in fact the aircraft is correctly oriented. Comparable problems arise from prolonged rolling maneuvers in which illusory sensations may be accompanied by nystagmus and blurring of the retinal image and hence apparent movement of visual targets. A common form of spatial disorientation is characterized by a sensation of banking when the aircraft is in level flight.

Motion sickness

Angular acceleration of one or more canals is the primary stimulus for motion sickness in airborne individuals. This is often brought about in susceptible pilots by a head movement in order to operate a control when the aircraft is spinning. The incidence of motion sickness is likely to be greater as flights become longer and, despite the possibility of adaptation, it must be considered as a serious hazard in long-term operations envisaged in space travel.

SELECTED BIBLIOGRAPHY

Adrian, E.D. Discharges from vestibular receptors in the cat. *J. Physiol.* 101:389-407, 1943.

Brodal, A., and Pompeiano, O. (Eds.). Basic aspects of central vestibular mechanisms. *Progr. Brain Res.* 37, Amsterdam: Elsevier, 1972.

Dohlman, G.F. The attachment of the cupolae, otolith and tectorial membranes to the sensory cell areas. *Acta Otolaryngol.* 71:89-105, 1971.

Hallpike, C.S. The caloric test: A review of its principles and practice with especial reference to the phenomenon of directional preponderance. In Wolfson, R. (Ed.): *The Vestibular System and its Diseases.* Philadelphia: University of Pennsylvania Press, 1966.

Naunton, R.F. (Ed.). *The Vestibular System.* New York: Academic Press, 1975.

Walsh, E.G. Role of the vestibular apparatus in the perception of motion on a parallel swing. *J. Physiol.* 155:506-513, 1961.

CHAPTER VII

Organ of Hearing

When sound reaches the external ear, it sets into motion the tympanic membrane with its attached chain of ossicles. The last bone in this chain is applied to the oval window of the cochlea, a fluid-filled spiral structure containing the receptor elements—the organ of Corti. Pressure waves, which are set up in the cochlea spiral by movements at the oval window, cause corresponding movements of the basilar membrane on which the organ of Corti is situated. Upward and downward movements of the basilar membrane stimulate the hair cells of the sensory epithelium, thus exciting the terminals of the auditory nerve fibers. In this way sound waves in the air are converted into pressure waves in fluid.

As the waves travel through the spiral cochlea, they reach a peak at specified points determined by the frequencies of the sound. At these points the basilar membrane vibrates more strongly than elsewhere, and a particular set of nerve fibers become active. The manner in which the basilar membrane responds to the pressure waves is important for the discrimination of pitch, loudness, and quality of different sounds. Pitch is determined by the site at which maximal vibration occurs, causing specific nerve fibers to discharge. Loudness is determined by the extent of the basilar membrane thrown into vibration, whereby the rate of discharge is increased and more nerve fibers are recruited. Quality of sound is determined by the various patterns of impulse discharge representing the total input of the individual auditory nerve fibers.

Like many other receptors in the body, the organ of hearing has certain related functions because of the complexity of its central connections. Apart from the main stream of impulses to the cerebral cortex, a proportion of the auditory input is consigned to lower levels of the nervous system, where collaterals from the auditory tracts spread diffusely into other projection systems. Thus, the reticular activating system provides background information for sharpening or attenuating the primary auditory signals. The motor systems activate reflex mechanisms for eliciting muscular reactions to sound; an example of such reactions is the instantaneous response to a sudden noise. Thus, impulses transmitted by the cochlea are essential not only for the sense of hearing, but also for many associated functions of behavior.

FUNCTIONAL PARTS OF THE EAR

The ear is described in three parts—external, middle, and internal (Fig. 58).

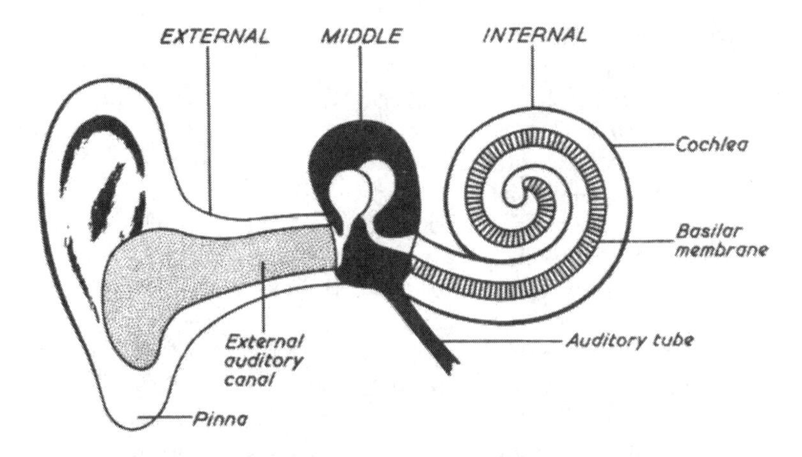

Fig. 58. Functional parts of the ear. The external ear comprises the pinna, external auditory canal, and tympanic membrane. The middle ear contains the ossicles, tensor tympani and stapedius muscles, and opening of the auditory tube. The internal ear consists of the spiral cochlea, which is divided into two compartments by the basilar membrane.

1. *The external ear* consists of the pinna, the external auditory canal, and the tympanic membrane.

• The pinna, a cartilaginous projection from the side of the head, is encircled by muscles and leads to the external auditory meatus. The purpose of the pinna is to direct sound waves toward the tympanic membrane. The muscles are used for very little action in human beings but are most useful in animals as detectors of sound direction.

• The external auditory canal, an oblique passage of cartilage and bone, is lined with skin that contains wax-secreting glands. The canal first passes backward and upward, then forward and downward, anterior to the mastoid process.

• The tympanic membrane, or ear drum, is a funnel-shaped membrane set at an acute angle in relation to the floor of the canal. The membrane separates the canal from the middle ear. When examined with reflected light through a speculum, the membrane has a glistening, blue-grey appearance with a whitish streak extending obliquely from the center upward and forward; the streak is formed by the handle of the malleus, the first bone of the ossicular chain, which is attached to the inner side of the membrane.

2. *The middle ear* is an air-filled cavity within the temporal bone. It contains a chain of three ossicles, two small muscles, and the auditory (Eustachian) tube.

The chain of three ossicles connects the tympanic membrane to the membrane of the oval window. The general form of the ossicles is illustrated in Figure 59.

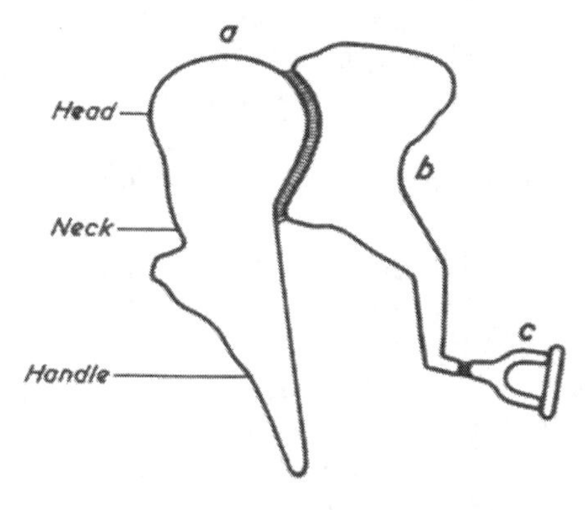

Fig. 59. Ossicles of the middle ear. a, malleus; b, incus; c, stapes.

The malleus (hammer) has a long process or handle attached, as mentioned, to the tympanic membrane and a rounded head projecting above the membrane where it articulates with the incus. The two bones are firmly bound together by ligaments. The incus (anvil) has a compressed body and a long process running parallel to the handle of the malleus but turning inward at its tip to form a nodule for articulation with the head of the stapes. The stapes (stirrup) has a head and two limbs supporting a footplate, which is fixed to the edges of the oval window. The three bones move as a unit, inward and outward, following the movements of the tympanic membrane. In this way, the vibrations of sound are communicated to the membrane of the oval window.

The two small muscles of the middle ear are the tensor tympani and the stapedius. The tensor tympani muscle arises from a bony canal in the anterior wall of the middle ear above the auditory tube; its tendon is inserted into the handle of the malleus. The muscle is innervated by a motor branch of the fifth cranial nerve and its action is to draw the handle of the malleus inward, thereby increasing the tension in the tympanic membrane.

The stapedius muscle arises from the posterior wall of the middle ear in front of the canal for the facial nerve; its tendon is inserted into the neck of the stapes. The muscle is innervated by a branch of the seventh cranial nerve, and its action is to pull the stapes away from the oval window.

As the two muscles function reflexively in response to sound, they serve together as a device for controlling transmission of vibrations to the internal ear. The two actions oppose each other to increase the system's sensitivity to weak sounds and to protect the cochlea from very loud sounds, especially sounds of low frequency. In the absence of the ossicular chain and its muscular control, sound waves can travel directly through the air of the middle ear and can enter the cochlea at the oval window. The sensitivity of the ear, however, is then considerably diminished.

The auditory (Eustachian) tube is a short canal connecting the middle ear to the nasopharynx. Since the orifice can be opened by yawning or swallowing, air can be renewed from the atmosphere as required. The advantage of this arrangement is that the pressure in the middle ear can be kept at the same level as the pressure outside. Otherwise the tympanic membrane might bulge inward as the air in the middle ear would be absorbed. This would result in impaired transmission.

If the mucous membrane of the nasopharynx is affected by catarrh, the orifice of the tube may be occluded. This may lead to much discomfort when flying in an aircraft. On ascending to high altitudes, air is readily expelled from the middle ear because its pressure is atmospheric; on descending, however, equalization of pressure becomes difficult if the orifice is occluded, and under these conditions temporary hearing loss can be expected.

3. *The internal ear,* or cochlea, lies in the petrous part of the temporal bone. In shape, it resembles a spiral staircase winding for two turns and three quarters round a central pillar or modiolus. The broad base of the spiral is directed toward the bottom of the internal auditory meatus while the apex is inclined toward the cavity of the middle ear. In vertical section, the cochlea appears as a coiled tube with a bony outer wall and a membranous inner wall, as shown in the diagram of Figure 60. The bony portion is filled with perilymph, which acts like a cushion for the delicate receptor elements; the membranous portion contains a special fluid, the endolymph.

Structure of the cochlea

The spiral tube is completely divided into two compartments by a partition called the basilar membrane. The epithelium covering the upper sur-

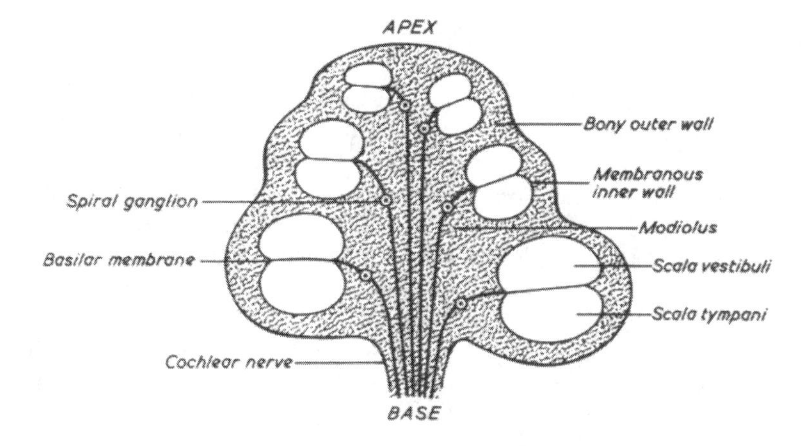

Fig. 60. Vertical section through the cochlea. The spiral tube and basilar membrane are seen in sections extending from base to apex. Terminal fibers emerging from the organ of Corti pass to bipolar cells in the modiolus to form the cochlear nerve.

face of the basilar membrane forms the organ of Corti, which contains the hair cells responding to sound and the terminals of the auditory nerve.

Compartments of the cochlea

The upper compartment, or scala vestibuli, lies between the roof of the cochlea and the basilar membrane. The compartment contains some perilymph, but a portion of it is shut off by Reissner's membrane, thus forming a spiral tube filled with endolymph. This tube is known as the duct of the cochlea, and its upper extremity is closed. Reissner's membrane is so delicate that the upper compartment behaves as a single tube in the transmission of sound. Pressure waves initiated at the oval window travel through the fluid from the base of the cochlea toward the apex.

The lower compartment, or scala tympani, lies below the basilar membrane. At the apex of the cochlea it communicates with the upper compartment by means of a small opening, the helicotrema, which serves to equalize pressure differences in the fluids. At the base of the cochlea it communicates with the round window of the middle ear. This aperture, however, is closed by a membrane which bulges outward when the pressure in the lower compartment is increased. Thus the membrane of the round window responds to the vibrations imparted to the basilar membrane by helping to take up the increased pressure.

Basilar membrane

This is the spiral partition that winds around the modiolus from base to apex, dividing the cochlea into upper and lower compartments. When viewed in cross-section, the membrane is seen to be attached medially to the spiral lamina of the modiolus and laterally to a bony projection on the outer wall (Fig. 61). The width of the membrane gradually increases from the basal turn, where it is narrowest, to the apical turn, where it is widest. This increase in width is accompanied by a corresponding increase in thickness. The auditory receptor is attached to its upper surface. The basilar membrane is a flexible structure responding to pressure changes in the fluid surrounding it by varying degrees of movement in its different segments.

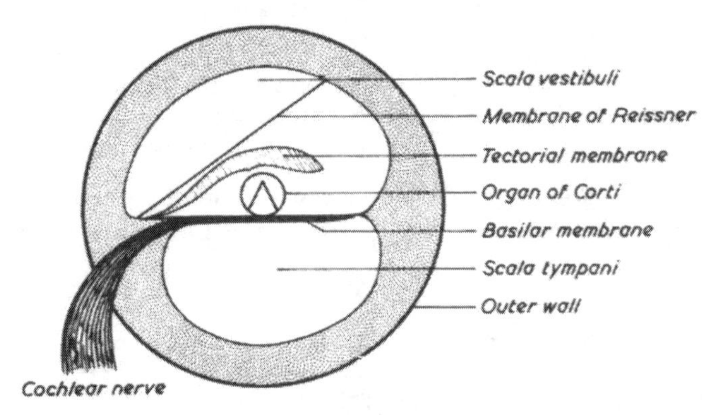

Fig. 61. Internal structure of the cochlea. The organ of Corti lies on the basilar membrane, which is attached medially to the spiral lamina of the modiolus and laterally to a bony projection on the outer wall. The tectorial membrane floats in endolymph and is separated from the perilymph of the scala vestibuli by the membrane of Reissner.

Organ of Corti

This is a collection of hair cells, supporting cells, and auditory nerve terminals (Fig. 62). Three rows of outer hair cells are separated from a single row of inner hair cells by two cartilaginous pillars inclined toward each other to form the tunnel of Corti. The hair cells are nucleated columnar epithelium with short hairs protruding from their upper ends. These hair cells are arranged in the form of a concave arch; they are affected by movements of the basilar membrane caused by sound waves, and, in turn, generate nerve impulses in the nerve fibers distributed to them.

Supporting cells are located between the rows of outer hair cells and the

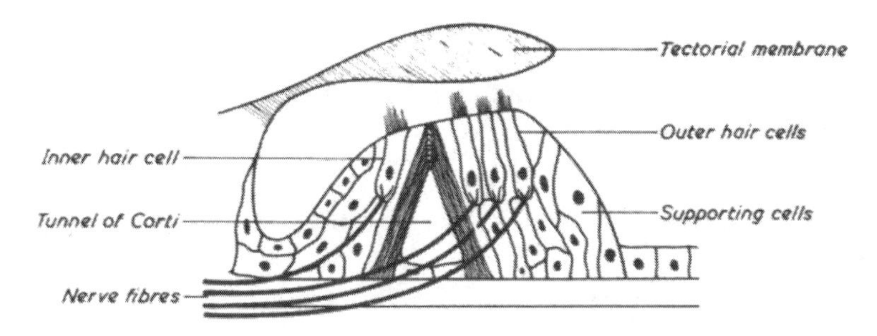

Fig. 62. Organ of Corti. The cartilaginous tunnel separates a single row of inner hair cells from three rows of outer hair cells. The tips of the hairs are embedded in the tectorial membrane. Numerous supporting cells surround the hair cells, which are supplied by terminal filaments of the cochlear nerve.

basilar membrane. Above the organ of Corti, the hair filaments make contact with the tectorial membrane, a thin structure anchored to the modiolus at one end and free at the other end. The auditory nerve terminals are distributed to the base and sides of the hair cells. They are the peripheral processes of the bipolar cells located in the spiral ganglion of the modiolus. The central processes of the bipolar cells emerge from the internal auditory meatus as the auditory nerve and reach the brain stem in company with the vestibular nerve. The main trunk also contains efferent fibers to the cochlea from the superior olive.

PROPERTIES OF SOUND

Definition

The adequate stimulus for the organ of hearing is sound. It is produced by changes in pressure set up by a vibrating object. A simple vibrator is a tuning fork which releases energy, when struck, in the form of rapid back and forth movements. Particles of air displaced by the movements alternately condense and expand as they speed away from the vibrating source. Sound is propagated in all directions as an advancing wave, each back and forth movement representing one cycle. If the advancing wave meets an object in its path, it can be scattered, absorbed, or reflected back as would an echo. The speed of a sound wave depends upon the density and elasticity of the medium through which it travels. Sound waves in air travel at 344 m/sec (769 m.p.h.).

Physical properties

Sound waves can vary in amplitude, frequency, waveform, and phase (Fig. 63).

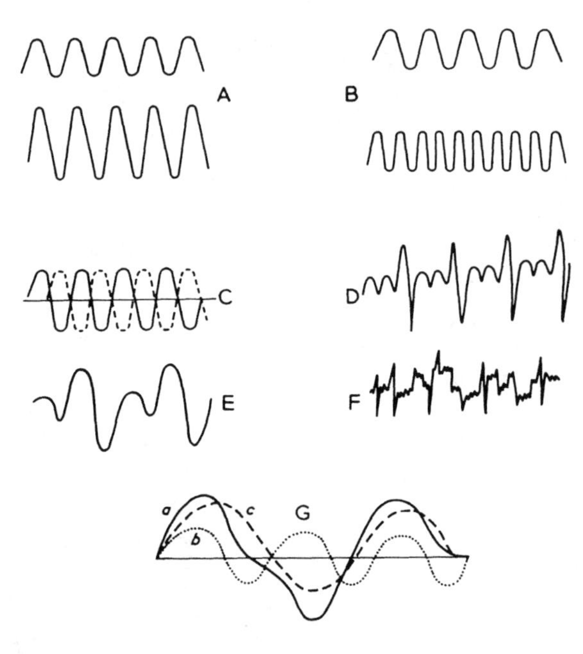

Fig. 63. Physical dimensions of sound. A represents sinusoidal waves of the same frequency but different amplitudes, B, sinusoidal waves of different frequencies, C, phase differences of two sinusoidal waves, and D, a musical note from a violin. E is a complex waveform regularly repeated, F, a typical noise signal, with varying amplitude at many frequencies, and G shows analysis of a compound wave. The frequency of wave b is twice that of c, and the amplitude of wave b is half that of c. Wave a is the result of b and c added together.

1. Amplitude. This is a measure of the strength of the pressure variations that create the sound. When represented graphically, amplitude is a measure of the distance between the peaks of the pressure waves.

2. Frequency. This is a measure of the number of times the pressure variations occur in one second. The frequency of a sound wave is measured in Hertzian rays (Hz).

3. Waveform. The simplest kind of waveform is a sine wave, usually emitted by a tuning fork. Most vibrating objects produce compound waveforms made up of a number of simple waves which, however, can vary in amplitude and frequency. For each compound wave there is a fundamental frequency, while the rest of the waveform is made up of variable frequen-

cies known as overtones or harmonics. No matter how complex the waveform, if its pattern has a definite rhythm or period, the sound produced is musical; if its pattern is irregular or non-periodic, the sound is termed "noise."

4. Phase. When two tuning forks of the same frequency are struck one after the other, the two waves may combine to form complex waves out of phase with each other. The phase difference can be measured between corresponding points on the waveforms. If two tuning forks of nearly the same frequency are sounded together, they may reinforce each other to produce periodic waxing and waning, which are heard as beats. A combination of rapid beats sounds unpleasant and is termed a "discord."

Perceptual properties

The sensations aroused by sound can vary in loudness, pitch, quality, and direction.

1. Loudness. The range of audibility is very extensive. The faintest sound that can be heard with one ear is called the threshold of hearing. At the other end of the intensity scale, loudness is limited by the tympanic membrane itself, which can be ruptured if the pressure waves are too severe. Sound intensity is correlated with the amplitude of the sound wave.

The decibel scale

The unit of sound intensity is the decibel (one tenth of a bel). It is a logarithmic unit in terms of which differences in sound intensitites can be expressed. If the faintest audible sound corresponds to 0 db, the intensity of a particular sound is the number of decibels above this threshold. The decibel unit is therefore a ratio between two sound intensities–E_1, the energy of the sound, and E_2, the energy of the reference sound. The formula, in terms of sound energy, that expresses an intensity difference is:

$$\text{Ndb} = 10 \log \frac{E_1}{E_2}.$$

The reference point is calculated from the minimum audible sound in a group of individuals with normal hearing; it is therefore the average threshold of hearing. It must be emphasized that the decibel is not an absolute measure of loudness, but a method of stating a difference between the physical intensities of two sounds.

Since the full range of audibility from the threshold to the loudest sound tolerable to the normal ear is very extensive, a logarithmic scale is useful. Thus if a sound is 10 times more powerful than the threshold, it has an intensity difference of 10 db; if 100 times, the intensity difference is 20 db, and if 1000 times, the intensity difference is 30 db. The decibel scale can be used to express the results of hearing tests. the normal ear being capable of detecting a change in sound intensity of 1 db. The intensities of some familiar sounds heard by a person with normal hearing is shown in Table II.

TABLE II. Intensities of Familiar Sounds

Sound	Scale of decibels
Threshold of hearing	0
Whisper	10
Quiet voice	20
Ordinary conversation	30
Noisy conversation	40
Loud voice	50
Traffic noise	60
Shouting	70
Full orchestra	80
Pneumatic drill	90
Overhead thunder	100
Discotheque music	110
Jet aircraft sounds	120

Audibility curve

The human ear complicates the problem of determining a scale of loudness, since it does not respond equally to all frequencies of sound. Hearing is best for the frequencies used by the ordinary voice (500 Hz to 2000 Hz) and falls off in the lower and higher frequency ranges. This means that the faintest sound a subject can hear in a quiet room depends upon the frequency of that sound. A plot of the relationship between the threshold of hearing at different frequencies is known as an audibility curve (Fig. 64). It will be evident that the frequencies most important for the components of speech are those which the ear hears best.

2. Pitch. The sensation of pitch depends upon the frequency of sound vibrations; the pitch of compound sounds is determined by the frequency of the fundamental wave. Sounds of low frequency produce notes of low

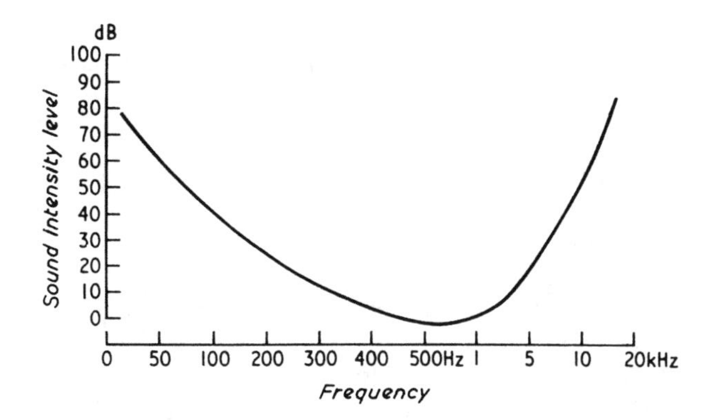

Fig. 64. Audibility curve plotting level of a sinusoidal tone as a function of its frequency. The threshold of hearing in a quiet room has its most sensitive range between 500 Hz and 2000 Hz, with the threshold dropping to 0 dB. Above 5000 Hz the threshold rises again to reach the frequency limit for hearing.

pitch or low tones, while sounds of high frequency produce notes of high pitch or high tones. The human cochlea is capable of detecting frequencies ranging from 16 Hz to 20,000 Hz. The lower and higher ranges, however, tend to suffer the greatest energy losses, with the result that the discrimination of pitch is much less acute at those frequencies, except perhaps to the trained ear of a musician. The pitch also changes slightly as the intensity of the sound changes; increasing the intensity lowers the pitch of a low note and raises the pitch of a high note.

3. Quality. As mentioned previously, most sounds produce compound waves made up of a series of smaller waves or harmonics, which impart to the sound a characteristic quality. The harmonics help to distinguish the voice of one person from that of another and enable the ear to detect the nature of a musical instrument, for example, whether a given note is being played on a piano or violin. In music, the frequencies of the harmonics may bear a simple ratio to the fundamental frequency. Thus, a violin string can vibrate as a whole and at the same time vibrate in sections. It is the combination of these frequencies that determines the quality of the note that is heard.

When several notes are played together on the same instrument, the various harmonics combine to compose a chord. A musical sound is heard only when certain intervals or consonants are used; otherwise the sound is discordant. The most important consonant interval in Western music is the octave, which links two notes together by a similar set of harmonics. The octave is the foundation of most classical compositions, but some modern forms of music deliberately employ combinations of discordant intervals.

4. Direction. The direction from which a sound is coming can be judged with a fair degree of accuracy and seems to depend upon a number of factors:

Intensity. Any difference in the intensities of sound reaching the two ears can be detected; the sound coming from a nearer source will tend to be a more powerful stimulus.

Time lag. Sound coming from the front or from behind an individual will reach both ears simultaneously; if the sound comes from a source on one side of the person only, the ear on that side receives the stimulus first. The nerve impulses from the two ears will therefore be out of phase, and the brain is able to appreciate the difference in time.

Frequency. The intensity factor is more important than the time factor for high frequency sounds, because the head acts as a sound barrier at these frequencies. On the other hand, frequencies below 3000 Hz are best localized by the time lag between the entry of sound into one ear and into the opposite ear.

Head movements. Faint sounds are made more audible by turning the head toward the source of the sounds. Animals locate sounds by scanning movements of the ears to find the direction of maximal intensity.

EXCITATION OF THE COCHLEA

Sound waves absorbed by the tympanic membrane are transmitted by the ossicular chain in the middle ear to the membrane of the oval window. Movements of this membrane cause pressure changes in the cochlear fluid, with resultant displacement of the basilar membrane and its attached organ of Corti. The vibratory motions stimulate the hair cells, thus exciting the auditory nerve fibers. Maximal excitation occurs in the part of the basilar membrane that undergoes the greatest movement.

Conversion of sound waves into fluid waves

Sound waves in air pass through the auditory canal to reach the tympanic membrane, which is thrown into vibration. An inward movement takes the handle of the malleus with it, and the force is transmitted through the incus to the footplate of the stapes. The three bones move as a single unit. With each vibratory cycle, the inward movement is followed by a movement outward, thus imparting similar motions to the membrane of the oval window. This allows sound vibrations to reach the fluid of the cochlea. The ossicular chain has the action of a lever that increases the force of movement and concentrates the energy of the sound. The important factor in the transfer of energy is the difference in area of the relatively large tympanic membrane from that of the smaller oval window (Fig. 65).

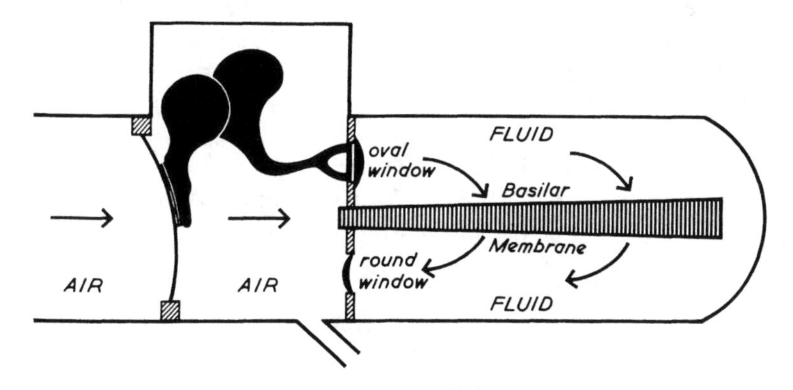

Fig. 65. Diagrammatic representation of the transmission of vibrations from the outer ear to the basilar membrane. Inward movement of the tympanic membrane is transmitted through the chain of ossicles to the oval window. Traveling waves are set up in the upper compartment of the cochlea, causing downward displacement of the basilar membrane and compression of the fluid toward the round window. The distance traveled by the waves depends upon the frequency of the note.

Movements of the basilar membrane

When the footplate of the stapes rocks inward, a pressure wave is set up in the fluid of the upper compartment of the cochlea. The wave travels along the spiral from base to apex, causing a downward displacement of the basilar membrane into the lower compartment; as a result of this increased pressure, the membrane of the round window bulges outward. The movements are reversed when the stapes rocks back again.

The basilar membrane does not vibrate as a whole because some parts of it are stiffer than others. In fact, maximal displacement occurs where the pressure wave is strongest and this in turn is related to the frequency of the sound wave. The driving pressure of a high frequency sound damps out only a short distance from the oval window; the basilar membrane has its greatest displacement here, in the basal turn of the cochlea. A medium frequency sound wave travels about halfway along the spiral and, finally, a low frequency sound wave travels the whole distance of the spiral, shifting the point of maximal displacement toward the apex. These characteristics of the traveling wave allow the cochlea to act as a mechanical analyzer of auditory frequencies.

Stimulation of the organ of Corti

As mentioned previously, the receptor elements of the organ of Corti consist of sensory hair cells held firmly in a reticulum of supporting cells. The free ends of the hairs project into the fluid spaces within the cochlea duct, and many are embedded in shallow grooves underneath the tectorial membrane. The oscillations of the organ of Corti result in a sliding back and forth movement of the tectorial membrane, described as a "shearing action" in which the protruding hairs are bent upon their attachments (Fig. 66). This action occurs whenever the basilar membrane vibrates and can be regarded as the final mechanical event leading to excitation of the auditory nerve terminals.

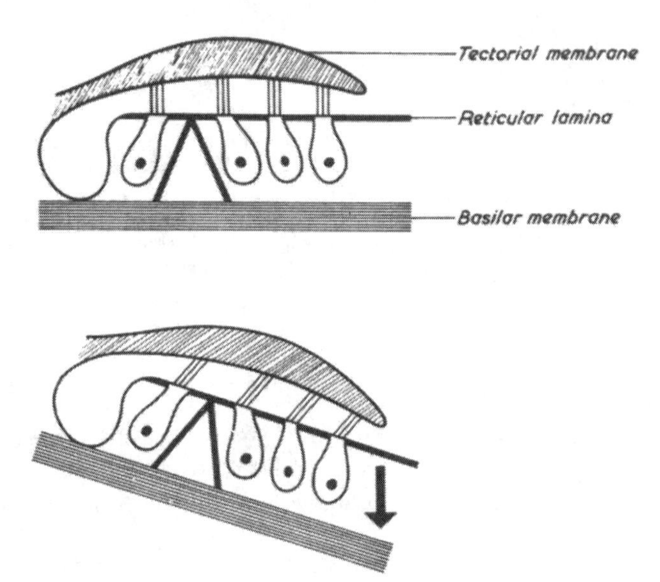

Fig. 66. Shearing action of tectorial membrane. The organ of Corti and adjacent structures all move as a unit. Downward movement of the basilar membrane (indicated by the arrow) pulls the reticular lamina downward and outward. The motion of the tectorial membrane bends the filaments of the hair cells, thus exciting the cochlear nerve fibers.

Cochlear microphonics

Electrical potentials, caused by bending the hairs, are sometimes referred to, with the names of their discoverers, as the Wever and Bray effect. The potentials can be recorded by placing an electrode on the auditory nerve or on the cochlea itself. They are produced in response to sound stimulation, but should not be confused with auditory nerve impulses. The amplitude of the potentials rises with increasing intensity of the sound but otherwise these potentials show none of the properties of action potentials. They are generated by displacement of the basilar membrane and bending of the hair filaments against the resistance of the tectorial membrane.

DISCRIMINATION OF PITCH

The ear distinguishes one sound from another by detecting the various tones of which each is composed. There are several theories regarding the way in which this is done but all of them center on the physical properties of the basilar membrane itself. The earliest theory, postulated by Helmholtz, regarded the basilar membrane as a series of tuned resonators–i.e., instruments capable of vibrating in tune to a particular sound frequency. It was supposed that the basilar membrane vibrated at different sites along the cochlear spiral according to the frequency of the sounds, high frequencies being represented at the base of the cochlea and low frequencies at the apex. The idea that tonal localization in the cochlea was a function of resonance has received a good deal of support:

1. Histological evidence. The basilar membrane is not a uniform structure. It contains fibers that progressively increase in length, elasticity, and thickness from the base of the cochlea to the apex. At the base, the fibers are short, stiff, and relatively thin and thus tend to vibrate to high frequencies; at the apex, the fibers are longer, more elastic, and thicker, characteristics that favor low frequency vibrations. These differences in the structure of the basilar membrane are represented diagramatically in Figure 67.

2. Experimental evidence. In one form of experimentation, holes are drilled through the wall of the cochlea to destroy the basilar membrane at various points. Then, by testing the animal's hearing of sounds of different frequencies, an audibility curve is constructed. The tonal gaps indicating hearing loss are related to the sites of the lesions, which can be verified afterward by histological examination.

In another type of experiment, animals exposed to a loud continuous sound suffer degeneration of the organ of Corti. The position of the damaged area in the cochlea can be correlated with the band of frequencies employed.

Finally, recordings from single fibers of the auditory nerve following tonal stimulation reveal that at threshold a fiber is excited only by a narrow band of frequencies: The one to which the fiber is most sensitive is called the "characteristic frequency."

3. Clinical evidence. Many people are tone deaf. In the absence of any

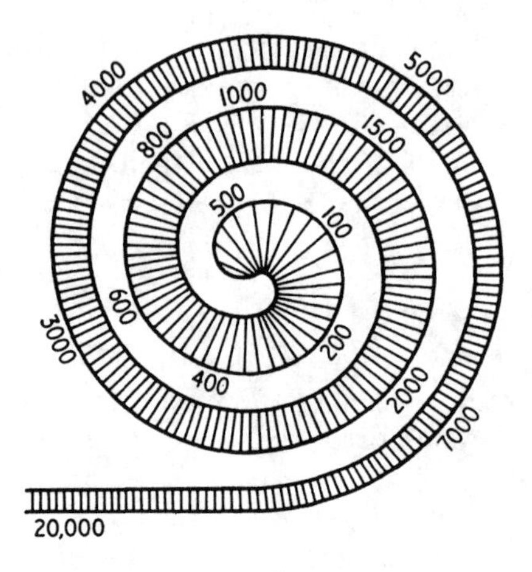

NOTE. The spiral is 2¾ turns. It increases uniformly in width.

Fig. 67. Diagram of basilar membrane showing range of audible frequencies. The middle range of frequencies occupies the greatest extent of the basilar membrane. High frequencies are located at the basal turn, and low frequencies at the apical turn. Total length in the human being is about 32 mm.

history of middle ear disease, the hearing loss is attributed to defects of the organ of Corti. It can be shown by audiometric tests that workers exposed to very noisy machinery often have tonal gaps in the high frequency range. Such persons may be unaware of a hearing disability since the frequencies used in speech remain unaffected.

Traveling wave mechanism

Pitch is a sensation for recognizing the different tones of sound. It appears from the above evidence that different sound frequencies preferentially vibrate selected portions of the basilar membrane. At these sites the hair cells receive a maximal stimulus, a particular set of nerve fibers is excited, and the corresponding tone identified. The problem is how best to explain the preferential movement of the basilar membrane. It is doubtful if the membrane behaves like a series of tuned resonators because its movements are complicated by (1) the shearing action of the tectorial membrane, and (2) the mass of moving endolymph. The displacment of the basilar membrane by a pressure wave traveling in the fluid of the cochlea is a more satisfactory explanation, since the peak of the wave varies with the driving frequency of the sound (Fig. 68). This means that a wave beginning at the oval window will reach its peak amplitude at a fixed distance away and then cease to travel. The peak of the wave is the point of strongest stimulation because at this point the vibration has its greatest effect. The discrimination of pitch therefore depends upon the part of the basilar membrane that has the maximal displacment. As the frequency of the sound changes, the peak of the traveling wave and point of maximal displacement are shifted to another part of the membrane.

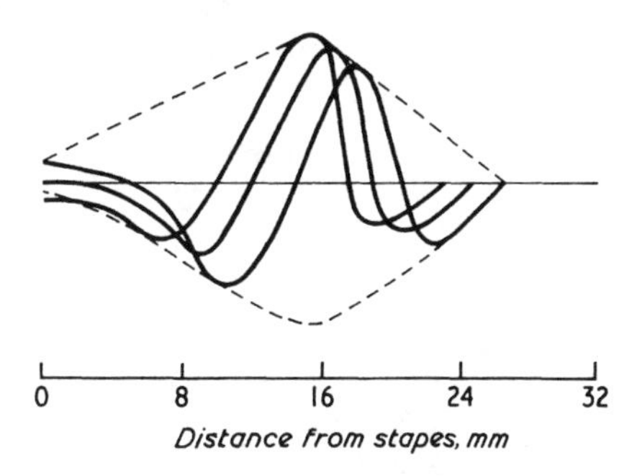

Fig. 68. Traveling wave along the basilar membrane. The solid lines represent the wave for a tone of 2000 Hz at successive intervals of time. The interrupted lines show the amplitude pattern formed by the wave peaks for this frequency. Maximum amplitude of movement is a function of distance from the stapes. Points of maximum amplitude occupy a particular position for each frequency.

DISCRIMINATION OF LOUDNESS

The sensitivity of the normal ear ranges from the faintest possible sound, or threshold of audibility, to a sound so loud that it becomes intolerable or even painful. Between these limits, the intensity of any sound is determined by the size of the pressure waves transmitted to the internal ear. The effect on the organ of Corti of increasing the intensity of a sound is to cause greater stimulation of the auditory receptor and, therefore, a greater discharge rate in the individual auditory nerve fibers. The maximum rate of discharge at the onset of the loudest tolerable sounds may be as high as 1000/sec, but the response to a continuous sound rarely exceeds 200/sec. Some hair cells reach their maximum discharge rate at relatively low stimulus levels.

A second factor in determining loudness is the frequency of the sound stimulus. Since each part of the basilar membrane is influenced by a particular band of sound frequencies, the discharge rate of a single nerve fiber might be increased by changing the tone without changing the intensity of the sound. Generally, there is less energy in the higher frequency components.

A third factor, which is illustrated in the curves of Figure 69, shows that increasing the loudness of a sound throws a longer stretch of the basilar membrane into vibration. This will have two effects: (1) A longer stretch of the organ of Corti will be stimulated and, as a result, a larger number of nerve fibers will become active. (2) Each nerve fiber will respond to a wider band of frequencies. The reason for this increased response is that a hair cell is tuned to a particular frequency only at threshold intensity. But it can discharge to other sound frequencies the more the intensity is increased.

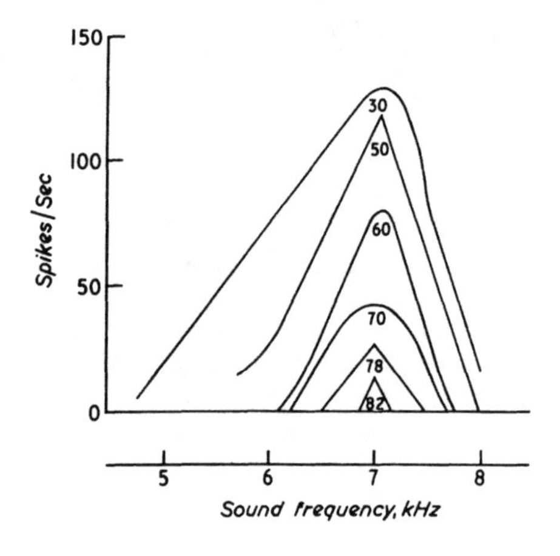

Fig. 69. Relation between intensity of sound and discharge rate of a single auditory nerve fiber. The ordinate represents nerve response in spikes/sec. The abscissa shows sound frequencies in kHz. The number of dB on each curve is the intensity level below a reference level. Note that as the level of sound intensity is increased, a wider range of frequencies excites the fiber. (After Galambos and Davis, 1943.)

In short, loudness is detected by the character of the impulse discharge in the auditory nerve fibers. As the intensity of sound increases, each hair cell excited discharges at a greater rate and responds to a wider band of frequencies, while the hair cells involved occupy a greater length of the basilar membrane, including those on the fringes of the vibrating portion.

Theories of hearing

A number of proposals have been put forward to account for the manner in which the basilar membrane responds to different sound frequencies:

1. The Telephone Theory assumes that vibrations of the basilar mem-

brane set up corresponding variations in waveforms that are transmitted by the auditory nerve to the brain. Discrimination of pitch and loudness is a function of the cerebral cortex and the cochlea does not behave as an auditory analyzer. With this theory it would be difficult to explain the capacity of the auditory nerve to conduct waveforms set up by high frequency vibrations, since the refractory period limits conduction to 1000/sec.

2. The Volley Theory accepts the limitation of the refractory period on nerve conduction, but suggests that higher frequencies are transmitted to the brain by supposing that several nerve fibers operate together in a group. Thus a sound frequency of 10,000 Hz could be transmitted by 10 nerve fibers, each conducting 1000 impulses/sec. The theory assumes that the basilar membrane vibrates as a whole to all frequencies of sound, but this seems highly improbable.

3. The Resonance Theory states that the basilar membrane operates as a series of tuned strings with different parts of it vibrating to certain notes. The basal part is adapted for high notes, and the apical part for low notes. Although there is good evidence for tonal localization in the cochlea, it seems that the strands connecting the entire length of the basilar membrane do not allow it to vibrate as do independent strings.

4. The Traveling Wave Theory is based on the motion of the cochlear fluid, which allows the peak of the wave to determine the position of maximal stimulation. For this reason it is commonly referred to as the Place Theory. The pitch of a sound is identified by excitation of a particular set of nerve fibers, while loudness is signaled by the rate of impulse discharge and by the recruitment of additional nerve fibers.

In summary, the internal ear appears to be endowed with the properties of an acoustic analyzer. Pitch is determined by the position of maximal movement of the basilar membrane, and loudness by the extent of vibration.

THE AUDITORY NERVE

Functional anatomy

The auditory nerve contains both afferent and efferent fibers. The *afferent* fibers arise from the hair cells of the organ of Corti. They consist of unmyelinated axons from bipolar cells in the spiral ganglion. The inner hair cells have a 1:1 innervation ratio, the outer hair cells about 1:10. The central processes of the bipolar cells emerge from the internal auditory canal and pass to the brain in company with the fibers of the vestibular nerve. The *efferent* fibers arise from the superior olive on both sides of the brain stem; they are distributed to the base and sides of the hair cells as vesiculated endings. Each auditory nerve, therefore, contains both crossed and uncrossed efferent fibers that are capable of modulating the afferent discharges.

Spontaneous activity

Microelectrode recordings from single auditory nerve fibers show that spontaneous discharges occur in the absence of sound stimulation. The rate of discharge varies from fiber to fiber, but the rate remains low and the interspike intervals constant. This electrical activity constitutes an important background signal or carrier wave for auditory impulse traffic.

Response to click stimulation

The response consists of a multiple spike discharge often showing two peaks. The latency of the first peak varies with the sound intensity but does not alter with repetitive stimulation. When identical clicks are sounded, each nerve fiber responds with a fixed latency for the first peak of the discharge. However, the latencies differ from fiber to fiber. These findings almost certainly reflect mechanical events in the cochlea and indicate that each auditory nerve fiber is excited in a specific way by the disturbances induced along the basilar membrane.

Response to tonal stimulation

When a particular tone is sounded, a certain part of the basilar membrane gives a maximal vibration. As the latency at the basal turn will be shorter than that at the apical turn, the auditory nerve response will have relatively short-latency peaks for high frequency tones and long-latency peaks for low frequency tones. The characteristic frequency of any individual fiber is, therefore, determined by its point of innervation along the stretch of basilar membrane from base to apex. Since each nerve fiber has a characteristic frequency, the basilar membrane may be regarded as an analyzer of sound frequencies, capable of resolving small differences in pure tones.

A continuous tone gives a sharp burst of impulses at the onset followed by gradual adaptation to a steady rate of discharge. The level of activity remains well below 200/sec, but is always significantly higher than the level of spontaneous activity. When the tone is switched off, there is usually a period of depression before a slow return to the spontaneous discharge.

Response to sound intensities

In the majority of auditory nerve fibers, the rate of impulse discharge generally rises with increased loudness of a continuous sound. Maximal discharge rate is reached with intensity levels 20 db to 50 db above threshold. The rate does not change with further increase of sound intensity. Some auditory nerve fibers tend to produce a fall in discharge rate with very high sound intensities. Since electrical stimulation of the cochlea can elicit discharges up to 1000/sec, the limiting factor must be attributed to the mechanical properties of the basilar membrane and not to the conduction properties of the auditory nerve.

SUMMARY

The ear responds to mechanical vibration of the sound waves in air. Sound energy is transmitted from the air to pressure waves in the cochlear fluid which in turn activate the sensory hair cells of the organ of Corti. The ossicles of the middle ear behave as a mechanical lever to communicate the pressure changes to the basilar membrane.

The discrimination of pitch is attributed to differences in the physical properties of the basilar membrane. The relatively short, stiff fibers near the base of the cochlea vibrate to high tones, while the more elastic and longer fibers of the apical region favor low tones. Pressure waves set up at the oval window travel through the upper compartment of the cochlea and reach their peak at sites relative to their driving frequency. At these sites the basilar membrane has its maximal displacement. Thus sound frequencies are discriminated from each other by the site of maximal stimulation of the nerve fibers supplying the organ of Corti.

The discrimination of loudness is attributed to differences in the pattern of impulse discharge to the brain. As the sound becomes louder, the amplitude of the pressure wave increases and a larger stretch of the basilar membrane is thrown into vibration. This results in a stronger stimulus being applied to the hair cells, so that the nerve endings discharge at more rapid rates. In addition, an increasing number of nerve fibers is excited.

The function of the cochlea is best studied in terms of the discharge patterns in the auditory nerve fibers. The studies indicate that there is a characteristic frequency for each fiber, according to its point of origin along the basilar membrane. At threshold intensity, a response is signaled when the rate of discharge rises significantly above the level of spontaneous activity. If the intensity of a sound is increased, the "tuning curve" for the appropriate fiber is extended in range, and there is a spread of energy to other frequencies. Finally, the presence of an efferent component in the auditory nerve suggests the possibility of central nervous control of the hearing mechanism.

SELECTED BIBLIOGRAPHY

Carmel, P.W., and Starr, A. Acoustic and nonacoustic factors modifying middle-ear muscle activity in waking cats. *J. Neurophysiol.* 26:598-616, 1963.

Davis, H. Biophysics and physiology of the inner ear. *Physiol. Rev.* 37:1-49, 1957.

Evans, E.F. The frequency response and other properties of single fibres in the guinea-pig cochlear nerve. *J. Physiol.* 226:263-287, 1972.

Fex, J. Auditory activity in uncrossed centrifugal cochlear fibers in cat. *Acta Physiol. Scand.* 64:43-57, 1965.

Galambos, R. Studies of the auditory system with implanted electrodes. In *Neural Mechanisms of the Auditory and Vestibular Systems,* 137-151. Springfield, Ill.: Thomas, 1960.

Galambos, R., and Davis, H. Responses of single auditory nerve fibers to acoustic stimulation. *J. Neurophysiol.* 6:39-57, 1943.

Klinke, R., and Galley, N. The efferent innervation of the vestibular and auditory receptors. *Physiol. Rev.* 54: 316-357, 1974.

Rupert, A. Moushegian, G., and Galambos, R. Unit responses to sound from auditory nerve of the cat. *J. Neurophysiol.* 26:449-465, 1963.

Tasaki, I. Nerve impulses in individual auditory nerve fibers of guinea pig. *J. Neurophysiol.* 17:97-122, 1954.

Tonndorf, J., and Khanna, S. The role of the tympanic membrane in middle ear transmission. *Ann. Otol. Rinol. Laryngol.* 79:743-753, 1970.

CHAPTER VIII

Organ of Vision

When light enters the eye it must first pass through an optical system to focus an image on the retina. Here the light initiates a chain of chemical reactions resulting in the breakdown and resynthesis of visual pigments. Specialized photoreceptor cells, the rods and cones, utilize the energy released from these reactions to set up electrical potentials in a network of retinal neurons.

The conversion of light energy into nerve impulses is the essential step in the process of vision. It amounts to bringing the outside world into the eye with attendant details of intensity, color, movement, etc., and to follow as accurately as possible all the features of a continuously changing scene. What is virtually a photochemical process becomes a pulse-coded message in a multi-channel system connecting the eye through the optic nerve fibers to the visual centers of the brain.

In studying the organ of vision, the eye will first be considered as an optical instrument adapted to giving an image under varying conditions of illumination. Second, an account will be given of the photoreceptor mechanisms responsible for visual acuity. The remainder of this chapter describes some of the complex interactions that occur in the neural organization of the retina itself.

OPTICAL PROPERTIES OF THE EYE

Properties of lenses

A lens consists of a piece of glass or other transparent medium with one or both surfaces representing a portion of a sphere. The surfaces may be either convex or concave. When light passes through such a medium, the rays emerge on a different course and continue in the new direction. This deviation of light rays from their original path is know as *refraction.* The rays are deviated toward the thickest part of the lens. Hence, rays passing through a convex lens will tend to converge and meet at a point called the principal focus of the lens, whereas rays passing through a concave lens will be made more divergent and will not meet (Fig. 70). Any ray of light passing through the optical center of a lens is not deviated when it emerges. A convex lens converges parallel rays and makes a divergent beam less divergent. A sharp image will be formed on a screen placed at a distance corresponding to the principal focus of the lens; if the screen is moved away from the principal focus in either direction, the image will be blurred.

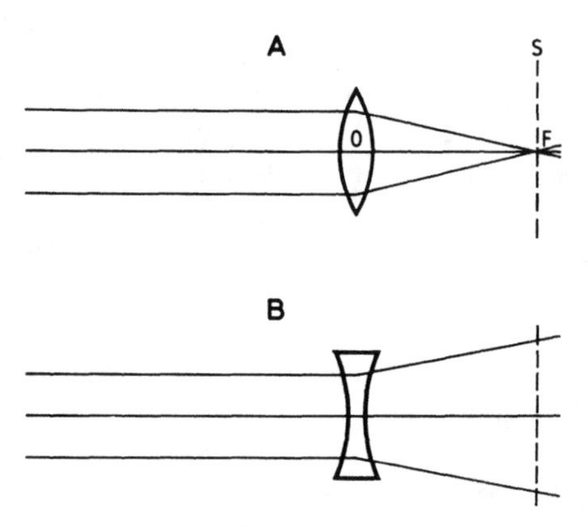

Fig. 70. Diagrams illustrating the passage of light through lenses. A, convex lens. Parallel rays of light converge to point F, the principal focus, where a sharp image is formed on the screen (S). If the screen is moved away from this point, the image will be blurred. B, concave lens. Parallel rays of light diverge as they pass out of the lens, thus not forming an image. Rays of light passing through the optical center (O) emerge without deviation.

All lenses used in optics are numbered according to their refractive power in units called *diopters.* A lens of 1 diopter has a focal length of 1 meter. Stronger and weaker lenses are measured as the reciprocal of the focal length. Thus a lens twice as powerful has a focal length of 50 cm; a lens five times as powerful has a focal length of 20 cm. A weak lens, e.g., 0.25 diopters, has a focal length of 4 meters. A positive sign (+) is used for a convex lens and a negative sign (–) for a concave lens. The two kinds can be distinguished by looking at the image while moving the lens across the field of view. If the lens is convex, the image moves in the opposite direction; if it is concave, the image moves in the same direction. When two lenses of equal power but opposite in sign are superimposed, they cancel each other out and have the effect of plane glass.

Refracting media of the eye

The eye possesses a compound optical system comprising four structures of different optical density:

1. The cornea is a transparent membrane, convex anteriorly, that forms a segment of a sphere smaller than the rest of the eyeball. It is the first part of the eye to affect the rays of light by convergence.

2. The aqueous humor is a watery fluid occupying the anterior chamber of the eye between the cornea and the iris diaphragm.

3. The lens, situated immediately behind the iris, is enclosed in a capsule and held in position by the suspensory ligament. It is a transparent elastic body, biconvex in shape, the convexity of its anterior surface being less than that of its posterior surface in the relaxed eye.

4. The vireous humor fills the space between the lens and the retina and its jelly-like consistency gives a firmness to the eye, keeping the retina at a fixed distance from the lens.

On entering the eye, light is refracted as it passes from one medium to the next. The amount of refraction is different at each of the interfaces since these have different curvatures and the density of the media also varies. Because the air has a low index of refraction, the rays are more sharply bent by the surface of the cornea than by the lens itself. Indeed, the lens contributes only about 15 diopters to a total refractive strength of about 65 diopters. The reason for this is that the difference of refractive index between the lens and the fluids surrounding it is small. Hence, light rays are not particularly bent at the lens interfaces.

In the normal eye, rays of light from an object come to a focus on the retina, about 15 mm behind the lens. Rays from each point of the object come to a separate focus on a common plane where an image is formed. Figure 71 shows that this image is inverted and reduced in size. The object

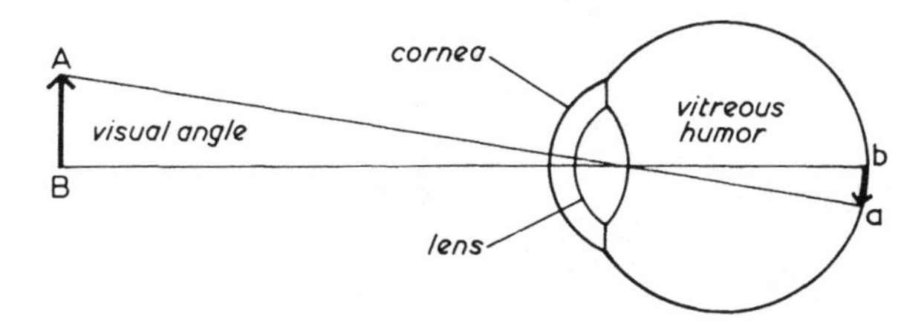

NOTE: anterior surface of lens is slightly more convex than the posterior surface.

Fig. 71. Formation of inverted image. The object is represented by the arrow AB. Light from A passes through the optical center to point a on the retina. Light from B passes along the principal axis to point b. Image ba is reduced and inverted. The visual angle is formed at the nodal point where the two axes cross.

is represented by the arrow AB. A ray from point B passes through the optical center to focus on the principal axis at b; a ray from point A passes through the optical center on a secondary axis to a focal point at a. The image is represented by the inverted arrow ba. The optical center where the rays are crossed is called the nodal point, and the angle formed by them is called the visual angle. The angle becomes wider as the object comes nearer to the eye.

Errors of refraction

The eye is considered to be normal, or *emmetropic,* when light from a distant object forms a sharp image on the retina. In many people, the optical system is not perfect; objects are not clearly brought to a focus on the retina and vision is blurred. The following errors of refraction are commonly found:

1. Astigmatism. This is caused by unequal curvature in any axis of the cornea or lens. As a result, rays of light are brought to a focus unequally; one set may converge to a point on the retina while those at right angles may focus in front of or behind the retina. The effect is that the rays do not come to a common focal point, and the subject sees a line instead of a spot. To test for astigmatism, he looks at a chart of radiating lines or letters and, if astigmatism is present, some lines will appear more distinct than others. These give the direction or axis of greatest curvature; the lines at right angles that appear indistinct correspond to the axis of least curvature. Astigmatism may be corrected by a cylindrical lens to make the lines appear identical in all meridians. In practice, the cylindrical lens is adjusted on the frame until the axis of least or greatest curvature is found. The curvature of the cylinder enables the subject to see all the meridians on the chart with equal clarity. If there is any accompanying error of refraction, this must also be corrected by using lenses of appropriate dioptric strength.

2. Myopia. In this condition the eyeball is too long, or the refraction of the lens too high, for light rays from a distant object to come to a focus on the retina. Objects beyond a certain distance will be out of focus and appear blurred; this is the myopic far point. All rays from more distant objects come to a focus in front of the retina and diverge again to form an indistinct image. Objects nearby, on the other hand, are seen clearly because the rays entering the eye are sufficiently divergent to focus accurately on the retina. The constant use of the eyes for close work may cause considerable strain and has led to the term *nearsightedness* or *shortsightedness.* Glasses should always be prescribed. Myopia is corrected by using a concave lens, which will diverge the light rays (Fig. 72A). The power of the lens required

to restore normal visual acuity is a measure of the degree of myopia present.

3. Hypermetropia. In this condition the eyeball is too short, or the refraction of the lens too weak, to focus rays from distant objects on the retina. The eye can overcome the disability by muscular effort, and this might be sufficient for seeing distant objects clearly, though at the expense of eyestrain. This has led to the use of the term *farsightedness* or *longsightedness*. It follows that the subject has more difficulty in seeing objects nearby since the rays come to a focus behind the retina and appear blurred. Hypermetropia is corrected by using a convex lens for near vision. Divergent rays are then made less divergent or made convergent to focus accurately on the retina (Fig. 72B). The power of the lens required to restore near vision for close work is a measure of the degree of hypermetropia present.

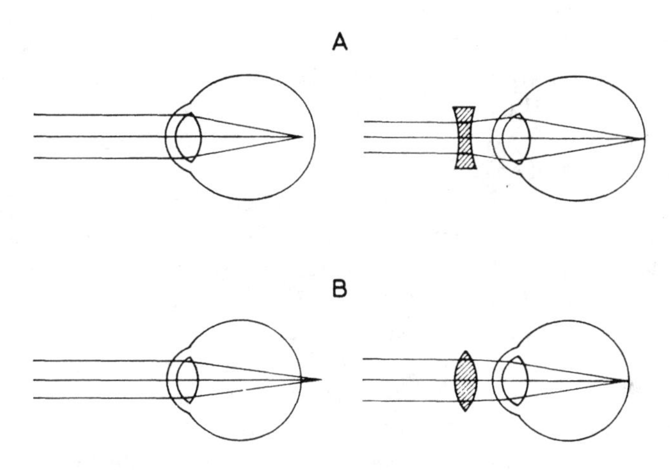

Fig. 72. Correcting lenses for errors of refraction. A demonstrates myopia. Rays of light come to a focus before reaching the retina (eyeball too long). The condition is corrected by a suitable concave lens to give a degree of divergence. B represents hypermetropia. Rays of light have not come to a focus when they reach the retina (eyeball too short). The condition is corrected by a suitable convex lens to give an additional amount of convergence.

Tests for visual acuity

Various tests are used to estimate errors of refraction and to find out whether vision can be improved by prescribing glasses:

1. Test charts. The letters typed on the chart subtend an angle of one minute when placed at a standard distance away. Each row of letters is

marked with the distance it can be read by a subject with normal vision. At a distance of 6 meters, the subject should be able to read the row of letters marked for that distance; his visual acuity is expressed as 6/6. A person who has defective visual acuity might be able to read only the larger typed letters marked 36; his visual acuity is expressed as 6/36; this means he can read from a distance of 6 meters letters that a person with normal vision can read at 36 meters. Trial lenses of different refractive power are inserted into the spectacle frame until visual acuity is brought close to normal. Each eye is tested separately, and the appropriate glasses prescribed for the individual to wear.

2. Retinoscopy. This is an objective method of estimating the refraction of the eye and, therefore, very suitable for small children. A beam of light is reflected into the subject's eye from a plane mirror, which is tilted from side to side. The observer looks through a hole in the mirror and sees the reflected beam on the pupil. If the reflection moves in the same direction as the mirror, the subject is hypermetropic; if it moves in the opposite direction, he is myopic. Lenses of different refractive power are placed in front of the eye until one is found to stop the movement. The appropriate glasses can then be prescribed.

3. Ophthalmoscope. This instrument consists essentially of a battery torch, a prism and a system of lenses. A beam of light is directed by the prism into the subject's eye, and the emerging rays are caught through a sight-hole to reach the observer's eye. In this way a view of the retina, optic disc, and blood vessels is obtained. If both the observer and subject are emmetropic, the retina will be sharply focused. If the observer is emmetropic but the retina appears blurred, trial lenses are inserted into the sight-hole until the retina is in proper focus. From the power of the lens used, the error of refraction is estimated. The ophthalmoscope contains a series of convex (+) and concave (−) lenses mounted in a rotary fashion and appropriately numbered in diopters. If the observer is not emmetropic, he must first correct for his own refractive error.

Accommodation

The eye is capable of altering its refractive powers through the process of accommodation. As mentioned previously, light from a distant object reaches the eye as parallel rays and a sharp image is formed on the retina. This is because the retina coincides with the principal focus of the optical system. The greatest distance at which an object can be clearly seen is called the Far Point when the eye is relaxed and no accommodation is necessary. For practical purposes, the rays are regarded as parallel for any distance beyond 6 meters from the eye. When an object is brought closer to the eye, the rays become more divergent and will reach the retina before the object comes into focus. As a result, the image will be blurred unless the refractive system of the eye can be increased. In fact, objects nearer than 6 meters away are kept in focus by muscular effort so that blurring does not occur. The shortest distance at which an object can be clearly seen is called the Near Point when the eye is fully accommodated. The range of accommodation is thus the distance between Far and Near Points. Since the curvature of the cornea and the fluids surrounding the lens remain unaltered, the change in the refractive system is brought about by alteration in the power of the lens.

Mechanism of accommodation. The lens is covered by an elastic capsule which in turn is attached by the suspensory ligament to the ciliary body. When the suspensory ligament is held taut, the capsule is stretched and the lens flattened; when the ligament is relaxed, the capsule becomes slack and the lens assumes a more spherical shape.

The release of tension on the suspensory ligament is brought about by contraction of the ciliary muscle. The smooth fibers of the ciliary muscle are arranged around the interior of the eyeball in the angle between the cornea and the iris. They are attached to this junction at one end, and the other end is inserted into the ciliary processes formed by the anterior edge of the choroid. Contraction of the muscle pulls on the ciliary processes, drawing the choroid forward toward the lens, thus relaxing the tension on the suspensory ligament. The lens, by virtue of its own elasticity, is allowed to become more convex. The greatest change in curvature occurs on the anterior surface where the lens bulges toward the pupil; the posterior surface of the lens remains unaltered owing to the greater resistance of the vitreous humor (Fig. 73).

Amplitude of accommodation. When the eye looks at a distant object, the lens is kept flat by the action of the suspensory ligament and its refractive power is at a minimum. In viewing objects nearby, the ciliary muscle contracts, diminishing the tension of the suspensory ligament and increasing the convexity of the lens until the Near Point is reached. The lens is now at its maximal refractive power. The difference between these two extremes is a measure of the amplitude of accommodation. In a young adult with normal vision the maximal amount of accommodation that can be elicited is equivalent to about 10 diopters. Thus the amplitude of accommodation can be expressed as the number of diopters added to the refractive power of the eye by muscular effort. The ciliary muscle is the principal agent for accommodating the eye to near vision. This muscle is supplied by the parasympathetic division of the oculomotor nerve.

Presbyopia. The amplitude of accommodation gradually diminishes as a

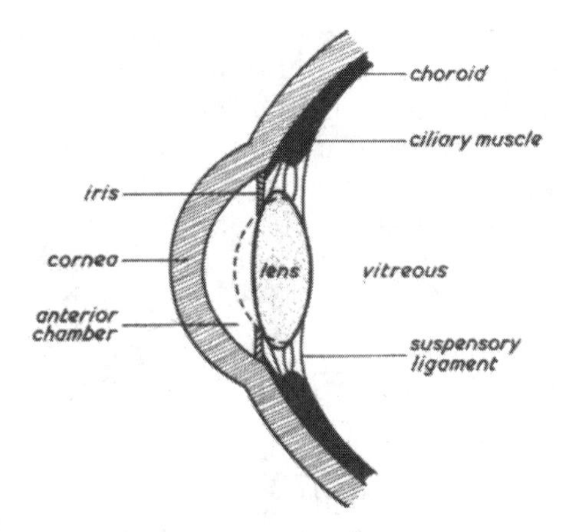

Fig. 73. Anterior segment of the eye illustrating mechanism of accommodation. Contraction of the ciliary muscle draws the choroid forward, relaxing the suspensory ligament. This allows the anterior surface of the lens to bulge into the anterior chamber (dotted line). The lens thus becomes more powerfully convergent.

person grows older, because of loss of elasticity and hardening of the lens. At birth, the amplitude is approximately 14 diopters, but decreases to 2 diopters in middle age. As a result the Near Point for clear vision recedes from about 7 cm in young children to about 30 cm or more in adults at the age of 45. At this age, there is difficulty in reading small print at a comfortable distance from the eye since the Near Point is then beyond the range of effective accommodation. This decrease in the amplitude of accommodation is called presbyopia. Apart from the difficulty in focusing near objects, the constant muscular effort involved in eye strain generally leads to symptoms of fatigue and headache. Presbyopia is treated by providing suitable convex lenses to increase the refractive power of the eyes. As no accommodation is required for distant vision, glasses need be worn for reading only or for very close work.

Role of the pupil

The amount of light entering the eye is determined by the size of the pupil. This is the aperture in front of the lens controlled by the sphincter mechanism of the iris diaphragm. Constriction of the pupil reduces the amount of light falling on the retina and eliminates scattered peripheral rays. Consequently, the definition of the image is improved. Constriction occurs automatically when the intensity of the light is high. When the light is dim, the pupils dilate to improve the ability to see in dark conditions.

Pupillary reflexes

1. Direct light reflex. If a torch light is shone into one eye, the pupil rapidly constricts and dilates again when the light is removed.

2. Consensual light reflex. Light shone into one eye causes pupillary constriction in the opposite eye.

The afferent pathway for both reflexes is via the optic tracts to the superior colliculus. The consensual light reflex is present because fibers from each eye pass into both optic tracts after decussation in the optic chiasm. From the superior colliculus, relays cross near the aqueduct of Sylvius to reach the small-celled component of the oculomotor nucleus, comprising parts of the Edinger-Westphal and anteromedian nuclei. The efferent pathway is via the parasympathetic outflow from the oculomotor nucleus to the ciliary ganglion; postganglionic fibers then proceed in the short ciliary nerves to the pupil.

3. Accommodation reflex. Pupillary constriction occurs when the eyes change their focus from a distant to a near object.

The afferent pathway is via the optic tracts to the lateral geniculate body and cerebral cortex. The reflex arc is completed by corticonuclear fibers to the oculomotor nucleus which supplies the medial rectus and ciliary muscles as well as the pupil. The oculomotor nerve is thus the common efferent pathway for both the light and accommodation reflexes. As syphilis of the nervous system may cause destruction in the region of the aqueduct and superior colliculus, the light reflexes may be lost while the reaction to accommodation is retained. This sign is known clinically as the Argyll Robertson pupil.

The diameter of the pupil is maintained by a tonic parasympathetic discharge until modified by some excitatory or inhibitory influence. The light reflex is an excitatory reaction causing further constriction of the pupil. Sympathetic stimulation, on the other hand, exerts an inhibitory influence on the oculomotor nucleus and dilates the pupil. Many spinal afferents, especially those carrying pain nerve impulses, may cause dilatation of the pupil. This is brought about by inhibition of the parasympathetic since dilator effects are present after section of the cervical sympathetic trunk.

Eye movements and their control

Although the sense of sight is normally binocular in function, the impression produced in consciousness is that of a single image. To accomplish this, it is necessary for the movements of each eye to be coordinated so that the two lines of vision meet at the object. Otherwise, two distinct images will be formed, resulting in double vision or diplopia.

Movements of the eye. Each eye is capable of movement about three axes—horizontal, vertical, and transverse. The point of intersection of each axis is termed the center of rotation. The movements are produced by six small striated muscles arising from the walls of the orbit and attached to the exterior of the eyeball. The actions and nerve supply of the extraocular muscles are outlined in Table III.

TABLE III. Actions and Nerve Supply of the Extraocular Muscles

Movements	*Muscles*
Inward	Internal rectus
Outward	External rectus
Upward	Superior rectus and inferior oblique
Downward	Inferior rectus and superior oblique
Inward and upward	Internal and superior recti, inferior oblique
Inward and downward	Internal and inferior recti, superior oblique
Outward and upward	External and superior recti, inferior oblique
Outward and downward	External and inferior recti, superior oblique

The oculomotor (third) nerve supplies four of the muscles: internal rectus superior rectus, inferior rectus, and inferior oblique. The trochlear (fourth) nerve supplies the superior oblique; the abducent (sixth) nerve supplies the external rectus.

Conjugate movements. When the two eyes are used in vision, the action of associated muscles is called conjugate movement of the eyes. This means the line of vision is directed to a common point in the visual field, while rays of light from this point will stimulate corresponding points of each retina. A movement of the eyes horizontally to the right is brought about by simultaneous contraction of the right external rectus and the left internal rectus. Similarly, when the eyes are moved upward, the superior rectus and inferior oblique contract together so that it is impossible to turn one eye up and the other down at the same time. These conjugate movements are made possible by simultaneous relaxation of the respective antagonistic muscles.

Convergence of the visual axes. When looking into the distance, the visual axes of the two eyes may be regarded as parallel, but as the distance is shortened, the eyes are turned inward to keep the object in focus. In movements of convergence the internal recti of the two eyes are associated. Convergence of the visual axes is accompanied by the accommodation reflex, since both adjustments are essential in order to follow an object moving closer to the eyes or further away from them.

Strabismus, or squint, is a condition in which the visual axes are not directed to the same point in the visual field; hence non-corresponding points on the retina are stimulated. The subject tends to use only one eye to prevent confusion from double vision, but suppression of one image leads to reduction of visual ability. The condition is due to lack of balance or of coordination between associated pairs of muscles, causing one eye to drift from its correct position. Ocular imbalance may be caused by overaction of one muscle or incomplete relaxation of its antagonist. Strabismus is usually a congenital defect, but may be acquired from lesions affecting the nerve supply to the muscles. Any attempt by the subject to use both eyes for near work results in considerable muscular strain. Treatment includes the use of eyeglass prisms adjusted to assist the weaker muscle or surgical intervention to limit the power of the stronger muscle.

Visual fields

1. Definition. The visual field is the extent of the outer world that can be seen when the eye is fixed on an object directly in front of it. The field of vision is limited by the nose, eyebrow, and cheek but can be extended without moving the head simply by using the extraocular muscles. The eye can then scan the field efficiently and rapidly, as occurs in reading a page of print.

2. Binocular vision. When both eyes are fixed on an object, the two fields of vision overlap to give an increased view extending about 120°. On both sides of this region is a field that can be seen by one eye only and is known as the monocular or temporal crescent.

3. Perimetry. The perimeter is an instrument used for mapping the visual fields. The subject closes one eye and keeps the other eye fixed on a central point, usually a small mirror. The observer moves a disc inward along the edge of a meridian, keeping it at the same distance from the eye, until the subject indicates he can see it. The angle is noted and plotted on a chart. This is repeated for each meridian until the entire field is mapped; the other eye can then be tested by the same procedure.

The perimeter can also be used for testing color sensitivity by plotting the fields with red, green, and blue discs. Figure 74 shows the shape of the normal visual field for the right eye. The area seen to the right of the chart is called the temporal field of vision, and the area seen to the left of the chart is called the nasal field of vision. It will be noted that the visual field is inverted upon the retina; objects in the upper half of the field produce an image in the lower half of the retina, and objects in the temporal field of vision produce an image in the nasal half of the retina. The field of vision can therefore be divided into four unequal quadrants. This may be useful in describing the localization of scotomas, or blind spots, in the retina. Perimetry is also helpful in the diagnosis of brain lesions in which contraction of the visual fields may result from damage to any part of the visual pathway.

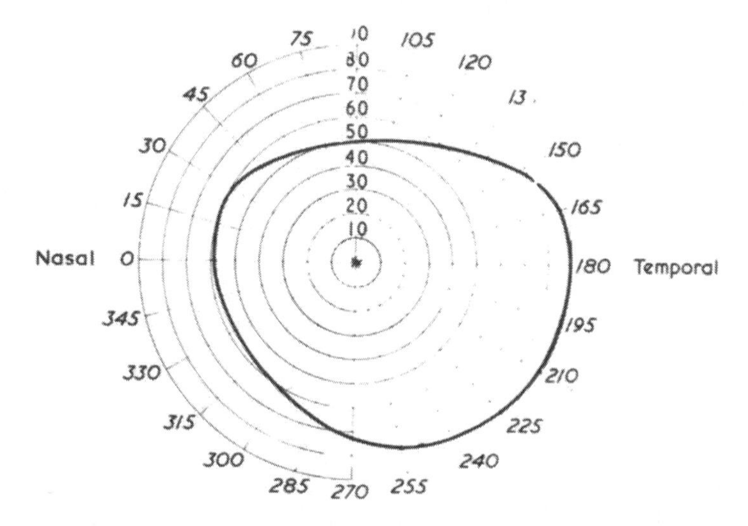

Fig. 74. Perimeter chart showing normal field of vision for right eye. The numbers along the meridians are degrees of visual angle measured from the fixation point at the fovea.

FUNCTIONAL PROPERTIES OF THE RETINA

In this section an account will first be given of the retinal structure since the retina is the light-sensitive portion of the eye, containing the specialized receptor elements, the rods and cones. Next, the nature of the light stimulus will be described and the suggestion made that the visual processes behave in a dual manner adapted for bright or dark conditions according to the intensity of the stimulus. Third, the photochemical basis of vision will be considered as reactions of the receptor elements to the bleaching of pigments. In the process these elements give rise to nervous excitation. Fourth, the effects of light stimulation on both rods and cones will be related to their remarkable adaptation abilities. Finally, it will be shown how the eye is capable of seeing small objects and discriminating fine detail in proportion to the conditions that bring the cone system into full operation. A summary of the factors that influence visual acuity will be found at the end of the section.

Structure of the retina

1. General features. The retina extends over the inner surface of the eyeball, lying near the choroid–a vascular, pigmented membrane which provides nutrients to the ocular tissues and serves as a light-absorbing lining. Opposite the center of the pupil there is a small area, the macula, and a central depression, the fovea, where the retina is thinnest and closely packed with cones (the depression is caused by the absence of nerve fibers and blood vessels). About 3 mm to the nasal side of the macula, the optic nerve leaves the eyeball at the optic disc or blind spot, since the disc is insensitive to light. The central artery of the retina and accompanying veins pierce the eyeball at the disc and immediately branch to divide the retina into four quadrants–upper and lower nasal, and upper and lower temporal.

The optic disc and retinal vessels can be observed directly through an ophthalmoscope (Fig. 75). The disc is pale yellow in color with a fairly sharp outline, sometimes partly ringed by pigment, and shows a slight depression, the physiological cup. If for any reason there is an increase of intracranial pressure, the outline of the disc becomes blurred because of back pressure on the retinal veins which appear engorged or choked. The condition is termed *papilledema* and is an important diagnostic sign.

2. Histological features. The human retina is a complex structure of neural and non-neural elements arranged in layers from the outside to the

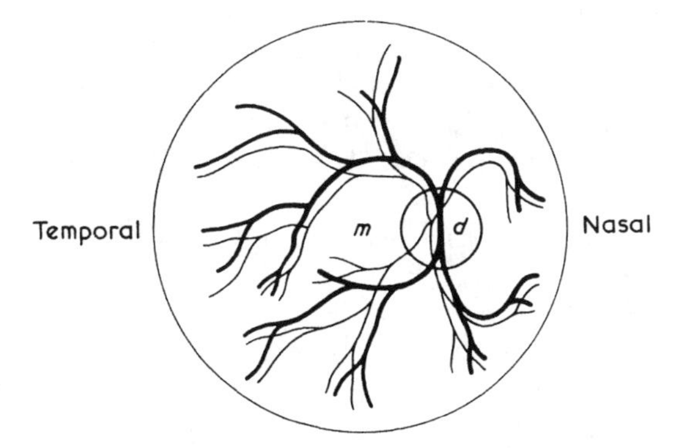

Fig. 75. Appearance of right fundus showing the main distribution of the blood vessels above and below the macula (m). The branches of the central artery emerge at the optic disc (d) and divide the retina into four quadrants–upper and lower nasal and upper and lower temporal. Arterial supply is shown as thick lines, accompanying veins as thin lines.

inside, as illustrated in Figure 76. The neural elements consist of (a) neurons linking the rods and cones to the optic nerve fibers, and (b) two intervening synaptic layers of horizontal and amacrine cells. The non-neural elements are the photoreceptors and their related pigment cells.

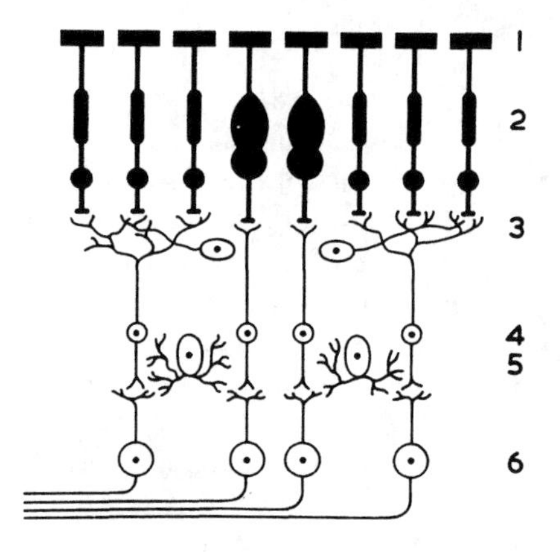

Fig. 76. Simplified diagram of retinal structures depicting neuronal arrangements—1, pigment layer; 2, rods and cones; 3, horizontal cells; 4, bipolar cells; 5, amacrine cells; and 6, ganglion cells. Note that many rods converge on one bipolar cell and that cones have their own pathway to ganglion cells. Axons of ganglion cells form the optic nerve.

(a) Outer pigmented layer. This consists of a single row of cells lying adjacent to the choroid and containing photosensitive pigment.

(b) Layer of rods and cones. Each rod is a cylindrical cell perpendicular to the surface. The outer part is in contact with the pigment layer; the inner part contains the nucleus and mitochondria and forms a synaptic body to connect with the neural elements in the next stage of the chain. Several rods terminate on the dendrites of one bipolar cell. Cones are similar in structure though less numerous. The nucleus is larger and the cell body wider in diameter; the terminal contact is usually on one bipolar cell with little or no sharing.

(c) Layer of horizontal cells. Processes from these cells connect at various points between the rods and cones.

(d) Layer of bipolar cells. These cells may be regarded as second order neurons connecting the rods and cones to the ganglion cells. The *diffuse* kind has a spreading dendritic tree embracing a large number of rods; the midget bipolar has dendritic contacts for single cones.

(e) Layer of amacrine cells. These cells appear to have no axons but their profuse dendritic expansions form a network between the bipolar and ganglion cells.

(f) Layer of ganglion cells. These are flask-shaped cells of various size and dendritic complexity, the dendrites interlacing with those of the bipolar and amacrine cells. They may be regarded as third order neurons whose axons converge on the optic disc and leave the eyeball in the optic nerve.

(g) Layer of optic fibers. This is the innermost layer of the retina through which light must pass to reach the rods and cones. Axons located in the nasal part of the retina take a fairly straight course to the optic disc, while those located in the temporal part sweep round the macula to avoid the fovea. The optic nerve fibers gain their myelin sheath as they leave the eyeball.

3. Distribution of rods and cones. The receptor elements in the peripheral region of the retina are mainly rods. At the macula, the cone-to-rod ratio steadily increases and in the fovea rods are entirely absent. In a transverse section through the fovea the cones are seen to be hexagonal in shape and closely packed, giving them the appearance of a "fine grain." As mentioned above, the nerve fibers and blood vessels of the retina sweep around the macula, away from the fovea, and thus obstructions are avoided that would cause the scattering of light. These features, as well as the one-to-one arrangement, which gives each cone a "private pathway," ensure that the central part of the retina is the region of greatest visual acuity.

Visual spectrum

Ordinary daylight consists of a narrow band of wavelengths ranging from 700 nm to 400 nm. If a beam of light is passed through a prism the rays are refracted in different degrees and appear as colored bands on a screen. The longer wavelengths appear red, the intermediate yellow-green, and the shorter ones blue. At the extreme ends of the visible spectrum, the longer wavelengths extend into the infrared region and the shorter ones into the ultraviolet. The amount of energy released by light falling on the retina is not the same for each wavelength. This is because the photosensitive pigments which absorb the light have different spectral sensitivities. Correspondingly, the amount of light required to produce a visual sensation also depends upon wavelength. The unit expressing the intensity of light reflected from an object is measured in millilamberts. Since the amount of light reaching the retina is determined by the size of the pupil, another unit which takes this factor into account, the photon, is used to express the level of brightness.

The distribution of energy changes among the different wavelengths of the spectrum may be plotted in the form of visibility curves (Fig. 77). The ordinate can be derived directly, in animals, from the frequency response of single optic nerve fibers. In the human being, the values are determined from the reciprocal of the energy required to match a standard light.

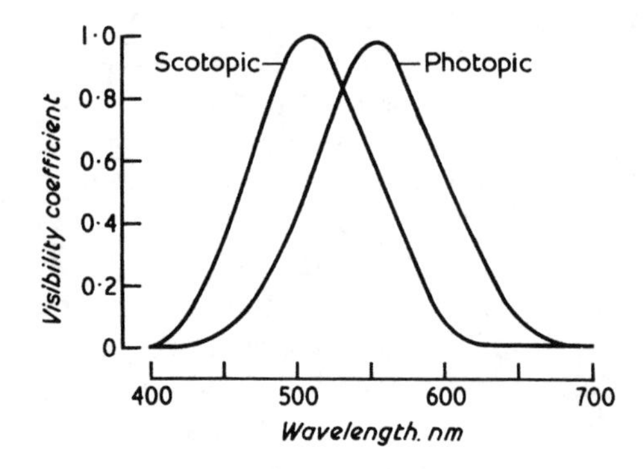

Fig. 77. Visibility curves for human retina. The scotopic curve due to rods has its peak at 510 nm; the photopic curve due to cones has its peak at 570 nm. For comparison, maximum sensitivity in each curve is taken as 1.0.

- The scotopic visibility curve is a function of the rod receptor system. It represents the spectral sensitivity of the dark-adapted eye and has a maximal response at 510 nm.
- The photopic visibility curve is a function of the cone receptor system. It represents the spectral sensitivity of the light adapted eye and has a maximal response at 570 nm.

A comparison between the two curves is made possible by plotting the ordinate on a common scale in which the peak of each curve is equal to 1.0. It will be evident that the intermediate wavelengths are more effective in stimulating the retina than the longer and shorter wavelengths and that the capacity to absorb light at any wavelength is dependent upon the brightness of illumination. The visibility curves show very definitely that the retinal mechanism is adapted for both twilight and daylight vision.

Photochemical basis of vision

All the photochemical substances of the retina have the capacity to absorb light and to initiate reactions that result in nervous excitation.

1. Scotopic vision. The rod receptor system is associated with the photochemical substance visual purple, or rhodopsin. Kuhne was the first to show that a solution of visual purple was bleached by the action of light.

Subsequently, an extract of frog rhodopsin prepared in the dark was exposed to different wavelengths of light and an absorption curve constructed as the one shown in Figure 78. The shape of the curve and degree of absorption at each wavelength are very similar to the visibility curve for rod function and therefore suggest that rhodopsin is the photochemical pigment bleached by light at low intensities.

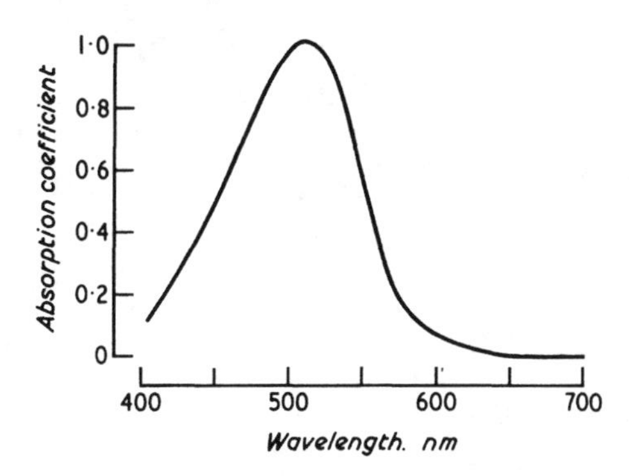

Fig. 78. Absorption spectrum for frog rhodopsin. The curve shows degree of bleaching at different wavelengths of light and closely corresponds to the scotopic visibility curve, with maximum absorption at 510 nm.

2. Photopic vision. The cone receptor system is associated with the photochemical substance visual violet, or iodopsin. It was first extracted by Wald from the retina of the chicken, which possesses relatively few rods. The spectral absorption curve of iodopsin is similar to the photopic visibility curve obtained in the human retina with high intensities of light.

Cycle of reactions. The visual pigments are chromoproteins, which rapidly decompose on exposure to light and re-form in a continuous process to maintain a steady level. The protein component is a colorless substance, opsin. The pigment with which it is combined is retinene, the oxidized form of the carotenoid, Vitamin A. The breakdown and synthesis of rhodopsin and iodopsin are similar. There are three main reactions—breakdown, rapid resynthesis, and slow resynthesis.

Breakdown

Rhodpsin → orange intermediates → opsin + retinene

The retinene is reduced to Vitamin A by the enzyme dehydrogenase and the coenzyme DPN, a derivative of yeast. The energy released from this reaction initiates the electrical changes in the receptor. As light continues to fall upon the eye, more and more retinene is reduced to Vitamin A and the level of visual pigment tends to fall. Rhodopsin, however, is continuously re-formed.

Rapid resynthesis

$$\text{Opsin} + \text{retinene} \xrightarrow{\text{dark}} \text{rhodopsin}$$

Rhodopsin is resynthesised in the dark by spontaneous action without the help of enzymes. Energy released from this process is used to oxidate Vitamin A to retinene, which then combines with opsin.

Slow resynthesis

$$\text{Vitamin A} \rightarrow \text{retinene (neo-b)} + \text{opsin} \rightarrow \text{rhodopsin}$$

Additional Vitamin A is supplied from the circulating blood, but only the correct isomer is effective. The Vitamin A formed by the bleaching of rhodopsin exists as the all-trans isomer; this must be converted into the cis-trans isomer to form neo-b retinene, which then combines with opsin.

The level of visual pigment is maintained by continuous recycling. The rapid process occurs spontaneously, the slower process requires enzyme mechanisms for conversion of Vitamin A to its specific isomer. A deficiency of Vitamin A in the diet may result in lowering of its level in the retina and therefore hinder the resynthesis of pigment. The condition is known as night blindness, or inability to see well in the dark.

Dark and light adaptation

It is a common experience to see very little on entering a dark room and soon have vision improve as the eyes become accustomed to the conditions. This change is known as dark adaptation. In a similar manner, the eyes are dazzled on passing from dark into daylight until they become less sensitive through light adaptation. The adjustment of the eyes to sudden changes of light intensity is a function of the cones as well as the rods.

To obtain a curve for dark adaptation, the threshold of illumination is plotted against time in the dark. The curve has two components, an initial, rapid portion due to adaptation of the cones, and a slower, but larger portion due to adaptation of the rods (Fig. 79). Cone adaptation is associated with the resynthesis of iodopsin and reaches a plateau after a few minutes; rod adaptation is associated with the resynthesis of rhodopsin and takes about 20 minutes to reach an effective level. It will be apparent that, although more slowly adapting, the contribution of the rods is quantitatively more important for securing the best results in night vision. The rate of dark adaptation can be improved by the avoidance of bright lights and by wearing red goggles, which allow cone vision to continue. If the eyes are suddenly exposed to a bright light, all visual pigments are rapidly depleted.

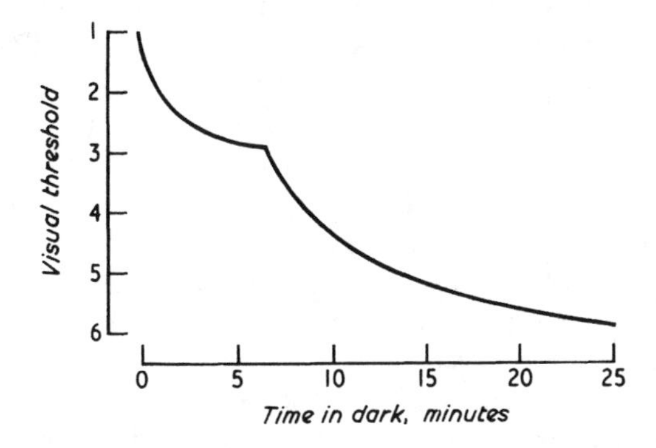

Fig. 79. Dark adaptation curve of human eye. Visual threshold is plotted against time spent in the dark. The first part of the curve shows rapid but limited adaptation due to the cones; the second part of the curve is more prolonged due to sensitivity of the rods. Dark adaptation is virtually completed in 20 minutes.

Visual acuity

The rods and cones belong to two receptor systems, each possessing photosensitive pigments that are bleached by light. The rod system is sensitive to low levels of illumination but is not capable of discriminating color or detail. As the illumination improves, the cone system comes into operation and fine, accurate work is then possible. At the same time the pupils constrict, sharpening the visual image. Since the cones are situated in the central part of the retina, only a small part of the visual field is utilized—the part where rays of light fall on the principal visual axis. Other factors influencing visual acuity are listed below.

Resolving power of the eye. Visual acuity is the ability to see very small objects with the unaided eye and to identify fine details clearly. Acuity is often measured by finding the greatest distance from which the eye can see two black dots or lines marked on white paper 2 mm apart. The angle subtended by the distance between the dots and the distance from the eye is called the minimal angle of resolution. Experimentation shows that the dots must be marked further apart as the distance from the eye is increased. In other words, if the angle subtended is less than a certain minimal value, the two dots will appear as one dot. The angle of resolution usually obtained is an angle of one minute, which corresponds to a distance between the images on the retina of about 3 μm, i.e., the diameter of a single cone. This suggests that two dots can be discriminated if the stimulated cones are separated by an unstimulated one. Maximal visual acuity, however, may be greater than this because light may spread to the intervening cones in the form of light and dark bands, the cones being sensitive to differences in the intensity of illumination at the edges of the bands.

Region of highest visual acuity. The fovea possesses the greatest capacity for detailed vision:

1. It lies opposite the center of the pupil, on the principal visual axis, and is therefore in the best position for a sharply focused image. Light passing through the periphery of the pupil is much less effective; visual acuity falls off as the rays strike the extra-foveal region of the retina. This characteristic of the light rays is referred to as "directional sensitivity."

2. The fovea contains densely packed cones and no rods. The ratio of cones to rods steadily decreases toward the margin of the macula and outside this zone the only receptor elements are rods. A curve of relative visual acuity is shown in Figure 80. Acuity is sharpest at the center of the fovea and falls off abruptly toward the periphery of the retina.

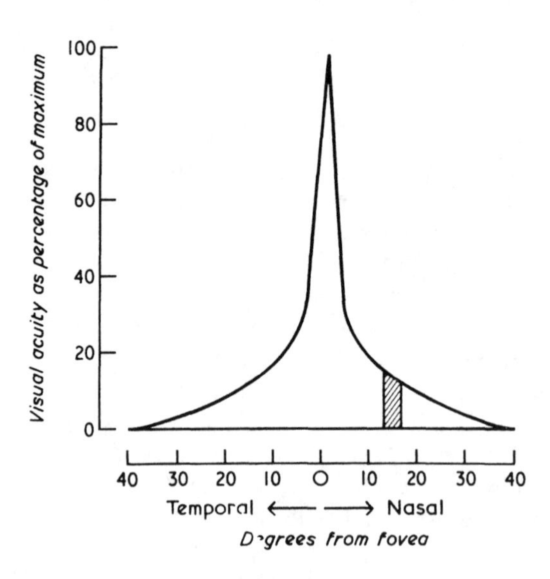

Fig. 80. Curve of relative visual acuity for cones. Acuity is greatest at the fovea and falls off steeply toward the edge of the macula, beyond which cones are absent. Shaded area is the blind spot, or position of the optic disc.

3. The nervous layers at the fovea are deflected to one side while the retinal blood vessels make a detour to avoid scattering of light.

4. Synaptic convergence on bipolar and ganglion cells is reduced to a minimum at the fovea where many cones have "private lines" to optic nerve pathways.

Influence of light intensity. Objects cannot be sharply defined in dim light, and a certain threshold of intensity is essential. Cones with the lowest threshold respond first, while those with the highest threshold respond only in very bright light. Accordingly, as brightness increases, the number of active cones increases and visual discrimination becomes possible. This relationship between the level of illumination and visual acuity is of great practical importance. Thus in bright daylight the closest work can be performed without difficulty, but when artificial light is employed, adequate illumination is required to achieve optimal working efficiency. Other factors affecting the level of illumination are the size of the object and background contrast.

- Details can be recognized more easily if the object is large in size. It is usual to adjust the lighting to suit the task. Small objects are brought closer to the eyes and in fine precision work it may be necessary to use a magnifying lens or low power microscope.
- A white object can be more easily seen against a dark background. It follows that if an object is brighter than its surroundings, the level of illumination required is less than when the surroundings are of equal brightness. Too great a contrast may be unsatisfactory, and the light must also be adjusted to avoid glaring, reflections, or shadows.

In short, acuity of vision requires ideal conditions of illumination. The discrimination of detail depends upon the ability of the eye to detect small differences of brightness. This can be accounted for if the intensity of light falling on one row of cones is noticeably different from the intensity falling on adjacent rows.

Summary of factors influencing visual acuity

Stimulus factors

1. Intensity of light. Adequate light is necessary for detailed vision, and definition improves with increasing illumination. Excessive light reduces visual acuity by producing glare.

2. Wavelength of light. The photopic visibility curve shows that wavelengths in the middle of the spectrum are more effective in causing visual sensation than those at the extremities.

3. Time of exposure. A brief glance may be sufficient to define the shape of an object; the details require more prolonged scrutiny.

4. Size of object. The larger the size, within limits, the more easily its detail can be seen.

5. Distance of object. The object must be within the effective range of accommodation; otherwise optical aids may be required.

6. Background contrast. The relative brightness of the object and of its surrounding is important for visual discrimination.

Optical factors.

1. Refraction. The optical system of the eye must be capable of focusing a sharp image on the retina or be corrected for errors of refraction.

2. Accommodation. Effective accommodation of the lens is required for all objects nearer than six meters from the eye.

3. Diameter of the pupil. The diameter must match the intensity of the light. Pupillary constriction eliminates stray rays passing through the peripheral part of the lens. Pupillary dilatation allows more light to enter the eye in dim conditions.

4. Binocular vision. Perfect ocular muscle balance is necessary to ensure that the visual axes are fixed on a common point in the field in order to prevent double vision.

Retinal factors

1. Macula vision. Visual acuity is a function of the central part of the retina. Acuity is maximal at the fovea, which has the greatest concentration of cones.

2. Directional sensitivity. Light rays falling perpendicular to the surface of the cones are more effective than angular rays passing through the periphery of the pupil.

3. Visual pigments. The photochemical cycle depends upon adequate availability of pigment and on the enzyme systems necessary for resynthesis.

4. Adaptation. The eyes must be adapted for a given level of illumination. Excessive brightness produces glare and leads to fatigue.

NEURAL ORGANIZATION OF THE RETINA

The eye is different from all other sense organs in that it possesses an intricate network of neural elements in addition to its receptor elements. While impulses from other sense organs have their first order relays within the central nervous system, the bipolar and ganglion cell layers of the retina represent two further levels of integration, where activity can be modified before the impulses reach the optic nerve fibers. Accordingly, the retina itself is capable of carrying out many complex interactions between rival afferent pathways in its attempt to deal with the constantly changing conditions of the visual field and varying levels of illumination.

Electroretinogram

If an electrode is placed in contact with the cornea and an indifferent electrode attached elsewhere on the body, a steady potential is recorded when no light is entering the eye. The steady potential is probably generated by neural elements in the retina. When light is admitted, a complex waveform known as an electroretinogram (ERG) is recorded. It is caused by the summated activity of many cells responding to the light stimulus. The complete potential has four components (Fig 81): the "a" wave is a small negative prepotential derived from the rising phase of the receptor potential. The "b" wave is a large positive wave resulting from activity of bipolar and ganglion cells. The "c" wave is a second positive wave generated by the pigment cells of the retina when the intensity of light is fairly high. And "d" wave is a third positive wave recorded on cessation of the stimulus.

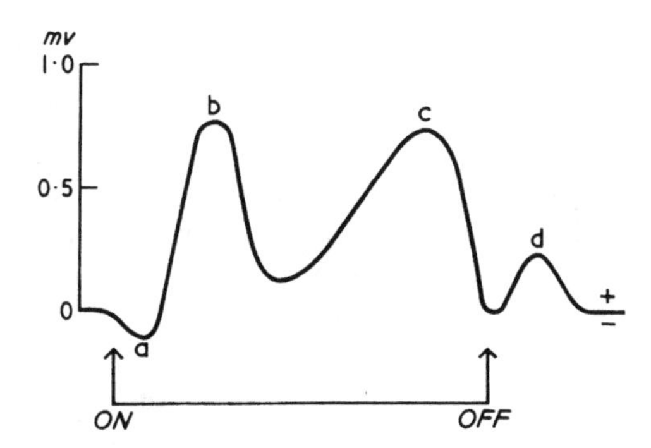

Fig. 81. ERG recorded from frog retina in response to stimulation by light. Upward deflections indicate positivity. Duration of stimulus is 2 sec.

The ERG recorded from the fovea shows a relatively large "a" wave and a greatly reduced "b" wave owing to the absence of neural elements in this region. The "a" wave represents the earliest potential in the chain of reactions leading to visual excitation.

Internuncial networks

The retina is organized on the basis of three principal neural layers—photoreceptors, bipolar cells, and ganglion cells—and two intervening plexiform layers composed of horizontal and amacrine cells. The individual cells in each layer can be identified by the injection of dye from a recording pipette. Many dyes have been used for this purpose, but Procion Yellow is favored because it reveals the fine structure of the cell without disrupting the electrical response.

1. External plexiform layer. This network intervenes between the receptors and the bipolar cells. It is derived from the processes of horizontal cells of which several types have been described. The usual form is that of a stellate body with a fine axon running horizontally, and lateral dendrites embracing the rods and cones. The arrangements suggests that horizontal cells act as a link between various groups of receptors, whereby the potential changes can be modified before being transmitted to the bipolar cells.

2. Internal plexiform layer. This network intervenes between the bipolar layer and the ganglion cells. It is derived from the processes of amacrine cells, many of which do not possess an axon. Electrical recordings from amacrine cells indicate that there are two main types giving transient or sustained responses. The arrangement allows surrounding information to be passed on to the ganglion cells.

Function of ganglion cells

The interactions that take place between the various neural layers result in discharges of the ganglion cells, the last stage of retinal processing. Early attempts to pick up the discharges were based on ERG deflections; later, recordings were obtained from single fibers of the optic nerve. More precise information came with the introduction of microelectrode techniques when the on/off character of the discharges was clearly demonstrated. The impulse patterns not only indicated how the rods and cones manipulated their respective ganglion cells, but also revealed the manner in which their receptive fields were organized.

Responses of ganglion cells. According to Hartline and Granit, three different kinds of response can be distinguished as illustrated in Figure 82.

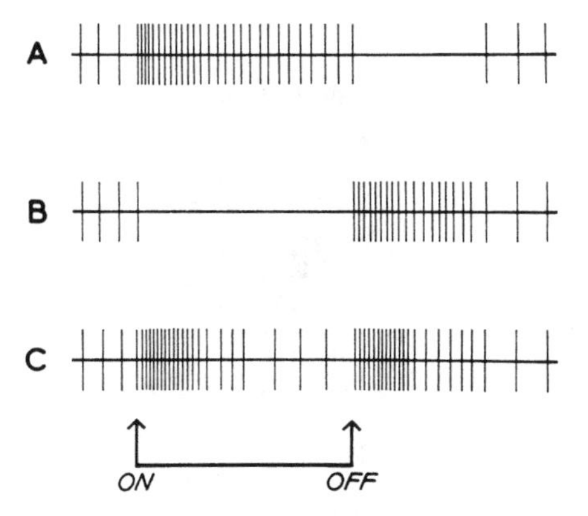

Fig. 82. Responses of retinal ganglion cells to illumination recorded from optic nerve fibers firing spontaneously. A shows "on" response inhibited when light is switched off. B is "off" response showing inhibition of cell during illumination. C represents mixed "on/off" response. The cell discharges when light is switched on and is excited again at off.

1. *On* response. The cell discharges at the onset of illumination and is silenced when the light is switched off.

2. *Off* response. The cell is inhibited during illumination and discharges when it is switched off.

3. *On/off* response. The cell discharges at the onset of illumination and discharges again when it is switched off.

All three response types occur in various proportions according to the level of illumination used and the state of adaptation, but the on/off response is the one most commonly seen in the cat's retina. Granit believed that the receptor elements operate by two antagonistic systems, one being excited by illumination and the other excited by dark. When the two systems converge on the same ganglion cell, the one inhibits the other. Thus if a cell responds to light, the pathway for dark is inhibited; if a cell responds to dark, the pathway for light is inhibited. In this way the response patterns of the ganglion cells can be explained by antagonism between the excitatory and inhibitory influences converging upon them.

Receptive fields. The receptive field of a ganglion cell can be mapped by studying the discharge patterns elicited in a single optic nerve fiber. The fields are roughly concentric and the discharge patterns observed generally conform to the principle of the on/off system. Thus a particular receptive field might consist of a central region giving on effects only, an intermediate zone giving on/off discharges, and the surroundings producing pure off discharges (Fig. 83). Another kind of receptive field might consist of a central zone of inhibition surrounded by a ring of excitatory elements. The center would give a discharge from the ganglion cell only when the light

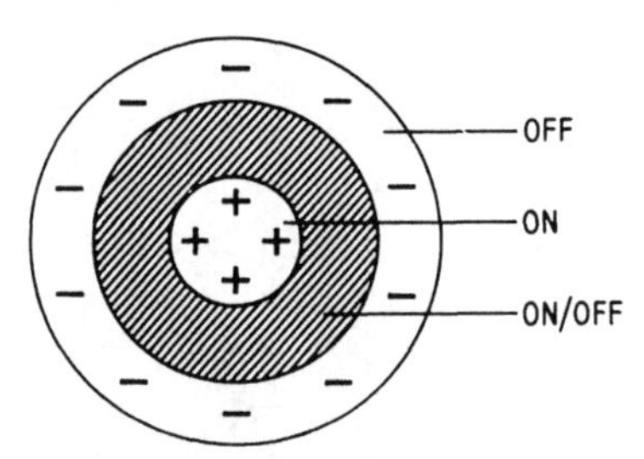

Fig. 83. Distribution of discharge patterns within receptive field of a ganglion cell. Central zone has "on" discharges (+); peripheral zone has "off" discharges (—); intermediate zone has "on/off" discharges (hatched).

would go off, and the periphery would remain silent until the light would go on.

Receptive field organization based on the interaction of antagonistic systems is obviously important for processing detailed information at the retinal level. Synaptic convergence leads to rejection of a proportion of the discharging pathways, and thus to a higher degree of sensitivity in one region at the expense of another region. If two pathways converge on the same ganglion cell, they cannot both operate it at the same moment. Inhibition in the surrounding zones of an active region improves the sharpness and contours of the image as well as providing a means for detecting shades of light and background contrast. The optic pathway and further stages in the processing of visual information will be discussed in Chapter XVII.

SELECTED BIBLIOGRAPHY

Baylor, D.A., and Fuortes, M.G.F. Electrical responses of single cones in the retina of the turtle. *J. Physiol.* 207:77-92, 1970.

Brown, P.K., and Wald, G. Visual pigments in single rods and cones of the human retina. *Science* 144:45-52, 1964.

Enroth, C. Spike frequency and flicker fusion of frequency in retinal ganglion cells. *Acta Physiol. Scand.* 29:19-21, 1953.

Granit, R. *Sensory Mechanisms of the Retina.* New York: Hafner Press, 1963.

Hecht, S. Rods, cones and the chemical basis of vision. *Physiol. Rev.* 17:239-290, 1937.

Hubel, D.H., and Wiesel, T.N. Receptive fields of optic nerve fibres in the spider monkey. *J. Physiol.* 154:572-580, 1960.

Marks, W.B. Visual pigments of single goldfish cones. *J. Physiol.* 178:14-32, 1965.

Marks, W.B., Dobelle, W.H., and MacNichol, E.F. Jr. Visual pigments of single primate cones. *Science* 143:1181-1183, 1964.

Rushton, W.A.H. Dark-adaptation and the regeneration of rhodopsin. *J. Physiol.* 156:166-178, 1961.

Wald, G. The chemistry of rod vision. *Science* 113:287-291, 1951.

Wald, G., Brown, P.K., and Gibbons, I.R. The problem of visual excitation. In Gross, C.G., and Zeigler, H.P. (Eds.): *Readings in Physiological Psychology,* 80-112. New York: Harper and Row, 1969.

CHAPTER IX

Organs of Taste and Smell

The special senses of taste and smell are usually classified together since the adequate stimulus for both organs is chemical. Taste receptors are stimulated by the substances present in food and drink, while smell is due to airborne particles exciting the olfactory epithelium. The association of taste and smell determines the palatability of food, pleasurable sensations adding to the appetite and enjoyment of the meal and assisting its proper digestion. On the other hand, unpleasant sensations clearly impair the appetite and harm the digestion. The chemical senses play a significant role in the commercial world and in the human social environment. In animals, these senses may be even more important for protection and survival. Taste may be critical for regulating the diet and for rejecting poisonous substances. An acute sense of smell can warn an animal of approaching danger and it also has a powerful influence on sexual behavior.

TASTE

The receptors for taste are scattered over the mouth, tongue, palate, and pharynx, the tongue being the chief organ. They are excited by a variety of soluble chemical substances belonging to one of four classes–sweet, bitter, sour, and salty. Nerve impulses generated by the receptors take complicated pathways to reach the brain stem and relay in the tractus solitarius to the thalamus. Their cortical destination is closely associated with that of other sensory modalities from the region of the face.

Taste buds

Each taste bud consists of two types of cell, gustatory and supporting. The gustatory cells are chemoreceptors. They are fusiform in shape with a large central nucleus and a terminal pore from which protrudes a number of fine, hair-like filaments. The supporting cells are loosely arranged around them, the whole being embedded in the epithelium of the mucous membrane. In the tongue, taste buds are found on thickenings of the mucous membrane or papillae. They are most numerous on the circumvallate papillae at the base of the tongue and on the fungiform papillae scattered over the sides and tip. Taste buds are absent over the mid-dorsal region where the papillae are mostly of the filiform type. The delicate filaments of the taste buds lie on the free surface and thus come into contact with the saliva. In children, the taste buds are more widely distributed over the oral cavity than in adults, but they do not reach full development until puberty. The nerve fibers supplying the taste buds form a plexus of small and large unmyelinated axons round the basement membrane and between the boundaries of the cells.

Methods of stimulation

The sense of taste may be evoked by direct contact, by injection of chemical substances, or by electrical stimulation.

1. Direct contact. Soluble particles clinging to the secretions in the mouth come into contact with the gustatory filaments of the taste buds. A strong taste requires a high degree of solubility. It is possible that enzymes in the epithelium of the mucosa act as an intermediary by combining with the particles before the gustatory cells are stimulated. The after-effects produced by certain substances may be due to the persistence of combined particles.

2. Systemic effect. Certain substances injected into the blood stream may arouse the sense of taste. For example, the injection of decholin into a vein at the elbow was used for estimating the circulation time. As soon as the drug reached the mouth it gave rise to an intensely bitter taste. The cravings for unusual foods often experienced in the early stages of pregnancy may be due to the action of circulating substances on the taste mechanism.

3. Electrical stimulation. A metallic taste is evoked when a weak current is passed through a pair of electrodes placed on the tongue. The effect is due to electrolysis resulting in the concentration of ions on the taste buds. It is a useful diagnostic test in facial paresis which may cause damage to the sensory fibers for taste.

Varieties of taste

Four distinct varieties are recognized:

1. Sweet. Most persons derive a pleasant sensation from sweet substances, which include the sugars, saccharin, alcohol, and many organic compounds.

All are characterized by high solubility in water and the capacity to mask other tastes in low concentrations. Sweetening agents are added to food and drink to increase palatability, and bitter pills are often coated with sugar. Excessive sweetness, however, can be distasteful, although individual tastes are extremely variable.

2. Bitter. The best known bitter substances are alkaloids like quinine and certain compounds like magnesium sulphate and sodium citrate. Mixed sensations of sweetness and bitterness occur according to the site stimulated, the back of the tongue being more sensitive to bitterness.

3. Sour. Many acids have a sour taste due to dissociation of H ions. Organic acids like acetic, tartaric, and citric acids seem more sour than do the inorganic ones, possibly because of the greater rate of penetration of the organic acids. The threshold of sensitivity is given as the minimum concentration at which the difference between the sour solution and water can be detected.

4. Salty. Substances producing a salty taste are typically the chlorides of sodium, potassium, calcium, magnesium, and ammonium. In water, the salts dissociate into positive and negative ions which bind directly with the receptors to produce depolarization without the assistance of enzymes.

Peripheral pathways

Impulses from the taste buds are conveyed to the brain in the sensory fibers of the seventh, ninth, and tenth cranial nerves.

1. The anterior two thirds of the tongue are innervated by the chorda tympani branch of the facial nerve. The fibers ascend under cover of the lateral pterygoid muscle to the geniculate ganglion, where they are joined by taste fibers from the palate. The central processes of the ganglion cells proceed in the nervus intermedius to the lower border of the pons.

2. The posterior third of the tongue is innervated by the glossopharyngeal nerve. The fibers pass to the petrous ganglion in the jugular foramen and their central processes form a number of filaments attached to the upper part of the medulla.

3. A few taste buds in the pharynx are supplied by branches of the vagus nerve. The fibers pass to the ganglion nodosum and from there by filaments to the dorsal vagal nucleus in the medulla.

Central pathways

All the taste fibers terminate in a continuous column of gray matter in the brain stem, known as the nucleus of the tractus solitarius. Second order neurons cross to the opposite side and ascend with the medial lemniscus to the posteroventral nucleus of the thalamus. Here, the taste fibers are located in a medial group of cells called the arcuate nucleus from which third order neurons project to the postcentral gyrus. The taste-receiving area of the cortex lies close to the area representing other sensory modalities of the face. Because of the sensory decussation, a cortical lesion produces loss of taste on the opposite side, while a tractus lesion affects the same side.

Single unit activity

Recordings from single ganglion cells reveal that the taste buds are spontaneously active, discharging usually at a very low rate. The discharge is enhanced on chemical stimulation, but adaptation occurs if the stimulus is prolonged. Several taste buds may converge on the same ganglion cell with the result that one stimulus can inhibit another. There is little to distinguish the electrical activity set up by different chemical substances, and the subjective response is no doubt very much an individual variant.

Loss of taste

Simple blunting of the taste quality is a common condition following a cold or excessive smoking. Taking heavily spiced foods often diminishes the taste of other substances, while the continuous repetition of the same taste renders the sensation less and less distinct.

Persistent loss of taste generally implies damage to the nerves supplying the taste buds or damage to any part of their central pathways. Bilateral lesions may occur after severe head injuries, but in the majority of cases the lesion is unilateral, causing loss of taste from one side of the tongue or palate. The site of the lesion may be supranuclear, pontine, or infranuclear.

1. Supranuclear lesions. A cortical or subcortical lesion causing the loss of taste may be associated with a facial paralysis. The upper part of the face, however, is not affected since the projections to the facial nucleus come from both cerebral hemispheres.

2. Pontine lesions. A lesion involving the tractus solitarius is often accompanied by an ipsilateral facial paralysis. Fibrillation of the facial muscles is a prominent sign, but it must be emphasized that in both types of lesion, upper and lower motor neuron, weakness of the facial muscles may be present while the sense of taste escapes impairments.

3. Infranuclear lesions. It happens that the facial nerve is particularly susceptible to toxic or inflammatory agents producing denervation on the

same side. Most cases originate in the bony canal containing the chorda tympani branch before its emergence at the stylomastoid foramen. Interruption of conduction in the chorda tympani nerve leads to loss of taste in the anterior two thirds of the tongue. If the nerve is not completely denervated, good functional recovery can be expected.

SMELL

The nose is the organ of smell. Odorous particles reaching the upper part of the nasal cavity excite the specialized olfactory epithelium in which the receptors are situated. Other parts of the nasal cavity are sensitive to touch, temperature, and pain but do not serve the sense of smell. Nerve impulses generated by the olfactory receptors are conducted by unmyelinated axons through the cribiform plate of the ethmoid bone to synapse in the glomeruli of the olfactory bulb with the dendrites of the mitral cells. These axons constitute the first cranial nerve.

The central processes of the mitral cells are collected together in the olfactory tract, a narrow white band lying on the orbital surface of the frontal lobe. The tract divides into lateral and medial roots through which the impulses gain direct access to the cerebral cortex. There does not appear to be any thalamic relay as with other sensory systems, but the existence of extensive intracortical connections must be assumed on the basis of the role that the sense of smell evidently plays in many behavioral reactions.

Olfactory epithelium

The mucous membrane of the nasal cavity has both respiratory and olfactory epithelium lining the septum and the three turbinate bones, or conchae, on the lateral wall. In normal respiration, air is conducted through the nasal passages between the turbinates but does not reach the roof of the nose above the superior turbinate. The olfactory epithelium is restricted to the uppermost passage and adjacent part of the septum. It can be distinguished from the respiratory epithelium by its yellowish color.

There are three layers of cells—nerve cells, supporting cells, and stellate cells. The nerve cells are bipolar nucleate receptors, which also serve as ganglion cells distributed between the supporting cells. Arising from the superficial part of each receptor are minute hair-like filaments, numbering up to 1000 in the human being, and greatly increasing the surface area exposed on the nasal mucosa. The axon arises from the deeper part of the receptor and passes through openings in the cribiform plate to reach the olfactory bulb. Here it breaks up into a terminal plexus, making synaptic contact with the dendrites from the mitral cells.

Olfactory bulb

The internal structure of the olfactory bulb is arranged in a number of layers (see Fig. 84)—(1) a layer of unmyelinated fibers, which are the terminals of the olfactory nerve; (2) a glomerular layer formed by synapses between the olfactory terminals and the dendrites of mitral cells; (3) a layer of mitral and tufted cells embedded in a matrix of neuroglia; and (4) a granule cell layer which also contains the myelinated axons of mitral and tufted cells.

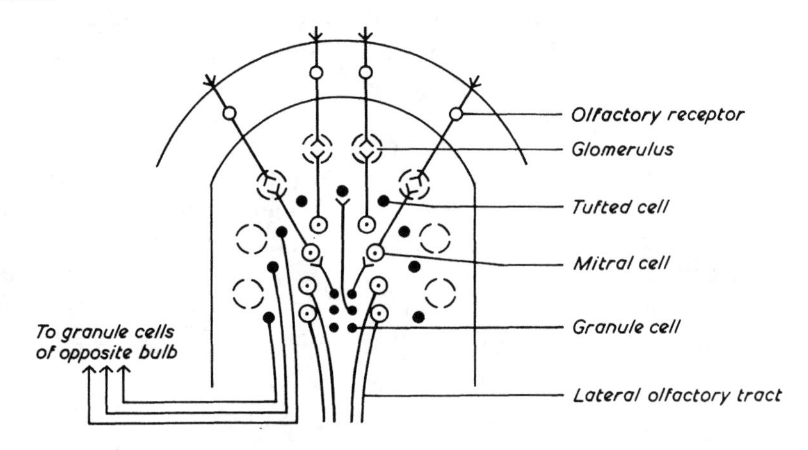

Fig. 84. Histological layers of olfactory bulb. The terminals of the olfactory nerve make synaptic contact in a glomerulus with the primary dendrites of a mitral cell. Axons from mitral cells form the bulk of the lateral olfactory tract. Axons from tufted cells cross to the opposite bulb where they terminate mainly in the granular layer. (Further explanations are given in the text.)

Olfactory tract

1. Axons from the mitral cells form second order neurons conveying impulses from the olfactory epithelium and bulb to the cerebral cortex. The tract emerges from the bulb as a white band which divides into a lateral and medial root. The lateral root, containing the axons of the mitral cells, enters the uncus and piriform area of the cortex; the medial root passes to the paraolfactory area on the medial surface of the hemisphere. Between the

two roots there is a small area of gray matter comprising the olfactory pyramid and anterior perforated substance.

2. Axons from the tufted cells enter the lateral olfactory tract and then course through the anterior commissure to reach the opposite olfactory bulb, where they make synaptic contact with the granule cells.

3. Axons from the granule cells do not pass through the olfactory tract but are directed backward to end on the dendrites of the mitral and tufted cells.

Olfactory cortex

The part of the human brain associated with the sense of smell is limited to a region known as the olfactory cortex. This is essentially the primary receiving area for the terminals of the lateral olfactory tract. It comprises the piriform area, anterior perforated substance, and olfactory tubercle, but the anatomical limits are uncertain since the gray matter of the cortex appears to be continuous with the gray matter of the bulb. In animals, the cortical representation for smell is more extensive and includes third-order neurons, which reach the anterior part of the hippocampus. It is unlikely that the hippocampus is directly concerned with discriminatory functions, although it may be important in some species, such as the badger, whose feeding habits are closely dependent upon the sense of smell.

Stimulation of the olfactory receptors

1. Animal experiments. A number of methods are employed, some more refined than others, to provide information about the olfactory system and its central mechanisms.

Controlled stimulation. The animal breathes through a tracheal cannula, while air containing the odorant material is blown through one nostril and collected from the other at a controlled rate. The olfactory epithelium possesses the property of physical adsorption, i.e., odorant particles are rapidly taken up by the surface membrane. The process is reversible since it can be shown that after one substance has been adsorbed, it can be displaced by introducing another substance into the air stream. Selective adsorption could be the basis of odor discrimination, depending upon the receptor site where the adsorbed molecules are most concentrated.

Discrimination method. Animals can be trained to distinguish between two containers, one of which holds food. If the olfactory organ has been previously destroyed, the choice is made purely by chance. Normal animals select their food to regulate their diet; the ones with destroyed olfactory organs show no preferences.

Conditioning method. An animal will lift its paw in response to an electric shock. If at the same time an odorant substance is inhaled, the animal learns to lift its paw in response to the odor alone. After placing a lesion in the olfactory pathway, the conditioned reflex is lost.

2. Human experiments. During quiet breathing, air does not usually pass above the superior turbinate, and any sense of smell aroused is due simply to diffusion of odorous particles into the upper nasal cavity. The subject must sniff in order to excite the olfactory receptors directly. Quantitative methods for testing smell use various substances of gradually increasing concentration. Sensitivity is measured in terms of the minimum concentration required to produce a perceptible sensation. To avoid the uncertainties of sniffing, a known quantity of the odor is blown into the nostril by a single puff of air and the threshold concentration of the substance is then determined.

Clinical application. The sense of smell is usually tested by asking the patient to identify familiar odors like camphor, peppermint, coffee, and many others. Each nostril is tested separately. Loss of the sense of smell, or true anosmia, must be distinguished from the failure to identify the odor, which may be due to local conditions in the nasal passages. Unilateral loss usually indicates a lesion of the olfactory pathways. In these patients the olfactory threshold on the two sides should be compared. Bilateral anosmia may be present in a number of intracranial disturbances, the commonest cause being a severe head injury.

Electrical activity of the olfactory system

The transfer of information from the olfactory receptors to the higher centers of the brain is accompanied by a series of electrical events, beginning with a slow wave generator potential and ending in a complex spatiotemporal pattern of impulse discharges. The response patterns are further complicated by bilateral corticofugal influences that limit or increase the excitability of the receptors.

Spontaneous activity. Irregular waves, often of high frequency, have been recorded from the isolated olfactory bulb of the frog and also in the rabbit under light anesthesia. Odorant stimuli were excluded by using filtered air. Adrian attributed these waves to intrinsic action of neuronal circuits within the bulb formed by short-axon cells, tufted and granule cells, none of which send axons into the lateral olfactory tract.

Slow potentials. If an electrode is placed on the olfactory epithelium, a slow negative wave is recorded in response to an odorant stimulus. The wave increases in size with increasing concentrations of the stimulus. If two or more odors are introduced into the air stream, the resulting waveforms may be very complex because of an additional positive-going potential. They are believed to represent true generator potentials initiated by the olfactory receptors.

Nerve fiber responses. Neural activity in response to odor stimulation can be recorded directly from the olfactory nerve fibers where they enter the bulb. Or the activity can be recorded by implanting electrodes into the cribiform plate. Many fibers show a low rate of discharge at rest. The introduction of an odorous substance into the air stream causes prolonged bursts of asynchronous discharges which increase in amplitude and duration with increasing concentration of the substance. When recordings are taken from individual nerve fibers it is found that some responses are increased and others inhibited by the same odor, suggesting the possibility of a discriminatory function existing at this level.

Responses from the olfactory bulb

1. When a microelectrode is inserted into the bulb, single unit activity can be demonstrated from all layers, the most prominent discharges being derived from the mitral cells. The initial effect of an odor is usually a reduction in the level of spontaneous activity and the eliciting of the discharge of discrete spikes. Different odors may have opposite effects on the same unit, some tending to increase the rate of mitral discharge and others tending to depress it.

2. Orthodromic responses of individual mitral cells may be obtained by electrical stimulation of the olfactory nerve. A single shock evokes a single spike discharge followed by a depression of excitability. Thus repetitive discharges are typically absent. The suppression of firing is brought about by hyperpolarization of the membrane as a result of the inhibitory synaptic action exerted by tufted cells and granule cells. It appears that axon collaterals of the mitral cells can excite the tufted and granule cells and so provide an effective inhibitory feedback mechanism.

3. Antidromic responses of individual mitral cells may be obtained by electrical stimulation of the lateral olfactory tract. Typical responses take the form of positive-negative spikes, 20 mv in amplitude peak-to-peak, with brief unvarying latencies; giant spikes over 50 mv in amplitude are also frequently observed. These spike potentials represent the invasion of excitable regions in the mitral cell membrane. If a second stimulus is delivered to

the lateral olfactory tract in a conditioning-testing sequence, the response to the testing stimulus is inhibited. Long-lasting suppression of mitral cell excitability can be produced by very weak conditioning shocks.

All the evidence suggests that different types of cells in the olfactory bulb are organized into neural circuits that control the excitability of the mitral cells and limit their frequency of discharge. The results are in agreement with the histological findings that axon collaterals of mitral cells operate through an inhibitory system which is clearly important in the processing of the afferent input.

Centrifugal influences

The excitability of the olfactory organ may be influenced by impulses originating in the brain itself. Two efferent pathways have been described: One connects the granule cells by way of the anterior commissure to the opposite bulb, and the other consists of fibers that run in the lateral olfactory tract from the pyriform cortex. Electrical studies indicate that the discharges of mitral cells can be modified by cortical stimulation, although it is difficult to exclude the possibility of antidromic excitation of these neurons. Further evidence suggests that each bulb can be influenced by activity originating in the opposite bulb. Thus stimulation of one olfactory organ may produce inhibitory effects on the opposite side. If the anterior commissure is sectioned, the amplitude of the evoked potentials is increased, suggesting that tonic inhibitory influences may be exerted on bulbar excitability by impulses from the opposite hemisphere.

Olfactory discrimination and fatigue

Although the sense of smell is less acute in the human being than in lower animals, most persons can distinguish one odor from another at remarkably low concentrations. Odorants like musk, camphor, acetates, and merceptan are commonly used for testing. The discriminating ability depends to some extent upon regional differences at the receptor sites; but a higher degree of selectivity can be seen in the patterns of nervous excitation set up by neurons in the olfactory bulb. Finally, it must be remembered that smell is a subjective sensation in which individuals show differences in perceptive power and may describe differently the sensation derived from the same odorous substances. One person may be insensitive to an odor that another person finds agreeable or intolerable. The cause of this difference is probably cerebral in origin.

1. Receptor sites. Adrian believed that the receptor sites differed in their sensitivity to odors. He showed that some sites responded to a certain group of compounds–for example, aromatics–while neighboring sites were unaffected when the same concentrations were used. It is obviously important to apply the odorant to as small an area as possible in order to locate the receptor sites with greater precision, and an olfactometer is now generally employed. This instrument permits delivery of a known concentration of the odor in a purified stream of air flowing at a controlled rate. The existence of differential sites in the olfactory epithelium is reflected by changes in bulbar activity in which the rise and fall times of a discharge may correspond to a specific site stimulated.

2. Patterns of response. The intricate network controlling the excitability of mitral cells in the olfactory bulb allows for odor discrimination. Some mitral cells have a lower threshold of excitability than others; these will tend to discharge at lower concentrations. An odor that excites one group of mitral cells may inhibit another group. This is illustrated schematically in Figure 85. It will be observed that the presentation of an odor (marked by an arrow) causes a different pattern of response in different units. The response may remain unchanged, show an increase in spike frequency or a complete suppression of firing lasting for the duration of the stimulus. When two odors are presented simultaneously, the resultant discharge may be increased by facilitation or decreased by inhibition, mediated by neuro-

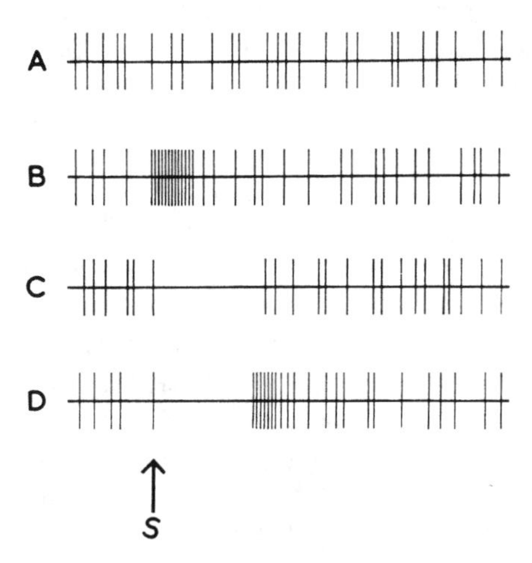

Fig. 85. Various patterns of response recorded from single mitral cells on presentation of an odor, indicated by arrow(s). A shows no change in spontaneous activity; B, excitatory response (increase of spike frequency with return to spontaneous activity); C, suppressive response with return to spontaneous activity; and D, initial suppression followed by increase of spike frequency.

nal mechanisms already described. It is also possible that the centrally directed input patterns are subjected to finer control by corticofugal influences from both cerebral hemispheres. It is clear from these studies that the olfactory organ possesses a temporal mechanism of differentiation analogous to that of other sensory systems.

It is a common experience that sensitivity to a particular odor tends to diminish over a period of time. This is probably due to adaptation of the olfactory receptors, which occurs when they are exposed to continuous stimulation. The olfactory sense is easily fatigued. If, however, another odor is introduced, the organ responds to the new odor. The observation has been used as a basis for classifying odors, since odors of the same class would mutually reinforce each other, whereas odors of a different class would not do so. Recovery from fatigue is usually rapid. For this reason, the view is held that sensitivity is not entirely peripheral, but is controlled by the level of spontaneous activity maintained by the bulb and its central connections.

BIOLOGICAL SIGNIFICANCE OF THE CHEMICAL SENSES

Human enjoyment of food and drink is derived to a great extent from their flavor. The smell of cooking arouses the appetite and stimulates the flow of saliva and gastric juices. The sense of taste influences the selection of food and its ingredients. This, of course, is an important consideration for the diet of hospitalized or convalescing patients. Bitter tastes can be masked by sweetening agents and unpleasant odors can be avoided. The aromas of fruit and wine, the scent of flowers and synthetic perfumes all tend to produce exciting and pleasurable sensations. On the other hand, offensive odors are clearly intolerable. Public health laws are designed to limit pollution of the atmosphere, improve the ventilation of buildings, and generally to create a purer environment.

In animals, the chemical senses play a far more critical role than they do in human beings. Most wild animals seek their food by the sense of smell, and, at the same time, smell is a warning to the quarry of approaching danger. The direction of the wind is important in tracking since odorant particles are transmitted in a current of air by the action of sniffing. Animals with their nose to the ground recognize smells that are not perceptible to man. Their behavior toward each other and their sexual habits are to a large extent determined by their olfactory impressions. Smell and taste are also important for selecting edible foods and for rejecting what may be poisonous. Thus animals kept in a laboratory on a vitamin-deficient diet are found to select foods with a high-vitamin content, whereas those deprived of the sense of taste show no preferences for different foods.

SUMMARY

Taste and smell are known as the chemical senses because the exciting agents are chemical substances. Taste buds are specialized sensory nerve cells with protruding hair filaments scattered over the tongue, mouth, and pharynx. The anterior two thirds of the tongue are supplied by the chorda tympani branch of the facial nerve with cell stations in the geniculate ganglion; the posterior third of the tongue is supplied by the glossopharyngeal nerve, while taste fibers from the pharynx travel in the vagus nerve. All the taste fibers terminate in the nucleus of the tractus solitarius. The rest of their course follows the projections of the somatosensory system to the cortical area representing the sensory innervation of the face. Taste buds are stimulated by direct contact of soluble particles clinging to the secretions in the mouth. Diminished sensation is therefore found in mild infections, but persistent loss of taste generally implies damage to the nervous pathways and frequently accompanies facial paralysis.

Olfactory receptors are located in the upper recesses of the nasal cavity. They are simple bipolar cells ending in a tuft of hairs. Their axons pierce the cribiform plate of the ethmoid bone to enter the olfactory bulb. The principal cell of the olfactory bulb is the mitral cell, whose dendrites form synapses with the olfactory nerve terminals in the glomerular layer and whose axons pass into the lateral olfactory tract. Mitral cell excitability is controlled by an inhibitory feedback mechanism operated through tufted and granule cells in the bulb. Olfactory receptors are stimulated by the physical process of adsorption. The ability to discriminate between different odors depends upon the concentration of odorant molecules that have been adsorbed on the olfactory mucosa. Specific receptor activity has been envisaged, but variations of impulse patterns between the bulb and the olfactory cortex provide an adequate mechanism of differentiation. Loss of the sense of smell may have serious consequences in the animal kingdom, but in the human being, it is usually temporary and not very important except as a symptom in the clinical evaluation of a head injury or brain disease.

SELECTED BIBLIOGRAPHY

Adrian, E.D. The action of the mammalian olfactory organ. J. Laryngol. Otol. 70:1-14, 1956.

Békésy, G. Von. Taste theories and the chemical stimulation of single papillae. In Gross, C.G., and Zeigler, H.P. (Eds.): *Readings in Physiological Psychology,* 128-146. New York: Harper and Row, 1969.

Douek, E. *The Sense of Smell and its Abnormalities.* Edinburgh and London: Livingstone, 1974.

Kerr, D.I.B., and Hagbarth, K.E. An investigation of olfactory centrifugal fibre system. *J. Neurophysiol.* 18:362-374, 1955.

Kimura, K., and Beidler, L.M. Microelectrode studies of taste receptors of rat and hamster. *J. Cell. Comp. Physiol.* 58:131-139, 1961.

Moulton, D.G. Spatial patterning of response to odors in the peripheral olfactory system. *Physiol. Rev.* 56:578-593, 1976.

Moulton, D.G., and Tucker, D. Electrophysiology of the olfactory system. *Ann. N.Y. Acad. Sc.* 116:380-428, 1964.

Pfaffman, C. (Ed.). *Olfaction and Taste; Proceedings of Third International Symposium.* New York: Rockefeller University Press, 1969.

Phillips, C.G., Powell, T.P.S., and Shepherd, G.M. Responses of mitral cells to stimulation of the lateral olfactory tract in the rabbit. *J. Physiol.* 168:65-88, 1963.

Powell, T.P.S., Cowan, W.M., and Raisman, G. The central olfactory connexions. *J. Anat.* 99:791-813, 1965.

Yammoto, C., Yammoto, T., and Iwama, K. The inhibitory system in the olfactory bulb studied by intracellular recording. *J. Neurophysiol.* 26: 403-415, 1963.

PART III

Central Nervous System

CHAPTER X

Reflex Activity

A reflex may be defined as a response to a stimulus that does not require the intervention of consciousness. A simple reflex involves only a small part of the central nervous system, perhaps just one segment of the spinal cord or one small nucleus in the brain. The majority of reflexes, however, occupy many levels of the nervous system, all combining in a coordinated manner to produce some desirable action.

The labyrinthine and oculomotor reflexes have already been described. Other examples are to be found in almost every major system of the body, including the alimentary, cardiovascular, and respiratory systems. The involuntary nature of reflex action had been recognized by the early part of the nineteenth century from observations of the frog and other cold-blooded animals. But it was not until the end of that century that the nature of the central mechanism was analysed, mainly from the research of Sherrington and his associates. Much of their work was based on spinal mechanisms controlling posture and movement and how these could be modified by descending influences from the brain. They developed the concept of synaptic states of excitation and inhibition, elucidating the principles of the transmitting process and demonstrating the dependence of reflex function on the character of receptor discharges. All varieties of reflexes serve a common purpose—to carry out an organized function of a particular organ for the protection and well-being of the body. In higher animals and in the human being, the spinal reflexes are under so much control of brain mechanisms that their own individual role is apt to be overlooked. Nevertheless, many actions are performed involuntarily with perfect ease and coordination when there is no obviously conscious direction. This is only too evident in diseases of the spinal cord or after injuries when muscular coordination is impaired and the reflexes become exaggerated or depressed.

The reflex arc

All reflexes have three essential components—an afferent limb, a reflex center and an efferent limb (Fig. 86).

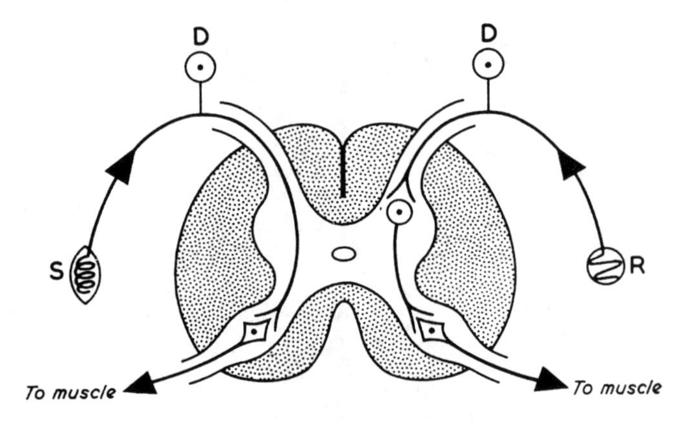

Fig. 86. Diagram of spinal reflex arcs. The left side of the figure illustrates a monosynaptic reflex showing the afferent neuron from a muscle spindle (S) making direct synaptic contact with an anterior horn cell. The arc is completed by the efferent neuron innervating the muscle. The right side of the figure illustrates a polysynaptic reflex showing the afferent neuron from a receptor (R) making synaptic contact with a posterior horn cell, which in turn connects with an anterior horn cell. D is the dorsal root ganglion.

1. Afferent limb. This comprises a receptor and its sensory axon. The nutrient cell lies in a dorsal root ganglion (or cranial equivalent), and the central process passes into the spinal cord via the dorsal root.

2. Reflex center. This lies in the gray matter of the anterior horn, connecting the two limbs of the arc. A two-neuron arc reflex is one in which the afferent limb makes a direct connection with the center and only one synapse is involved. Such reflexes are also termed *monosynaptic*. If the afferent limb ends in the dorsal horn, one or more interneurons link it to the center to form a multi-neuron reflex arc in which many synapses may be involved. Such reflexes are therefore called *polysynaptic*.

3. Efferent limb. This consists of an anterior horn cell, its axon or motor nerve fiber, and the muscle fibers it supplies. The skeletal muscle that responds to reflex stimulation is termed the effector organ.

The knee jerk is an example of a simple monosynaptic reflex. When the patella tendon is tapped, the quadriceps muscle is stretched and impulses generated by the muscle spindles enter the spinal cord by the lumbar dorsal roots. As a result of this the motoneurons supplying the quadriceps are excited and the muscle contracts. Most other reflexes, whether spinal or cranial, involve at least one relay cell or interneuron. The afferent impulses entering the dorsal horn may pass up or down the spinal cord through many segments before reaching their motoneurons. In such circumstances, where the reflex pathway is a complicated one, there is a greater opportunity for modification of the reflex by other pathways converging upon it.

Internuncial system

This is a chain of interneurons forming an intricate network between the dorsal and ventral roots of the spinal cord. Impulses entering the dorsal horn can take many alternative pathways, some shorter, some longer than others. As there is a delay of about 0.5 msec at each synaptic junction, impulses spreading through the internuncial system will take a longer time than those using the shorter routes. Eventually, all the impulses arrive at the motoneurons which form the "final common path" to the muscles. Figure 87 shows two possible arrangements of the internuncial network:

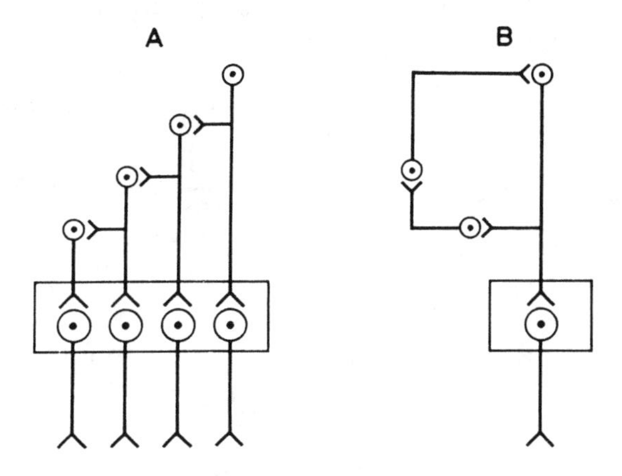

Fig. 87. Diagram of internuncial circuits. A is multiple chain showing how impulses spread through a number of synapses before arriving at the motoneuron pool. B depicts closed chain in which impulses complete a circuit re-exciting the neuron. In either case, activation results in bombardment of the motoneurons in the pool.

1. The multiple chain allows impulses to spread diffusely through a varying number of synapses before arriving at the motoneurons at successive intervals and thus cause a continuous discharge.
2. The closed chain shows how a neuron can be re-excited by impulses completing a circuit and thereby cause a repetitive bombardment of the motoneurons.

Most spinal reflexes can be modified by descending influences from the brain. These may be excitatory or inhibitory. The effects are sometimes brought about by direct synaptic action on the anterior horn cells, but more usually through the internuncial system, especially if the effect is an inhibitory one. In clinical terms, the descending pathway from the brain to its terminals in the cord is referred to as the upper motor neuron. The anterior horn cells and the peripheral nerves are called the lower motor neuron.

Motoneuron pool

The nerve cells of the anterior horn are arranged in columns or pools, each pool sending its motor nerves to a particular set of muscles. Only a proportion of the cells in a pool need be active in a simple reflex, the remainder of the cells being uninfluenced or kept below threshold by inhibition. The pool is then said to be fractionated into active and inactive parts, the latter constituting the subliminal fringe. The number of cells actively discharging is related to the size of the incoming volleys. Thus a weak stimulus produces only a small rise of tension in the muscle. As the intensity of the stimulus is increased, more motoneurons are excited and the size of the reflex response is increased. A maximal response results when threshold is reached in all the cells of the pool including those in the subliminal fringe. In polysynaptic reflexes, the response may not be related directly to the stimulus intensity, since much depends upon the state of the internuncial neurons and their constantly changing levels of excitability.

SPINAL REFLEXES

There are four classical spinal reflexes—stretch, flexor, crossed extensor, and long spinal.

Stretch reflex

The response of the quadriceps muscle to a known degree of stretch was first demonstrated by Liddell and Sherrington in a decerebrate preparation. They attached the tendon to a myograph and recorded the total isometric tension resulting from a sudden stretch. After cutting the motor nerve, the small amount of tension developed was caused by the elastic component of the muscle. The reflex character of the response was established by cutting the dorsal roots conveying afferent impulses from the muscle, but leaving the motor nerve intact. Only the elastic tension, comparable to the paralyzed muscle, was then recorded (Fig. 88).

In the human being, stretch reflexes elicited by the tendon jerks are important clinical signs. The knee jerk can be recorded by placing a pair of silver cup electrodes on the skin overlying the quadriceps and connecting the leads to an oscilloscope. The hammer for tapping the patella tendon contains a contact switch and battery to provide the trigger current. The ankle jerk is recorded similarly by placing the electrodes on the calf and

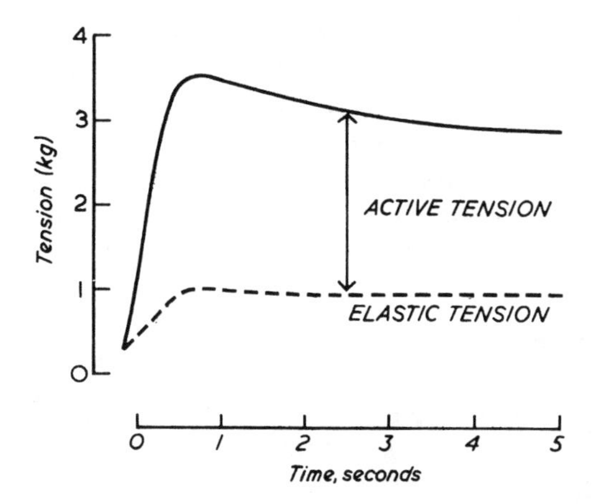

Fig. 88. Reflex response of quadriceps muscle to stretch. Tension developed reaches 3.5 kg and is then sustained at a slightly lower level. Active tension is the difference between this and the elastic tension indicated by the dashed line. After denervation only elastic tension is recorded. (After Liddell and Sherrington.)

tapping the tendon Achilles (Fig. 89). The latent period of the tendon reflexes is relatively short since the internuncial system is not involved. The total reflex time can therefore be calculated from (1) conduction time in the afferent nerves (Group I with conduction velocity 100 m/s); (2) central delay (0.5 msec); and (3) conduction time in the efferent nerves (motor nerve fibers with conduction velocity 100 m/s).

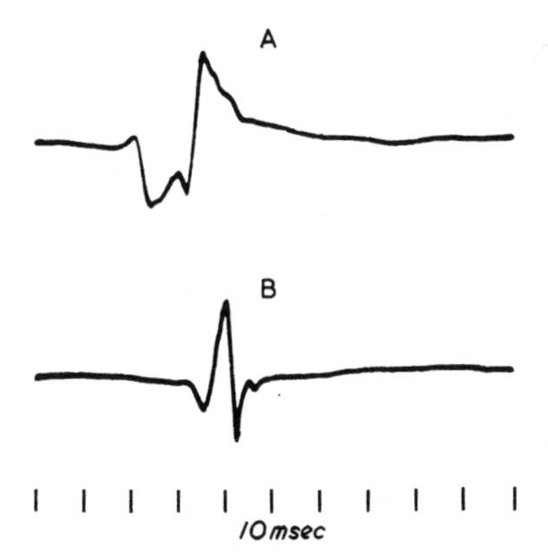

Fig. 89. Tendon reflexes recorded from human subject by placing electrodes on skin overlying the muscle. A represents knee jerk; reflex conduction time is about 18.0 msec. B shows ankle jerk; reflex conduction time is about 32.0 msec.

Simultaneous recordings of the electrical and mechanical events occurring during the stretch reflex show that the muscle action potential precedes the contraction, which is followed by a period of inactivity. The silent period is due partly to the refractory state of the motoneurons, which respond to the stimulus in a synchronized fashion, and partly to the loss of afferent inflow from the muscle spindles when the muscle contracts.

Characteristics of the stretch reflex

1. It is monosynaptic. Afferent fibers from the muscle directly excite the dendrites of the motoneurons supplying the muscle.
2. Central delay is minimum. Because no interneurons are involved, there is no after-discharge. Consequently, the reflex tension developed by the muscle falls rapidly when the stimulus ceases.
3. Stimulation of an ipsilateral afferent nerve promoting flexion of the limb inhibits the stretch reflex. This is because impulses from the stretched muscle are effectively blocked by afferents from the antagonistic muscle.
4. The stretch reflex has both phasic and static functions. The phasic component is illustrated by tendon jerks in which a synchronous afferent volley gives a brief contraction. The static component takes the form of a steady inflow of impulses when the muscle is under continous slight stretch; the motoneurons respond with an asynchronous discharge giving a prolonged contraction. The static stretch imposed on the muscular system by gravity assists in the maintenance of posture.

Flexor reflex

If the hand accidentally touches something very hot, it is rapidly withdrawn. This abrupt movement was called by Sherrington a *nociceptive reaction.* Flexor reflexes are evoked by any stimulus that is harmful or painful and therefore serve to protect the body from injury. The response mechanisms differ in nearly every detail from the stretch reflexes: The receptors are usually pain nerve endings in the skin; the afferents are carried in cutaneous nerves to the dorsal horn of the spinal cord; the connections to the motoneurons are rarely direct and internuncial involvement may extend over many segments; the muscles concerned are the flexors of the corresponding limb. The rise of tension is more gradual and relaxation prolonged because of the late arrival of impulses at the motoneuron pools.

Characteristics of the flexor reflex

1. It is polysynaptic. At least one interneuron lies between input and output. With stronger stimulation more interneurons are invaded and the reflex discharge prolonged.

2. It is multi-segmental. A weak stimulus applied to the sole of the foot causes flexion of the ankle. As the stimulus intensity is increased, the response spreads to the knee and hip.

3. The motoneuron pool is fractionated. One afferent nerve excites only a proportion of the motoneurons in the pool. If more than one afferent nerve is stimulated, summation effects occur in the subliminal fringe. As a result, the total reflex tension developed may be considerably increased.

4. The antagonistic muscle is inhibited. Afferents entering the spinal cord have their first synaptic relay in the dorsal horn. Excitatory fibers pass to the motoneurons of the flexor muscles and their synergists, while inhibitory interneurons supply the extensor group. In this way relaxation of the antagonistic muscles accompanies flexion of a joint.

5. It has protective and postural functions. The protective function, illustrated by the withdrawal response, is also an important clinical sign of underlying injury. On palpation, the flexed muscle feels rigid compared with muscle on the opposite side. In posture, the flexor muscles contribute to the stability of the body and react to changes in the position of the limbs when the center of gravity is shifted.

Crossed extensor reflex

When a flexor reflex is elicited in one limb, the extensor muscles of the opposite limb contract. The spinal mechanism accounting for this reaction suggests that impulses must cross the mid-line to synapse with the contralateral motoneuron pool. Thus an apparently simple reflex has a fairly complex organization. The afferent volley which excites the flexors and relaxes the extensors of a joint may spread to involve other joints in the same limb as well as those on the opposite side.

Reciprocal innervation. The crossed extensor reflex illustrates a basic feature of spinal organization whereby the contraction of protagonists is accompanied by relaxation of antagonists. This is known as the principle of reciprocal innervation. The arrangements are shown in Figure 90. A cutaneous afferent volley excites the flexors and inhibits the extensors on its own side, but on the opposite side, the extensors are excited and the flexors inhibited.

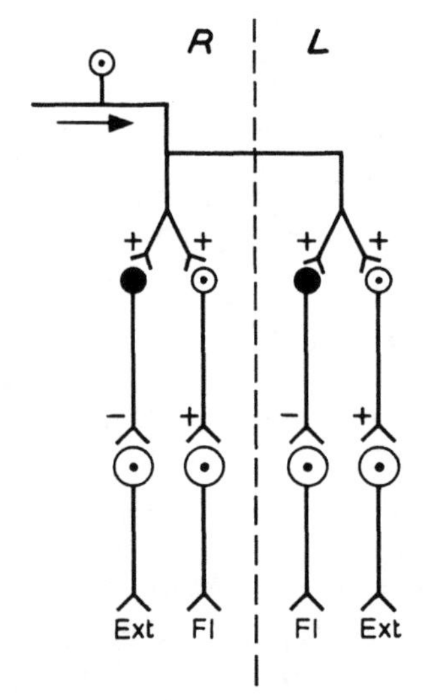

Fig. 90. Diagram illustrating the principle of reciprocal innervation. The afferent volley entering the cord through the dorsal root produces a flexor reflex on the right side and a crossed extensor reflex on the left side. Relaxation of the antagonistic muscles is produced by the inhibitory interneurons indicated by the closed circles.

Long spinal

The reflexes so far described indicate that the spinal cord does not behave as a series of isolated segments, but responds to incoming signals with an integrated pattern of activity. Sherrington's description of the scratch reflex is a good example of an integrated action involving long stretches of the cord. Assuming the skin over the shoulder region is irritated, impulses will enter the cord at the cervical and upper thoracic levels and leave the cord at the lumbosacral level. The propriospinal tract provides a conducting pathway for eliciting the appropriate movements of the leg. The scratch reflex is strictly an ipsilateral response, but other long spinal reflexes, such as those employed in posture and balance, may elicit movements in all four limbs.

PRINCIPLES OF REFLEX ACTIVITY

The principle of convergence

There is considerable overlap of the afferent nerve terminals in their central distribution. This means that an individual cell is shared by afferents from different sources converging upon it. Originally, Sherrington described convergence in relation to the motoneuron pool and showed that when two afferent fibers were stimulated together, the reflex tension developed was increased. Later, Eccles demonstrated the role of presynaptic convergence in transmission through sympathetic ganglia. Transmission may be modified by an increase or reduction of the postsynaptic discharge. An increased discharge is brought about by summation of the excitatory synapses causing a greater number of neurons to respond. This is called *facilitation.* A reduced discharge is brought about by neuronal sharing and is called *occlusion.*

Facilitation. Consider the diagram of Figure 91. Within a pool of neurons supplied by two presynaptic fibers, a and b, some are activated by a, some by b, while others are shared by the two fibers. Assuming that the shared neurons are in the subliminal fringe, stimulation of a alone will result in the discharge of three neurons; likewise, stimulation of b alone will result in the discharge of three neurons. When a and b are stimulated together, nine neurons discharge instead of the expected six because of summation in the subliminal fringe. The total response is thus enhanced. Facilitation occurs with converging afferents when their combined effects are greater than the sum of their individual actions.

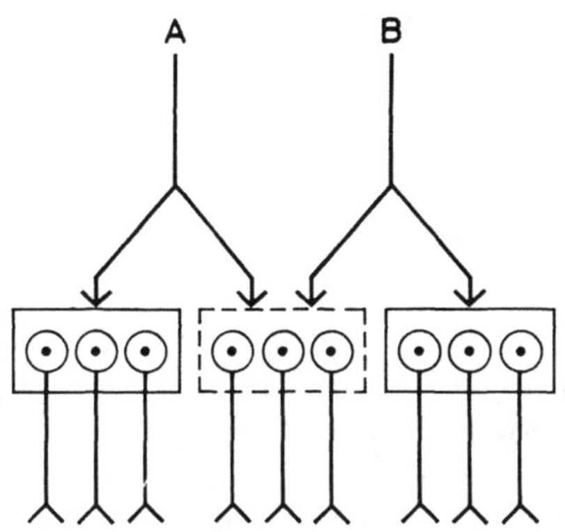

Fig. 91. The diagram shows three groups of motoneurons. The continuous lines are groups of three motoneurons excited by the presynaptic fibers a and b, while the broken lines represent the subliminal fringe. Stimulation of a alone or b alone excites only three motoneurons. Combined stimulation results in the discharge of nine motoneurons, an example of facilitation.

Occlusion. In the diagram of Figure 92, the presynaptic fibers, a and b, share a common pool of neurons. Assuming that the stimulus is strong enough to reach the threshold of excitability, stimulation of a alone will result in the discharge of six neurons; likewise, stimulation of b alone will result in the discharge of six neurons. When a and b are stimulated to-

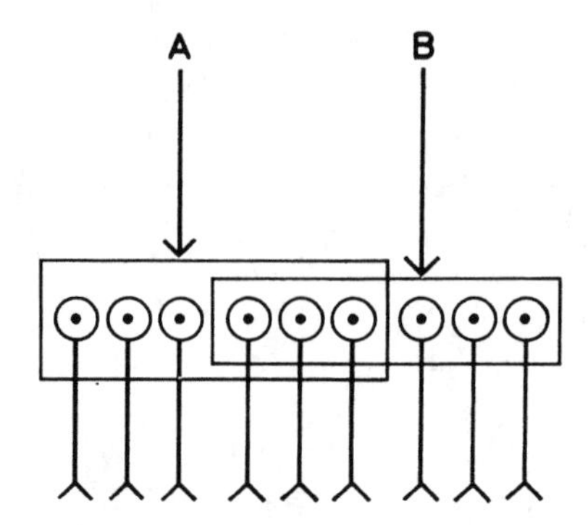

Fig. 92. The presynaptic fibers a and b each excite six motoneurons. Stimulation of a and b together excites only nine motoneurons, as three are common to both fibers in the overlapping field, an example of occlusion.

gether, only nine neurons discharge instead of the expected 12, because of neuronal sharing. The total response is thus reduced. Occlusion occurs with converging afferents when their combined effects are less than the sum of their individual actions.

Facilitatory and occlusive influences due to overlapping fields of activity are prominent features of synaptic transmission in the central nervous system.

The principle of divergence

Most dorsal root fibers break up into many branches that establish synaptic contact with a large number of spinal neurons. By this means, a single afferent alone can influence the discharge of several cells and, if the whole nerve is stimulated, the influence simply spreads from one segment of the cord to the next. Irradiation of impulses producing widespread effects is another important feature of central nervous organization. If the intensity or frequency of afferent stimulation is increased, the size of the reflex response is also increased as more and more motoneurons are excited.

After-discharge

The cause of after-discharge has already been discussed in relation to the internuncial system. It is seen characteristically in the flexor reflex because of the fact that impulses traveling to the motoneurons become delayed in a network of interneurons and thus prolong the motoneuron discharge. Comparing the flexor reflex with the stretch reflex, differences in the latency and duration of the ventral root discharges can be observed. The traces in Figure 93 are taken from the classical experiment of Lloyd in which action potentials were recorded from a ventral root following stimulation of muscle and skin afferents. In the former, the latency is short; only a large spike potential is recorded since the internuncial system is not involved and there is no after-discharge. In contrast, the lower trace shows a much longer latency; afferent stimulation of a cutaneous nerve evokes widespread activity in the internuncial system, resulting in delayed bombardment of the motoneuron pool and prolonged after-discharge.

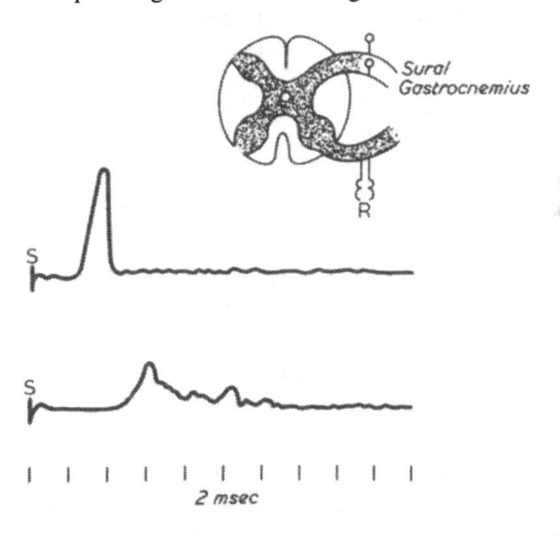

Fig. 93. Oscillographic tracings of monosynaptic and polysynaptic reflex discharges recorded from the ventral root of the cat. The upper trace represents the response to a single shock (S) delivered to the central cut end of nerve to gastrocnemius muscle. Note the short latency and lack of after-discharge. The lower trace shows the response to stimulation of sural (cutaneous) nerve exciting the internuncial system, and resulting in prolonged after-discharge. The experimental arrangement is shown in the inset. Time is 2 msec.

Rebound

In studying reciprocal innervation, Sherrington observed that if a muscle is reflexively inhibited by stretching the antagonistic muscle, its tension is sometimes increased above its resting level, i.e., inhibition is followed by a state of increased excitation. This "rebound" contraction has been attributed to over-compensation of the transmitting mechanism at synaptic junctions. It is comparable to other conditions of enhanced synaptic activity such as the response immediately following a conditioning tetanus, as explained below.

Post-tetanic potentiation

The experimental evidence is as follows. A single shock is delivered to a presynaptic fiber, and the response recorded from a ventral root. If the shock is preceded by a conditioning burst, the size of the test response is markedly increased. As the interval between conditioning and testing shocks is lengthened, the potentiation effect becomes progressively less. It is supposed that intense repetitive stimulation of a presynaptic fiber increases the output of transmitter substance from the synaptic vesicles, which is then available for the generation of EPSP. Excess transmitter increases the size of the postsynaptic response.

Renshaw inhibition

Renshaw cells are a type of interneuron located in the medial region of the anterior horn. They are excited by axon collaterals from motoneurons operating within the same segment of the cord. The axons of the Renshaw cells make inhibitory synaptic contacts on the soma and dendrites of the motoneurons and thus form an antidromic pathway like that depicted in Figure 94. Impulses in the motor-axon collateral liberate acetylcholine and depolarize the Renshaw cell. Impulses discharged by the Renshaw cell lib-

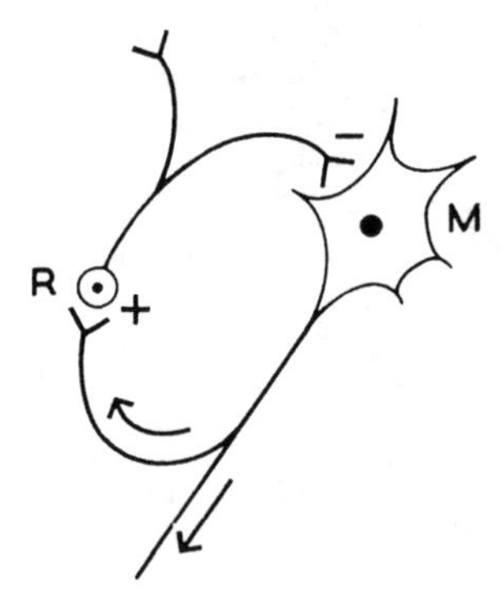

Fig. 94. Diagram of Renshaw cell inhibition. Collaterals of the motoneuron axon excite the Renshaw cell (R), which passes back to make inhibitory synaptic contact with the motoneuron (M).

erate an inhibitory transmitter substance at its synaptic terminals. Renshaw inhibition is an example of a negative feedback control whereby motoneuronal reflex discharges are effectively damped down.

Gate control

Although the output of the spinal motoneurons is regarded as the "final common path," their background excitability is subject to a constant stream of excitatory and inhibitory influences. Sherrington used the descriptive terms "central excitatory and inhibitory states" without, however, defining their mechanisms. It is now certain that the internuncial system plays a major role in generating impulse patterns in the spinal cord from the interplay of both local and descending pathways. According to the "gate" control theory, cells of the substantia gelatinosa in the dorsal horn modify impulse patterns before they are passed on to the next order of neurons. Extensive convergence of all types of fibers occurs in the dorsal horn where excitatory and inhibitory afferents can be balanced one against the other. Inhibition is turned on by increased activity in the large diameter fibers, which consequently close the gate to impulses in the slower-conducting small diameter fibers. On the other hand, inhibition is turned off and the gate opened when small fiber activity is intense. The theory of gate control has an interesting application to the mechanism of pain and will be discussed further in Chapter XVI.

ISOLATED SPINAL CORD

Spinal animal

A spinal preparation is one in which the cord has been severed from the medulla; if the head is also removed, the remaining portion of the body is known as a decapitate preparation. In either case the animal will stop breathing and must be artificially ventilated to survive. If transection is made below the fifth cervical segment, leaving the phrenic nerves intact, the animal will usually breathe on its own despite a fall of blood pressure due to isolation of the vasomotor center. Providing the blood pressure can be maintained above 40 mmHg, the reflex centers recover their excitability and the reactions of the spinal cord can be studied without interference from the brain.

In man, lesions of the spinal cord are not uncommon and complete transection occurs in many injuries. Examination of such patients gives insight into the important role exerted by descending pathways from the higher centers and reveals how the spinal cord can operate in a reflex manner when it is cut off from the influences that normally control it.

Spinal shock

This may be defined as a state of reflex depression that occurs immediately after complete transection of the spinal cord. It may not be noticeable in lower vertebrates, such as the frog, but in higher animals it may last for hours and, in man, several weeks. During this period, all movements supplied by segments below the level of the lesion and all sensations are lost. In other words, when motor and sensory activities are suddenly cut off from the brain, the muscles become paralyzed, the skin is found to be completely anesthetic, and all forms of reflex response are absent. Spinal shock may be caused by a combination of factors—loss of facilitatory pathways from supraspinal centers; deficiency of excitatory transmitter substance in spinal synapses; and unopposed activity of inhibitory interneurons.

It seems that the motoneuron pools are dependent upon a background of excitation normally supplied by the internuncial system, and when this is removed, the motoneuron discharges are temporarily depressed. In animals other than primates, the vestibulo-spinal tracts are considered to be essential for the prevention of spinal shock.

Complete transection in man

The results of acute transection of the cord vary to some extent with the level of the lesion. Thus the upper limbs are unaffected when the injury is below the second thoracic segment. At the level of the transection there may be some hyperesthesia due to irritation of neurons entering the spinal cord through the dorsal roots. In gunshot wounds and lacerated fractures, assessment of the extent of a spinal lesion may be a problem since a number of segments may be involved without, however, destroying all the pathways. Below the level of the transection, all activities of the spinal cord are absent throughout the period of shock.

1. Motor functions. Voluntary movement is lost and the muscles are paralyzed. The legs lie in any position and remain flaccid due to the absence of skeletal muscle tone.

2. Sensory functions. All conscious sensation is abolished; the skin is

completely anesthetic. It may be possible to localize the site of the lesion simply by testing the reaction to a pinprick and noting the segmental level at which a sensation is felt.

3. Reflex functions. As the spinal centers are depressed, no reflexes can be elicited. The superficial reflexes, tendon jerks, and the withdrawal reflex are all absent.

4. Visceral functions. Smooth muscle is paralyzed as well as skeletal muscle. Consequently, there is a fall of blood pressure due to widespread vasodilatation, unless the lesion is below the second lumbar segment. The vasoconstrictor fibers leave the cord above this level. The skin remains dry as sweating is abolished. Since both voluntary micturition and the bladder reflexes are absent, the state of the bladder presents a serious nursing problem and catheterization is essential to prevent retention of urine. The reflex evacuation of the rectum is also depressed during the period of shock.

Stages of recovery

Providing the patient is kept relatively free from infection, reflex function begins to return to the spinal cord after about three weeks.

1. The blood pressure is restored as vasomotor tone returns to the paralyzed blood vessels, and at the same time the bladder and rectal reflexes partially recover. The bladder responds to filling by reflex evacuation, but the force of contraction is feeble. Consequently, the bladder may not be completely emptied, leaving the residual urine prone to infection.

2. Spontaneous flexor movements occur after about six weeks when the withdrawal reflex can be elicited by strong stimulation. As the reflex becomes more brisk, it tends to spread to other spinal centers including those controlling the outflow to the bladder, rectum, and skin. The response then becomes a "mass reflex" in which both extremities are withdrawn, the bladder and rectum contract, and the patient perspires profusely. If the sole of the foot is stimulated, the big toe tends to turn upward during the response. Dorsiflexion of the toe, instead of the usual plantar flexion, is an important clinical finding known as the sign of Babinski.

3. Extensor reflexes return last of all. Some months after spinal transection, the knee and ankle jerks can be elicited and the crossed extensor response obtained. A degree of tone returns to the muscles. The legs are now able to take the weight of the body when the patient stands, and at this stage he is encouraged to walk by using the muscles of the hips for assistance. In some cases, the presence of a slight degree of spasticity in the muscles may be due to an incomplete transection. If some of the descending

pathways between the brain and the spinal cord escape the lesion and are left intact, these tend to reinforce the spinal centers with resultant reflex hyperactivity.

Importance of supraspinal control

While there is always a possibility of good functional recovery after a spinal injury, the reflex centers remain below their normal level of excitability, muscle tone is reduced, and the tendon jerks are correspondingly weak. It seems that the spinal centers are very much dependent upon facilitating influences from the brain and cannot operate in the same way when those influences are absent. This is made only too evident by the state of spinal shock that develops even when the lesion is incomplete. It is true that, assuming a general infection is prevented, a fair degree of recovery can be anticipated. This prognosis emphasizes once again that the isolated cord is capable of carrying out fully coordinated reflex actions. Emptying of the bladder and rectum is well-controlled, and the skin assumes a healthy color. Nevertheless, the affected muscles remain weak and hypotonic, the extensor groups in particular showing the need for reinforcement from higher centers.

SELECTED BIBLIOGRAPHY

Creed, R.S., Denny-Brown, D., Eccles, J.C., and Sherrington, C.S. *Reflex Activity of the Spinal Cord.* London: Oxford University Press, 1932.

Hunt, C.C., and Perl, E.R. Spinal reflex mechanisms concerned with skeletal muscle. *Physiol. Rev.* 40:538-579, 1960.

Liddell, E.G.T. Spinal shock and some features in isolation-alteration of the spinal cord in cats. *Brain* 57:386-400, 1934.

Lloyd, D.P.C. Reflex action in relation to the pattern and peripheral source of afferent stimulation. *J. Neurophysiol.* 6:111-119, 1943.

Lloyd, D.P.C. Functional organization of the spinal cord. *Physiol. Rev.* 24:1-17, 1944.

Lorente de Nó, R. Analysis of the activity of the chains of internuncial neurons. *J. Neurophysiol.* 1:207-244, 1938.

Renshaw, B. Central effects of centripetal impulses in axons of spinal ventral roots. *J. Neurophysiol.* 9:191-204, 1946.

Ruch, T.C. The spinal cord and reflex action. *Ann. Rev. Physiol.* 4:359-374, 1942.

Sahs, A.L., and Fulton, J.F. Somatic and autonomic reflexes in spinal monkeys. *J. Neurophysiol.* 3:258-268, 1940.

Sherrington, C.S. *The Integrative Action of the Nervous System.* 2nd ed. New Haven: Yale University Press, 1947.

Wilson, V.J., and Kato, M. Excitation of extensor motoneurons by group II afferent fibers in ipsilateral muscle nerves. *J. Neurophysiol.* 28:545-554, 1965.

CHAPTER XI

Skeletal Muscle Tone and Posture

Sherrington maintained that all movements and posture are superimposed on a background of muscle tone, the degree and extent of which depend upon afferents coming from the muscles themselves. Today, this view is generally held and remains of great practical importance to the neurologist, since alteration in the tone of skeletal muscle is a common result of pathological processes in human diseases. From what has been said in the last chapter, it seems fairly clear that the spinal cord is capable of maintaining reflex activity, yet muscle tone is deficient when deprived of reinforcing influences from the brain.

In order to determine which parts of the central nervous system are important for the regulation of muscle tone, sections are made at different levels through the brain of an experimental animal and the deficits resulting are observed. Sherrington demonstrated that an abnormally high degree of tone developed in the extensor muscles of the cat after decerebration through the brain stem; if, however, the lesion was above the level of the red nucleus, rigidity did not occur. Liddel and Sherrington established that the rigidity of the decerebrate animal was reflexive in origin and resulted from increased tonic activity of the stretch reflexes, for if the posterior nerve roots were sectioned, the limbs became flaccid. Their work was extended by Denny-Brown, who showed that stretch reflexes also occurred in flexor muscles, thus converting the limbs of the decerebrate animal into rigid pillars. Lloyd established the excitatory monosynaptic nature of the stretch reflex and its transmission through Group Ia afferent fibers, while Granit emphasized an inhibitory component contributing to the reflex by the Group Ib afferents.

All this led to the question about the various receptors in muscle responsible for maintaining tone. Muscle spindles were described by histologists more than 100 years ago and soon afterward Golgi discovered the tendon organ. A paper by Barker gave a particularly good account of the intrafusal muscle fibers and their innervation, both sensory and motor. Single fiber recordings of the frog's muscle spindle by Matthews gave a more detailed analysis of the reflex responses of a muscle to stretching. Following this came Leskell's pioneering work on gamma efferent discharges. Hunt and Kuffler elucidated the mechanisms of spindle control, and the Japanese school, led by Tasaki, contributed significantly to the role of the spindle in the regulation of posture.

Modern ideas on the functioning of muscle receptors are based on the concept of a servo-mechanism, described by Merton and others as a self-regulating closed loop with sensory feedback from the spindles. This mechanism is designed to maintain a constant muscle length and thereby automatically compensate for the changes imposed by external forces such as gravity and postural movements. Finally, it is evident that if the servo-loop operates as part of a spinal reflex, its performance will depend upon the activities of higher centers in the brain which modify and control it by facilitation and inhibition.

Definitions of tone in muscle

1. Clinical tone. When a healthy muscle is palpated at rest it appears firm to the grasp. On manipulating a limb, the examiner notes the degree of resistance to passive movement. Clinical tone is assessed by the amount of resistance offered in the normal limb compared with the resistance offered on the side of the lesion. Thus there is a marked increase of resistance to passive movement in a patient who has suffered an upper motor neuron lesion. Because part of the resistance is due to reflex contraction of the muscles, the sensitivity of the stretch reflex can also be estimated. When tone is increased, the tendon jerks are brisk; when tone is decreased, the tendon jerks are sluggish.

2. Physiological tone. This is due to the sensitivity of the muscle spindle reflex system. It is a state of reflex contraction brought about by intermittent motoneuron discharges firing at low rates down the ventral roots. The discharges are maintained by afferent impulses coming from the muscle spindles which, in turn, are controlled by the activity of the gamma efferent nerves. Tone can be abolished by destroying any part of the reflex arc. Correspondingly, tone can be modified by any disturbance that alters the excitability of the gamma efferent system.

3. Postural tone. Sherrington defined tone as the activity of muscle when functioning in posture. He stated that "reflex tonus is the expression of a neural discharge which maintains attitude. It arises in the proprioceptors and varies in different muscles, being most marked in the anti-gravity muscles." In the erect posture, the flexor and extensor muscles of the trunk and lower limbs contract slightly as soon as they are stretched. When a muscle is

stretched, its spindles are stimulated and a reflex is evoked; this reflex tends to return the muscle to its original length.

Actions of the muscle spindle

The stretch reflex has two components, phasic and tonic. The former is the reflex response to a brief, sudden stretch and the latter to a prolonged or maintained stretch. Both components have their origin in the sensory endings of the muscle spindle.

1. Role of intrafusal muscle fibers. The structure and innervation of the muscle spindle were described in Chapter V. The spindle contains two types of intrafusal fiber, nuclear bag and nuclear chain:

Nuclear bag and intrafusal fibers

· Contraction of the fibers is brought about either by stretch or by gamma efferent stimulation. Contraction is confined to the region of the end-plates and leads to a stretching of the primary endings. The secondary endings are also excited but not to the same extent.
· The predominant sensory ending is the annulo-spiral. The afferent fibers belong to the Ia group, which has an excitatory role in the stretch reflex, forming monosynaptic connections with the alpha motoneurons.
· The gamma efferent nerves supplying nuclear bag fibers have relatively low thresholds and terminate in discrete end-plates.
· The nuclear bag fibers are sensitive to the dynamic component of stretch. Their role is to resist sudden changes in muscle length by causing reflex contraction, which opposes the stretch and relieves the spindle of its stimulus. At the same time the discharge of the primary endings is silenced.

Nuclear chain intrafusal fibers

· Contraction of the fibers is brought about by stimulation of the static gamma efferent nerves. This leads to excitation of the secondary endings, causing an increased afferent discharge from the spindle.
· The secondary or flower spray endings lie predominantly on the intrafusal chain fibers. The afferents belong to the II group of fibers which produce their effects polysynaptically.
· The static efferent fibers have relatively high thresholds and tend to terminate diffusely as trail endings rather than as discrete motor end-plates.

• The roll of the nuclear chain intrafusal fibers is to maintain a tonic afferent discharge from the spindle. This compensates for the silence of the primary ending discharges during contraction of the muscle.

2. Efferent control of the muscle spindle. The small motor fibers entering the spindle are derived from anterior horn cells and form about one third of the outflow from a ventral root. They are called gamma efferent fibers since their diameters range between 3 μm to 9 μm. Stimulation experiments show that there are two kinds of motor supply to a spindle, dynamic and static (Fig. 95).

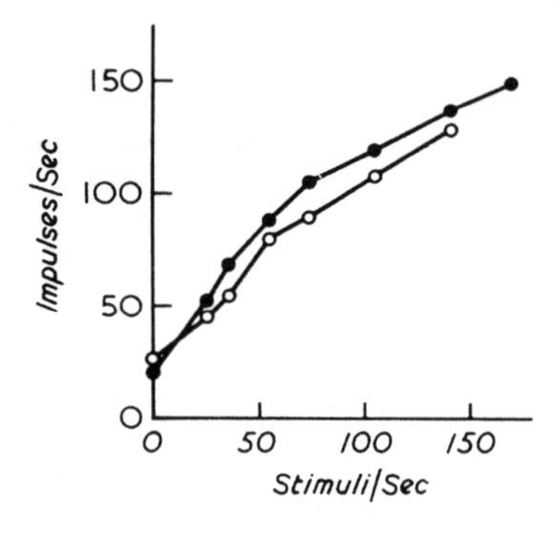

Fig. 95. Graph showing increasing discharge of primary ending in response to increased frequency of stimulation of the two kinds of gamma efferents. Open circles represent stimulation of dynamic fiber, closed circles, stimulation of static fiber. (After P.B.C. Matthews.)

• Stimulation of a dynamic fiber causes an increase of firing from the primary annulo-spiral ending, while the secondary ending is not affected.
• Stimulation of a static fiber causes an increase of firing from the secondary ending, while having little action on the primary afferent discharge.

These differences in action between the two kinds of fusimotor fibers are important from the point of view of the use made of the muscle spindle receptor by the central nervous system. The dynamic action of the spindle signals a change in length of a muscle when it is stretched. The spindle responds by operating the monosynaptic stretch reflex, then ceases to discharge when the muscle contracts. The static action of the spindle produces a sustained afferent discharge irrespective of the state of the muscle, i.e., whether it is stretched or slack or contracting. This sustained discharge accounts for the presence of tone in the muscle. Tone can obviously be modified, however, by altering the sensitivity of the spindle, either by increasing the gamma efferent discharge or by reducing it. In this way the central nervous system can be expected to exert its influence on the distribution of tone in the various muscles under its command.

Decerebrate preparation

The neural structures involved in the regulation of skeletal muscle tone were revealed in Sherrington's work on the decerebrate cat. An abnormally high degree of tone developed in the extensor muscles of the animal, converting the limbs into rigid pillars. It was supposed that the hypertonus occurring in human disease was dependent upon similar disturbances causing interruption of the flow of impulses from the higher centers in the brain.

1. Classical decerebration. Transection of the brain stem is performed above the level of the vestibular nucleus (A–A in Fig. 96), thereby disconnecting the spinal cord from the red nucleus and the structures above it, but leaving the cerebellum intact. Shortly after the operation, the animal develops an exaggerated posture due to an increase of tone in the anti-gravity muscles. The posture is described as a "release phenomenon" because the local reflexes responsible for tone are released from higher control.

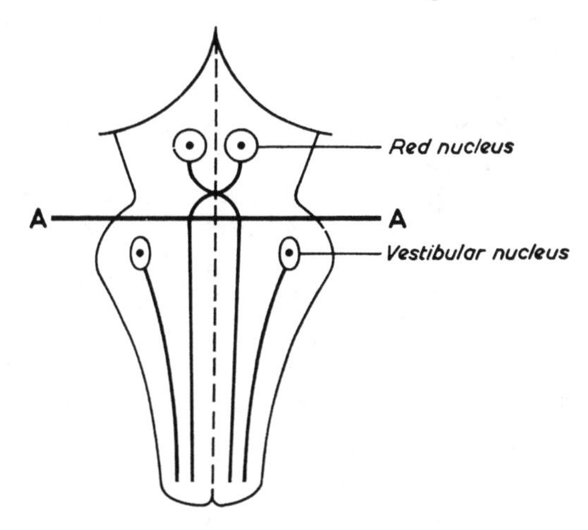

Fig. 96. Diagram of cat brain stem. A-A depicts the level of the transection used to produce decerebrate rigidity.

2. Anemic decerebration. This is performed by tying the rostral end of the basilar artery on the ventral surface of the brain stem and clamping the common carotid arteries in the neck. Branches of the basilar artery supply the brain stem and most of the cerebellum, but the anterior lobe of the cerebellum as well as the cerebral hemispheres are deprived of circulating blood. The rigidity slowly disappears if the carotid clamps are removed.

Reflex activity in the decerebrate cat

1. De-afferentation. The rigidity of a decerebrate animal can usually be abolished by cutting the appropriate dorsal roots. From this it can be concluded that the tonic activity required for maintaining rigidity originates in the muscles themselves. The spinal motoneurons, however, are still capable of being excited by impulses descending from higher levels, and rigidity may persist after de-afferentation if the balance of the converging pathways is in favor of facilitation. One of the pathways considered to be of the utmost importance in maintaining decerebrate rigidity is the vestibulo-spinal tract. It is supposed that vestibular influences left intact after decerebration are sufficient to excite the spinal centers, for if the vestibular nucleus is destroyed on one side, the rigidity promptly disappears on that side.

2. Autogenetic facilitation. The term *autogenetic* is used to describe the contributions made by the spinal cord itself toward the development of the excessive tone seen in decerebrate rigidity. Autogenetic facilitation is the result of the excitatory influences arising from muscle spindle discharges. Thus stimulation of Group Ia afferents elicits monosynaptic excitatory effects on the extensor motoneurons accompanied by inhibition of the antagonists. At the same time, motoneurons supplying synergistic muscles of the same and neighboring joints are facilitated. This suggests that purely spinal mechanisms do, in fact, contribute toward maintaining skeletal muscle tone by suitably facilitated reflex arcs.

3. Autogenetic inhibition. This is the result of the inhibitory influences elicited from tendon organ discharges. Thus stimulation of Group Ib afferents inhibits the extensor muscles of a joint while exciting the flexors of the same and neighboring joints. Since both types of muscle receptor are sensitive to stretch, the spinal centers are endowed with a significant amount of self-control. While the muscle spindles provide a tonic excitatory discharge, the tendon organs tend to counteract the effects by reflex inhibition. This has led to the concept of a "plastic" quality of reflex behavior, which, of course, is lost when the dorsal roots are sectioned.

Lengthening and Shortening reactions

These reactions demonstrate how the muscles maintain their hypertonic state when their length is suddenly changed by flexion or extension.

The *lengthening reaction* is the elongation reflex of a muscle produced by forcible stretching. It is elicited by attempting to flex the knee joint of a decerebrate animal. At first, resistance is felt owing to the high degree of tone in the quadriceps muscle. If the attempt is persisted, the resistance is overcome suddenly and the muscle relaxes, adjusting itself to the new longer length. Sudden relaxation of the previously rigid limb is brought about by autogenetic inhibition. Forcible stretching of the muscle stimulates the Golgi tendon organs, which have a higher threshold than do the muscle spindles and discharge when a certain level of stretch is exceeded. The excitatory side of the stretch reflex is switched off by the more powerful Group Ib afferent volleys. The lengthening reaction is thus a protective reflex designed to prevent injury from excessive stretching. A similar phenomenon in spastic patients is known as the "clasp knife" reflex, the name derived from the way in which a pocket knife behaves when it is opened.

The *shortening reaction* refers to the isotonic contraction of the muscle when it assumes its original length. It is elicited by extending the knee joint after forcible flexion; the muscle adjusts itself to maintain reflex tone at the shorter length.

Physiological tremor

This term refers to the fine oscillations seen in a muscle during contraction, for example, in the outstretched fingers of a human subject whose eyes are closed. Although tremor is an important clinical sign in many diseases, the responsible factor seems to lie in some part of the reflex arc. The most likely mechanism is to be found in the servo-loop controlling muscle length. It will be remembered that the dynamic component of the stretch reflex originates from the primary endings of the muscle spindles and that the rate of spindle firing is controlled by the gamma efferent system. The oscillations may be due to instability of the reflex arcs when the muscles are subjected to gravitational stretch.

Supraspinal control of muscle spindles

In the intact nervous system the higher centers can influence spinal reflexes in two different ways—by direct action on the alpha motoneurons or, indirectly, by action on the gamma efferent servo-loops. Since the majority of descending pathways terminate on interneurons in the cord, the changes imposed are mostly on the static stretch reflexes. Spindle activity is normally measured in terms of impulse frequency, a rise or fall of frequency being an indication of the bias that has been applied to the sensory endings. Thus any increase in spindle discharge will relate to an increase of tone in the muscle. Conversely, a fall of frequency will relate to a reduction of tone.

Electrical stimulation of certain points in the brain stem can excite or inhibit spindle afferent discharges, presumably by activation of pathways converging on the spinal motoneurons. These supraspinal influences are present after cutting the dorsal roots and therefore act independently of reflex stimulation. Descending facilitating influences acting on the spindle loop provide an intense driving force tending to produce hyperreflexia and hypertonia. Likewise, descending inhibitory influences have the opposite effects and tend to reduce or abolish the spindle discharge. Thus a balance struck between these opposing systems will account for the normal distribution of tone in the muscles. The development of rigidity in the decerebrate animal can be explained partly as an excess of facilitation on the spinal centers and partly as a deficit of descending inhibitory factors.

POSTURAL REFLEXES

Posture may be defined as the attitude assumed by the body under different circumstances, whether lying perfectly still, preparing to carry out a movement, or during movement itself. The entire muscular system may be involved, with flexors as well as extensors playing their part in supporting the body against gravity. In standing, the extensor muscles are particularly important. As the central nervous system is capable of integrating the various reflex mechanisms which form the basis of posture, some idea of the contributions from the different regions can be gained by studying the postural reflexes in experimental isolation. Three classical preparations are commonly described—decerebrate, thalamic, and decorticate.

Decerebrate

In this preparation postural reflexes are more readily elicited than they are in the spinal animal and affect the whole body, not only the limbs. The abnormal degree of tone in the muscles supports the weight of the body in the upright position, but the animal is unable to right itself or make walking movements.

1. Positive supporting reactions. When the pad of the toe is lightly touched, all the muscles of the limb (both flexors and extensors) contract. The limb is converted into a rigid pillar. The reflex is initiated by impulses from the pad, then reinforced by proprioceptive impulses from the stretched muscles. If the skin of the pad is anesthetized, the reaction can be elicited by stretching the muscles. If, on the other hand, the toes are flexed, the muscles are reflexively inhibited and the reaction abolished.

2. Tonic neck reflexes. These reactions produce coordinated movements

of the limbs in relation to the position of the head. To study them separately, it is necessary to remove both labyrinths. Afferent impulses arise from the muscle spindles in the neck muscles and from receptors in the vertebral joints.

- Rotation of the head to one side causes the limbs of that side to extend while the limbs on the opposite side relax.
- Bending the head forward causes flexion of both forelimbs and extension of the hindlimbs.
- Bending the head backward causes extension of both forelimbs and relaxation of the hindlimbs.

Tonic neck reflexes adjust the tone and position of the limbs to any alteration in the position of the head. The posture of the whole animal is prepared for carrying out certain automatic movements, such as looking up or crouching under a shelf.

Thalamic

In this preparation the cerebrum is removed to eliminate the higher centers, but those of the mid-brain remain intact. Rigidity does not develop, and the animal maintains a normal posture. By means of the righting reflexes, the head can be raised the right way up if the animal is placed on its back. The righting reflexes consist of a chain of reactions and can be readily demonstrated in the intact animal. A cat which is blindfolded and dropped with its legs uppermost will turn quickly in the air and land on its feet. The reaction begins with the labyrinthine reflexes causing the head to turn the right way up. As the head turns, the neck is twisted, and impulses from the neck muscles bring about rotation of the trunk so as to follow the head. Tonic neck reflexes increase extensor tone in the limbs to meet the force of landing. If the labyrinths are destroyed, the blindfolded animal cannot right itself, as cues about the position of the head in space are lost. It may be concluded that all the essential mechanisms are present in the thalamic animal for maintaining the normal distribution of tone in the muscles and for making appropriate adjustments to the limbs when the head is moved or when the weight of the body is shifted.

Decorticate

Removal of the entire cortex has the same effect as removing the cerebrum, but in primates, tone is gravely disturbed because of the loss of powerful inhibitory influences. Similar disturbances of posture are seen in the human being, especially in individuals with extrapyramidal disease, which causes the affected muscles to become spastic and the reflex centers to become hyperexcitable. Two further reactions may be mentioned here:

1. Grasp reflex. This is seen following an isolated lesion in the premotor cortex. If an object is placed in contact with the hand or foot, flexion of the fingers or toes occurs. The reaction is called *forced grasping* and is involuntary in character. It is due to a release of pyramidal influences on subcortical centers and is abolished if the lesion extends into the motor cortex.

2. Contact Placing. If the paw of an animal is placed in contact with the side of a solid object, the limb is immediately placed on top of it. The purpose of the reflex is to seek support for the body. Likewise, the foot seeks a position suitable for normal standing. If the animal is moved sideways, the leg hops in the same direction attempting to keep the body supported. All these reactions are dependent upon the integrity of the cerebral cortex.

SUMMARY

Tone in skeletal muscle is assessed by the amount of resistance offered to passive movement of a limb. A study of the physiological mechanism of skeletal muscle tone is essential to an understanding of posture. And the disturbances of tone so commonly seen in disease are most important to the clinician.

The static component of the stretch reflex is responsible for the maintenance of tone. Impulses arise from the muscle spindles, mainly from the secondary endings lying predominantly on the intrafusal chain fibers. The muscle spindles are under the control of a special system of motor nerve fibers, called gamma efferents, which modifies their sensitivity and drives the spindles independently of the stretch stimulus. Thus the muscles are kept in a tonic state irrespective of their dynamic actions. Tone is modified simply by altering the sensitivity of the muscle spindles. This is usually brought about by a combination of local and descending influences converging on spinal reflex arcs. Purely autogenetic factors arising from slight degrees of stretch ensure a steady motor output to the muscles.

Supraspinal influences on the reflex arcs operate by direct action on the alpha motoneurons or, indirectly, by influencing the discharge of gamma motoneurons. The parts of the brain involved can be studied in the experimental animal by mapping the position of excitatory and inhibitory points or by making selective lesions through the brain at different levels. The abnormalities produced and deficits of function are amply confirmed on clinical grounds. It seems that the distribution of tone in the muscles is dependent upon the overall effects of the various neural systems converging on the spindle loops.

Normal distribution of tone is the basis of posture. It accounts for the ability to stand up against gravity and for the attitudes taken up to support the body when its position is changed. Postural reactions are readily elicited in the decerebrate preparation, but righting reflexes require an intact midbrain; the grasping and placing reactions have their reflex centers in the cerebral cortex. Abnormalities of tone result from any condition that switches the balance between excitatory and inhibitory actions on the motoneuron. Thus hypotonia occurs in lesions of excitatory pathways, whether they be peripheral or central in origin. Such lesions are characterized clinically by flaccid muscles and diminished or absent reflexes. Likewise, hypertonia occurs in lesions of inhibitory pathways, but in this case, the hypertonia must be suitably facilitated. The exaggerated reflexes of the spastic subject represent a condition of over-activity of mechanisms released from inhibitory control.

SELECTED BIBLIOGRAPHY

Barker, D. The innervation of the muscle spindle. *Quart. J. Micros. Sci.* 89:143-186, 1948.

Boyd, I.A. The structure and innervation of the nuclear bag muscle fibre system and the nuclear chain muscle fibre system in mammalian muscle spindles. *Phil. Trans. R. Soc. B* 245:81-136, 1962.

Eldred, E., Granit, R., and Merton, P.A. Supraspinal control of the muscle spindles and its significance. *J. Physiol.* 122:498-523, 1953.

Grigg, P., and Preston, J.B. Baboon flexor and extensor fusimotor neurons and their modulation by motor cortex. *J. Neurophysiol.* 34:428-436, 1971.

Hunt, C.C., and Kuffler, S.W. Stretch receptor discharges during muscle contraction. *J. Physiol.* 113:298-315, 1951.

Matthews, P.B.C. *Mammalian Muscle Receptors and their Central Actions.* London: Edward Arnold, 1972.

McCouch, G.P., Deering, I.D., and Ling, T.H. Location of receptors for tonic neck reflexes. *J. Neurophysiol.* 14:191-195, 1951.

Pollock, L.J., and Davis, L. The reflex activities of a decerebrate animal. *J. Comp. Neurol.* 50:377-411, 1930.

Takano, K., and Henatsch, H.D. The effect of rate of stretch upon the development of active reflex tension in hind limb muscles of the decerebrate cat. *Expl. Brain Res.* 12:422-434, 1971.

Wiesendanger, M. Rigidity produced by de-afferentation. *Acta Physiol. Scand.* 62:160-168, 1964.

CHAPTER XII

Control of Posture

Normal posture depends upon the integrated activity of many different regions of the brain. Certain postures are assumed in repose; others are exhibited before any coordinated movement is carried out. The essential elements for most postural activities are present in the spinal cord. But muscular tone in the spinal animal is deficient, and the spinal reflexes cannot adapt themselves to ensure stability when body position is disturbed. The adjustments to restore the body to its correct attitude are highly complex and usually follow a succession of accurately timed events. Such events are largely guided by input systems capable of exerting continual control. It will therefore be appreciated that posture is a function of the nervous system as a whole; each component has a special contribution to make or imposes some kind of modification. Interference with any of these components by injury or disease will almost certainly cause a disturbance of posture. The type and severity of the disturbance obviously will depend upon the particular region involved and upon the extent of the damage.

Regions of the brain that will now be considered in terms of their role in regulating posture are reticular formation, cerebellum, basal ganglia, premotor cortex, and limbic system.

RETICULAR FORMATION

Functional anatomy

The reticular formation is essentially a network of interneurons occupying the mid-line of the brain stem between the medulla and the thalamus. Several recognizable nuclei are scattered throughout its substance. An important afferent source comes from the anterolateral columns of the spinal cord. It also receives collaterals from the ascending sensory systems as well as from the deep nuclei of the cerebellum. In addition, descending projections from the cerebral cortex and from many subcortical nuclei terminate in a diffuse manner among the reticular neurons. Within the network are many short synaptic chains where opportunities occur for convergence and interaction. But there are faster routes whereby the reticular formation can exert its influence both upward toward the cerebral structures and downward into the spinal cord.

Descending facilitatory influences

Electrical stimulation of certain points in the brain stem causes facilitation of cortically induced movements and enhances reflex activity. The effects were first described by Magoun and Rhines from experiments in the monkey; the facilitatory region lay in the pons and mid-brain and extended rostrally to the intralaminar nuclei of the thalamus (Fig. 97). When the brain stem was cut below this region, the effects were abolished. The facilitatory reticular-spinal tracts occupy the anterolateral columns and terminate on interneurons in the cord; they provide a background of excitation to the motoneurons, especially those supplying the extensor muscles. Figure 98 shows how the knee jerk is greatly enhanced by stimulating the reticular formation. The facilitatory effects are entirely independent of the vestibular-spinal tracts.

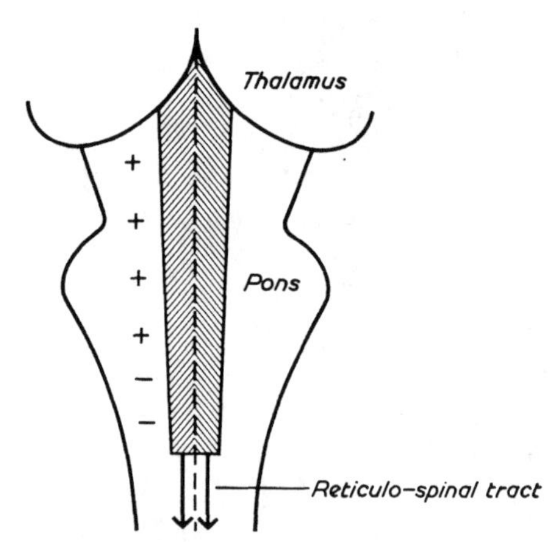

Fig. 97. Diagram of brain stem reticular formation showing extent of facilitatory (+) and inhibitory (−) regions giving rise to corresponding reticulo-spinal tracts. The inhibitory region is caudally situated in the brain stem.

From the evidence discussed in the previous chapter, it seems likely that facilitatory influences cause an increase in the rate of gamma efferent discharge which, in turn, results in an increased firing of the muscle spindles. The spindle response is still present after cutting the dorsal roots to prevent

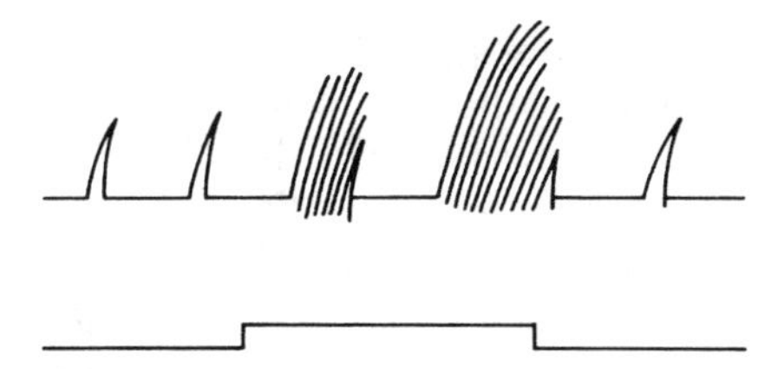

Fig. 98. Facilitation of the knee jerk during stimulation of the brain stem reticular formation is indicated by the signal in the lower tracing. (After Magoun and Rhines.)

the impulses from reaching the motoneurons, but it is lost if the ventral roots are cut. Descending facilitatory influences are therefore responsible for increasing muscle tone and enhancing reflex activity. If the facilitatory pathway is interrrupted, muscle tone is diminished and the reflexes are weakened.

Descending inhibitory influences

Electrical stimulation of this part of the reticular formation causes a reduction of cortically induced movement and abolishes reflex activity. The inhibitory points are more caudally situated in the brain stem and the spinal tracts are confined to the lateral columns of the cord. Hence they can be easily cut in the experimental animal to show the effects of loss of inhibition. Figure 99 shows how the knee jerk is abolished during the period of reticular stimulation. The descending fibers terminate on short interneurons in the cord through which spindle activity is reduced by means of the gamma efferent loop. Accordingly, if the inhibitory pathway is interrupted, muscle tone is increased and the reflexes become exaggerated.

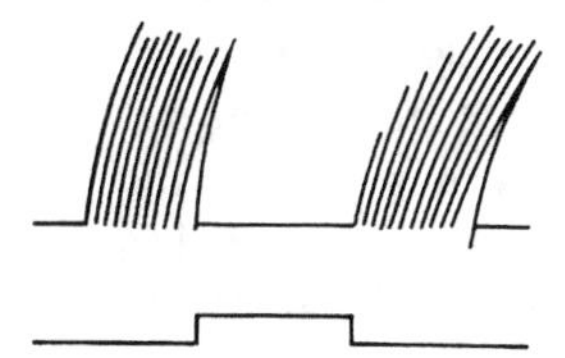

Fig. 99. Inhibition of the knee jerk during stimulation of the bulbar reticular formation is indicated by the signal in the lower tracing.

Effects on posture

The brain stem reticular formation is capable of modifying the distribution of tone in the muscles by influencing the discharge patterns of the motoneurons which supply the muscles. The facilitatory component increases the excitatory state of the spinal centers already subject to tonic reflex inputs from the muscle spindles. The level of tonic activity is, of course, determined by the gamma efferent mechanism to which reference has just been made. The inhibitory component opposes these actions, causing a reduction of spinal excitability and of spindle afferent discharge. The integrated activity of these two reticular systems accounts for a wide range of control extending between maximal excitation and maximal inhibition. Tone is thereby adjusted to meet all bodily requirements.

Damage to the reticular formation can affect both systems, with resultant changes in posture. Loss of reticular drive theoretically should give rise to hypotonia and diminished reflexes, but this is not commonly seen in practice, since many other facilitatory systems also contribute to the state of the spinal motoneurons and thus maintain their drive. On the other hand, damage to the inhibitory part of the reticular formation leads to a change in posture closely resembling that of the decerebrate animal. There develops a state of spasticity characterized by a marked increase of tone in the muscles, increased spindle discharges, and exaggerated tendon reflexes.

CEREBELLUM

Functional anatomy

The cerebellum lies in the posterior cranial fossa below the tentorium. It consists of a central vermis and two laterally placed hemispheres. The cerebellum may be divided into three functional parts (Fig. 100).

1. The anterior lobe has a major role in the control of posture. The various lobules, designated I to V in Larsell's classification, show a definite topographical organization in which the body is inversely represented. The lingula represents the sacral structures and is small in man. The centralis corresponds to the leg; the culmen corresponds to the trunk and arm, while the simplex or head region lies most caudally, below the primary fissure. Visceral afferent representation in the anterior lobe is described in Chapter XX.

2. The posterior lobe has a major role in the control of voluntary movement. It is the largest part of the cerebellum, extending from lobules VI to IX, and from its topographical arrangements, it appears to coincide with the development of the motor cortex. In addition to its motor projections, the posterior lobe, or neocerebellum, contains localized areas for auditory, visual, cutaneous, and visceral representations.

3. The flocculo-nodular lobe is involved primarily with the function of

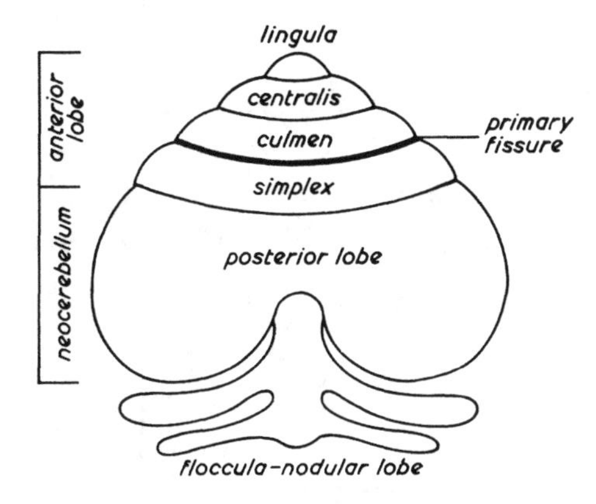

Fig. 100. Diagram of the cerebellum, illustrating its three functional parts. The physiological anterior lobe includes the simplex below the primary fissure. All parts are subdivided into a central vermis, an intermediate or para-vermal zone and a lateral zone. (Modified from Fulton.)

balance. It lies immediately over the fourth ventricle, developing as an extension of the vestibular nuclei and serving as a center for the labyrinthine reflexes, as described in Chapter VI.

Structural organization

The structure of the cerebellum is remarkably uniform. The cerebellar cortex covers the whole of the visible surface and dips into the various fissures of the interior. The white matter forms a central core, projecting into the cortex in the form of laminations. The deep nuclei are masses of gray matter lying in the substance of the cerebellum through which the output fibers are conducted to the brain stem.

1. Cerebellar cortex. There are three well-defined histological layers (Fig. 101):

(a) The molecular layer is densely packed with dendrites and axons. The dendrites belong to the branching processes of Purkinje and Golgi cells, while the axons are derived mainly from granule cells. In addition, there are numerous terminal filaments of climbing fibers distributed from the white matter. The principal cells of the molecular layer are basket and stellate. Each basket cell supplies about eight or 10 Purkinje cells. The axons of the basket and stellate cells make synaptic contacts with the Purkinje cells by means of spines; they are powerfully inhibitory.

(b) A single layer of Purkinje cells consists of flask-shaped bodies, each

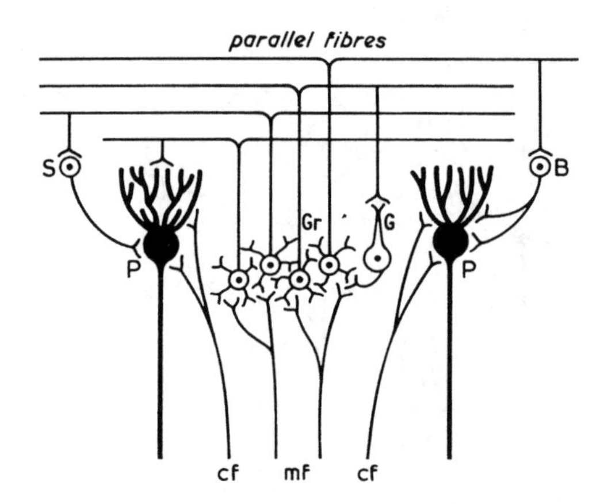

Fig. 101. Diagram of the principal cells and their synaptic connections in the cerebellar cortex. The parallel fibers are the bifurcated axons of granule cells (Gr) innervated by mossy fiber afferents. The terminals of the parallel fibers end on the dendrites of stellate (S), basket (B), and Golgi cells (G) in the molecular layer, which also contains the dendritic tree of the Purkinje cells (P). Climbing fiber afferents terminate directly on the Purkinje cells whose axons pass through the white matter to the deep cerebellar nuclei.

with a profuse dendrite tree branching in the transverse plane of a folium. The axon descends from the bottom of the flask and gives off abundant collaterals as it traverses the white matter to end in the deep nuclei of the cerebellum.

(c) The granular layer contains round, closely packed cells with radiating dendrites ending in claw-shaped tufts. The axons of the granule cells ascend through the molecular layer where they bifurcate to form the parallel fibers. In addition, the granular layer contains large stellate neurons or Golgi cells whose axons ramify on the granule cells with the terminals of a mossy fiber. Together they form a complex synaptic arrangement known as a "cerebellar glomerulus." The composition of a glomerulus therefore includes the following–the presynaptic terminal of a mossy fiber; the presynaptic terminal of a Golgi axon; and the postsynaptic dendrite tufts of a granule cell. The Golgi cell has a key position in this arrangement: Its dendrites are driven by the parallel fibers; its axon is influenced by the mossy fibers.

2. White matter. There are two systems of afferent fibers terminating in the cerebellar cortex. The first is the mossy fiber input and the second, the climbing fiber input.

(a) Mossy fibers originate from the spinal cord and brain stem. They carry impulses from a wide variety of sense organs. On reaching the granular layer, each mossy fiber innervates about 20 glomeruli in addition to supplying numerous collaterals.

(b) Climbing fibers originate mainly from cells in the contralateral inferior olive. Each climbing fiber makes extensive synaptic contact with the dendritic tree of a Purkinje cell. There are also many collaterals distributed to other cells in the cerebellar cortex.

3. Deep nuclei. These are three in number on each side and constitute the efferent pathway from the cerebellum.

(a) The fastigial nucleus is found close to the mid-line above the roof of the fourth ventricle. It receives the axon terminals of those Purkinje cells which are situated in the vermis and also receives a contribution from the mossy fibers.

(b) The intermediate nucleus is a relay station for Purkinje projections from the paravermian zones.

(c) The dentate nucleus appears as a double fold of gray matter in transverse section. It receives the terminals of Purkinje axons from the lateral zones of the cerebellum and also a contribution from climbing fiber collaterals.

The above account of the structure of the cerebellum describes the various types of cell and their principal connections. It will be clear that the cerebellar cortex has two distinct input systems—mossy fiber and climbing fiber—but only one output system—the Purkinje cell axons to the deep nuclei. A diagram of these relationships is shown in Figure 102.

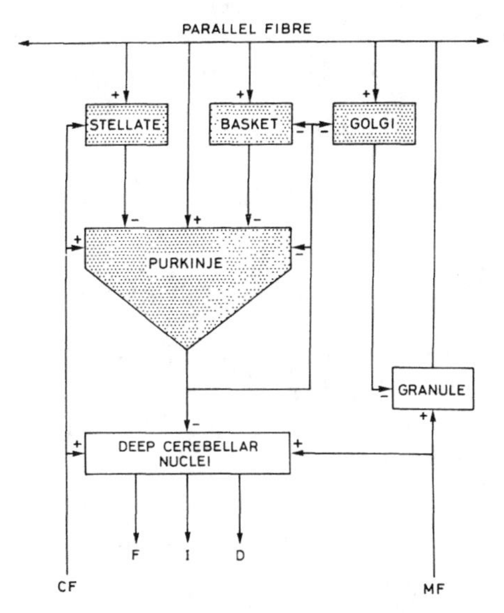

Fig. 102. Circuit diagram of input-output relations in the cerebellum. The two input pathways, mossy fiber (MF) and climbing fiber (CF), are shown to have both excitatory (+) and inhibitory (–) actions on the Purkinje cell. Stellate and basket cells give inhibitory synapses to the Purkinje cell, while the Golgi cell inhibits the granule cell. The output pathway along the Purkinje axon gives inhibitory synapses to the deep cerebellar nuclei, namely fastigial (F), intermediate (I), and dentate (D). Purkinje, stellate, basket, and Golgi cells are all shaded to indicate their inhibitory function.

Analysis of synaptic actions

The output of the cerebellum represents the final stage in the processing of impulse patterns due to the integration of opposing excitatory and inhibitory synaptic actions.

Action of mossy fibers

The various afferent systems terminating in the cerebellum as mossy fibers form excitatory synaptic contacts with the dendrites of granule cells. A single mossy fiber is capable of exciting several hundred granule cells. The axon of each granule cell bifurcates into a parallel fiber, which in turn makes excitatory synaptic contacts with about 300 Purkinje cells, thus achieving considerable divergence. The parallel fibers also excite Golgi, basket, and stellate cells in the same folium:

- The axons of Golgi cells terminate inside a glomerulus where they form inhibitory synapses with granule cell dendrites. Thus the granule cell has its own negative feedback control. This limits the output of the parallel fiber whereby weak "unwanted" signals are effectively blocked.
- Basket and stellate cells are inhibitory interneurons. Their axons terminate exclusively around Purkinje cells, on which they exert a powerful inhibitory action.

Action of climbing fibers

Each climbing fiber ends on the dendritic tree of a Purkinje cell where excitatory synaptic contacts are established. The pathway from inferior olive to Purkinje cell is monosynaptic. Climbing fiber collaterals extend to adjacent Purkinje cells and to basket and stellate cells through which the threshold of the Purkinje cells can be modified. In this way the Purkinje cells are brought more closely to or directed away from the mossy fiber influence.

Action of Purkinje cells

The vast number of excitatory and inhibitory synapses embracing the dendritic tree of a Purkinje cell provides excellent opportunities for interaction and selective propagation of impulse patterns. It can be envisaged that within any folium many Purkinje cells will be discharging while adjacent

Purkinje cells are inhibited, the total effect being a fluctuating pattern of impulses representing a computational analysis of all the inputs sampled. The Purkinje discharges are transmitted to the cells of the deep cerebellar nuclei on which they have an inhibitory action. Accordingly, an increased frequency of discharge produces a greater inhibition, while a decreased frequency has the reverse effect. It is interesting to note that the cerebellum operates by means of a negative control. This form of output has been described as "inhibition sampling."

Control of posture

Axons from the deep nuclei proceed through the cerebellar peduncles to the brain stem and so constitute the efferent pathway from the cerebellum. Many of them terminate in the excitatory and inhibitory regions of the reticular formation. By means of descending reticulo-spinal influences, the desired level of tone is attained in the muscles. It can be envisaged that for each part of the body there is an anatomical loop extending from muscle, joint, and skin receptors to discrete points in the cerebellum and back again to the appropriate spinal centers. The cerebellum provides the necessary instructions to maintain the correct distribution of tone in the body musculature and to make adjustments for any change in posture that may occur. It is capable of providing this service because it is continuously sampling receptor activity from all sense modalities. The evidence is as follows:

1. Effects of stimulation. Sherrington described inhibition of rigidity in the decerebrate animal when the vermis of the anterior lobe was stimulated. This inhibition of tone was seen predominantly in the extensor muscles and resulted in collapse of the limb. The effects are exerted through the fastigial nucleus and its outflow to the inhibitory reticular tract. Stimulation of the paravermian portion of the anterior lobe has the opposite effect, causing an increase of rigidity in the extensor muscles. This suggests activation of the facilitatory reticular-spinal tract. The anterior lobe can thus exert an influence on the spinal motoneurons in either direction. It can act on the alpha motoneurons or influence tone indirectly via the spindle loops. The facilitatory pathway increases afferent discharge from the muscle spindles; the inhibitory pathway decreases the discharge.

2. Effects of ablation. The cerebellum is not essential to life; its removal does not cause paralysis of muscles or loss of conscious sensation. Ablation of the anterior lobe increases the rigidity of a decerebrate animal, accompanied by hyperactive reflexes and the positive supporting reaction. These effects are due to the release of spinal centers mediating the stretch reflexes,

suggesting that inhibition of extensor tone is the predominant role of the anterior lobe. If the lesion is restricted to the culmen, only the forelimb is affected; if the centralis is ablated, the increased tone is limited to the hindlimb.

Ablation of the posterior lobe of the cerebellum interferes with the execution of voluntary movements while changes in muscle tone are of secondary importance. Generally, hypotonia is observed in the experimental animal. Presumably it is due to loss of facilitatory influences on gamma fusimotor neurons. Hypotonia is more pronounced following cerebellar injuries in man owing to the relatively large size of the posterior lobe. The resultant loss of facilitatory drive may account for the characteristic pendulum knee jerk and inability to support the weight of the body.

In conclusion it may be stated that the postural functions of the cerebellum are the outcome of opposing facilitatory and inhibitory influences distributed by the efferent pathways from the cells of the deep cerebellar nuclei. Their firing rate may be increased, reduced, or totally silenced according to the signals conveyed by the Purkinje axons which converge upon them.

BASAL GANGLIA

Functional anatomy

Within each cerebral hemisphere there are collections of subcortical grey masses bordering on the lateral ventricle and internal capsule. They include the corpus striatum (caudate nucleus and putamen), globus pallidus, and the subthalamic nucleus. The various connections between these masses form two important circuits linking the basal ganglia to the cerebral cortex on the one hand and the mid-brain on the other. The first is known as the cortico-strial circuit and operates through the thalamus. The second is known as the cortico-nigro-pallidal circuit and operates through the substantia nigra and reticular formation.

1. Cortico-strial circuit. This circuit is between the motor cortex, corpus striatum, globus pallidus, thalamus, and back to motor cortex. The corpus striatum receives myelinated and unmyelinated fibers from the motor cortex of both cerebral hemispheres, the contralateral projection being limited to the head of the caudate nucleus. A second source of afferents comes from the intralaminar nuclei of the thalamus. Efferent fibers from the caudate nucleus and putamen proceed to the globus pallidus, which is divided into external and internal segments. Fibers from the external segment form a secondary loop distributed to the subthalamic nucleus and back again to

end on cells in the internal segment. The primary circuit is continued from the internal segment by a well-known projection called the ansa lenticularis which terminates on the ventro-lateral, ventro-anterior and central nuclei of the thalamus. The diagram in Figure 103 shows how the circuit is completed by thalamo-cortical projections returning to the motor cortex. This basal ganglia circuit, with its own feedback mechanism (the secondary loop), may be regarded as a "damping" device controlling the excitability of the motor areas of the cerebral cortex. Its role in the organization of voluntary movement will be discussed in Chapter XIII.

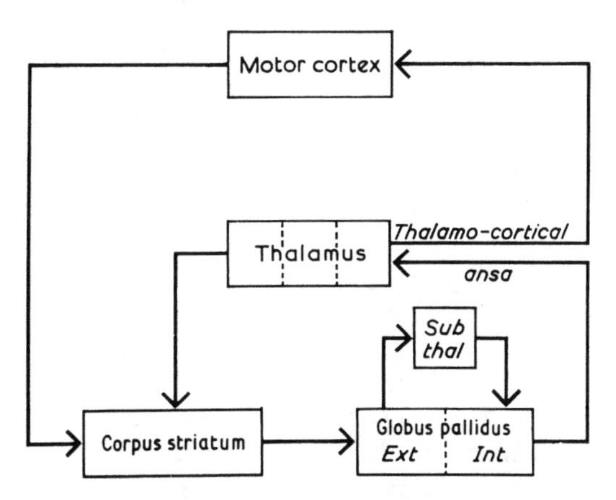

Fig. 103. Diagram of cortico-strial circuit. The corpus striatum receives fibers from the motor cortex and intralaminar nuclei of the thalamus and projects to the globus pallidus. Fibers from the globus pallidus proceed via the ansa lenticularis to the ventrolateral and central nuclei of the thalamus, which in turn give rise to thalamo-cortical projections extending back to the motor cortex. A secondary loop is shown between the globus pallidus and the subthalamic nucleus.

2. Cortico-nigro-pallidal circuit. The substantia nigra is a pigmented layer of gray matter found in the cerebral peduncle of the mid-brain. Its deeply pigmented cells contain the transmitter substance *dopamine,* a precursor of noradrenaline. A direct cortico-nigral system of fibers arising from the precentral motor cortex has been described by several authors, but others believe that cortical influences on the substantia nigra are relayed through the corpus striatum. The efferent pathway from the substantia nigra consists of dopaminergic fibers terminating diffusely in the caudate nucleus, putamen, and globus pallidus. The diagram in Figure 104 shows how the circuit is completed by projections from the ventro-lateral nucleus of the thalamus to the cerebral cortex. The globus pallidus is evidently concerned in both of these neuronal circuits, and dopamine must play an important role in the transmission process.

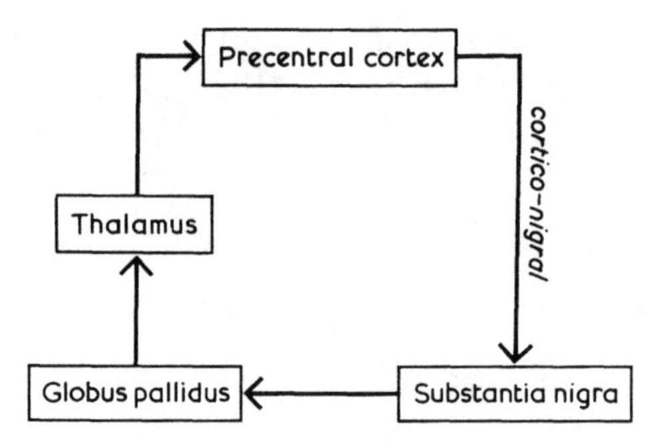

Fig. 104. Cortico-nigral-pallidal circuit. The substantia nigra receives fibers from the precentral motor cortex and projects diffusely to the caudate mucleus, putamen, and globus pallidus. These projections are relayed back to the cortex via the ventrolateral thalamic nuclei. A descending system of fibers, not shown in the diagram, connects the substantia nigra to the brain stem reticular formation.

Relation to posture

1. Extrapyramidal influences. The significance of the basal ganglia lies in their ability to exercise control over extrapyramidal pathways. This control is most likely exerted through the precentral motor cortex, since there does not appear to be any direct link between the basal ganglia and the skeletal muscles. Another pathway, however, has been suggested. It relates the subthalamic nucleus to the inhibitory zone of the reticular formation through which the globus pallidus can influence the spinal motoneurons. The abnormalities resulting from damage to this pathway may account for some types of postural disorders.

2. Rigidity. The most constant pathological finding in diseases of the basal ganglia, Parkinsonism, for example, is a loss of the pigmented neurons of the substantia nigra. Less commonly, there is a degeneration of small nerve cells in the globus pallidus. The consequent reduction in the dopamine content of these structures creates an unbalanced state of impulse transmission. Since dopamine is inhibitory in action, the opposing transmission systems become hyperexcitable, gamma efferent activity is exaggerated, and muscular rigidity ensues.

In the normal functioning of the basal ganglia, both ascending and descending influences are involved in the control of movement and posture. Ascending influences are primarily concerned with the execution of movements since these are initiated at the cortical level. Thus tremor and other involuntary movements are consistent features of the basal ganglia syndrome. Abnormalities of posture probably arise from deficits of extrapyramidal mechanisms, which reach their targets in the reticulo-spinal tracts. Muscular rigidity is therefore an expression of imbalance of opposing transmission systems and is dependent upon the integrity of spinal reflex arcs.

PREMOTOR CORTEX

Functional anatomy

The premotor area is a subdivision of the precentral motor cortex. Designated area 6 in Brodmann's cytoarchitectural map, it lies between the pyramidal cortex (area 4) and the frontal eye fields (area 8). The superior border extends on to the mesial surface of the hemisphere, while the inferior border lies close to the lateral sulcus or Sylvian fissure (Fig. 105). Efferent pathways from the premotor area take three principal routes:

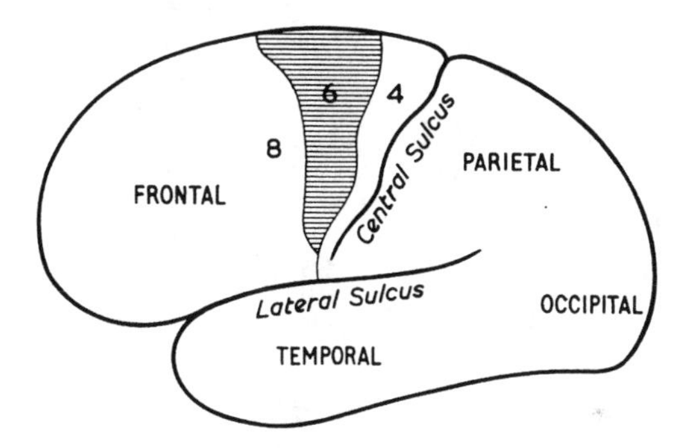

Fig. 105. Diagram of left cerebral hemisphere shows position of premotor area (shaded).

1. Projections to the basal ganglia. These have already been described. The premotor area is part of a neuronal circuit through which the basal ganglia exert their influence on the motor functions of the cerebral cortex.
2. Projections to the cerebellum. A proportion of the outflow from the premotor area is mediated by the cortico-spinal tracts. Some of the corticofugal projections relay in the pontine nuclei through the middle cerebellar peduncle; others descend to the inferior olive and contribute to the climbing fiber input.
3. Projections to the spinal cord. Extrapyramidal fibers accompany the pyramidal tract to the reticular formation of the brain stem. The pathway is completed by spinal relays to gamma efferent neurons acting on the spindle loops.

Suppressor bands

Marion Hines was the first to describe a strip of cortex lying along the anterior border of area 4 in the brain of the monkey. This strip was designated area 4s. On stimulation, muscle tone and movements on the opposite side were suppressed, while ablation resulted in the development of spasticity. Later, other suppressor bands were discovered on the surface of the cortex by the local application of strychnine. All were found to pass directly to the caudate nucleus. The suppressor bands form a well defined topographical projection. Thus area 8s projects to the head of the caudate, 4s to the body, 2s to the posterior third, and 19s to the tail of the nucleus (Fig. 106). These disclosures emphasize a close functional relationship between the cortex and the basal ganglia and explain how the excitability of the motor cortex is kept under continuous control. The existence of comparable mechanisms in the human brain may be responsible for the occurrence of epileptic attacks at the site of a focal cerebral lesion.

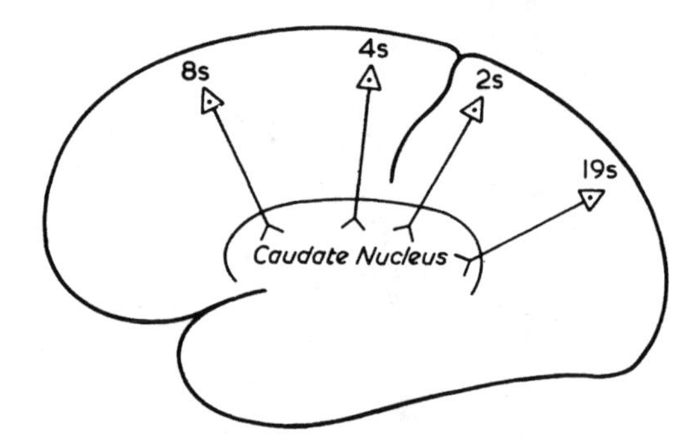

Fig. 106. Diagram of monkey brain shows the projections of suppressor areas 8s, 4s, 2s, and 19s on the caudate nucleus. (Modified from Garol and McCulloch.)

Spasticity

Spastic muscle implies a condition of hypertonia and exaggerated reflexes caused by the release of nervous structures from inhibition. Clinically, the term is applied to disturbances involving the premotor cortex itself or its descending neural pathways. Spasticity is therefore a sign of extrapyramidal disease. It affects mainly the muscles of the limbs, producing a feeling of heaviness and requiring increased effort for carrying out volitional movement. The patient is consequently disinclined to move, although the mechanism of muscular contraction may be intact. The problem of spasticity has been investigated in animals by lesion experiments, while a good deal of information has also come from clinical studies.

1. Animal experiments. Localized ablation restricted to the leg area of the primary motor cortex (area 4) causes paralysis of the limb on the opposite side. The muscles of the leg become hypotonic. This reduction of skeletal muscle tone is known as *flaccidity*. If the lesion is extended to include the upper part of the premotor area, the leg muscles become spastic as well as paralyzed. This evidence suggests that spasticity is not a disturbance of pyramidal function. On the other hand, a lesion confined to the premotor area does not cause paralysis if area 4 is left intact. Yet the muscles on the opposite side show signs of increased tone and exaggerated reflexes. When the premotor cortex is ablated to include the suppressor area 4s, the animal develops a clearly defined spasticity. The conclusion may be drawn that spasticity is encountered only when the extrapyramidal cortex or its projections are interrupted. The cause of the spasticity is removal of inhibitory influences passing through the reticular formation to the spinal motoneurons.

2. Clinical studies. Spasticity is a common clinical sign in many neurological disorders, particularly in hemiplegia. It is associated with brisk reflexes that occur whenever the affected side is displaced from the position of rest. The patient has a disordered gait. The difference between the spastic and the normal limb lies in the much greater sensitivity of the muscle spindle loop, which causes a great outburst of reflex activity even on the slightest stretch. The heightened muscle tone may produce a rhythmic series of reflex contractions in the ankle or knee referred to as *clonus*. In the hemiplegic patient there is also some loss of volitional movement. These findings indicate involvement of both extrapyramidal and pyramidal systems. A lesion confined to the internal capsule can be more discrete than a cortical lesion, but in either case the clinical picture is that of a spastic paralysis. The affected limbs show an increased resistance to passive movement, brisk tendon jerks, and clasp-knife rigidity.

LIMBIC SYSTEM

Functional Anatomy

Certain regions of the brain, often taken together under the broad term *limbic system*, serve many functions in common. They comprise the anterior cingulate and posterior orbital gyri in the frontal lobe and the hippocampus in the temporal lobe. Together with their associated subcortical nuclei, the principal target for their activity appears to be the hypothalamus and, for this reason, they are intimately concerned with the control of visceral functions (Chapter XXI). However, the diffuse character of their projections suggest that impulses may spread along a variety of pathways to structures beyond the hypothalamus including the brain stem reticular formation and cerebellum. Consequently, the limbic system is capable of exerting an influence on motor functions, particularly those involved in defense and attack. It may have no direct influence on the skeletal musculature, but changes in muscle tone and posture are essential physiological activities that accompany behavioral reactions.

Influence on posture

It has long been known that bilateral removal of the limbic region causes the appearance of rage reactions in the cat. The animal exhibits a marked increase of tone in the skeletal musculature, extension of the claws, raising the back, and other postural changes associated with aggressive action. Similar outbursts of rage can be evoked by electrical stimulation of discrete points through implanted electrodes in the conscious animal. Most of the postural changes are those concerned with organized responses such as defense reactions, flight or attack, restlessness and growling or searching for food.

Despite the vast amount of experimental work on emotional expression in animals and also in man, there is still a lot of conflicting evidence concerning its cerebral control. This is especially true when comparing results from different species. One of the difficulties appears to be that adjacent excitable points may serve different or even opposing neuronal systems. Another difficulty is that in performing a localized ablation, the effects may be obscured by interruption of fibers of passage which belong to a different network altogether. The contribution made by the limbic system to the regulation of posture is therefore far from settled. Present evidence suggests that it exerts a modulatory influence on the hypothalamus and the structures affecting the total response of the animal to an emotional situation.

SUMMARY

Normal posture is the attitude taken up by the body when the nervous system is intact. It allows the body to adopt characteristic positions according to circumstances, for example, the curled up position of a sleeping animal or the crouching, nose-down position of an animal searching for food. When standing still, the full weight of the body is taken by the limbs with the head upright and, as the animal moves, appropriate adjustments must be made to support the body when the weight is shifted from one limb to another. In the human being, the entire locomotor system may be involved simply in changing from the horizontal to the vertical position. When walking or running, the adjustments required are more complex in order to maintain stability and rhythm and to prevent the body from falling.

Postural tone is developed best in the extensor or anti-gravity muscles. It is a reflex phenomenon dependent upon the sensitivity of the muscle spindles, which, in turn, is regulated by descending influences from the brain. To get some idea of the role played by different parts of the central nervous system, a study has been made of the postural disturbances that develop in an animal after selective lesions and, in general, the results relate closely to clinical findings. Important contributions come from the following regions of the brain:

1. The reticular formation has both facilitatory and inhibitory roles. Lesions of the facilitatory system cause hypotonia and diminished reflexes; lesions of the inhibitory system cause hypertonia and exaggerated reflexes. Decerebrate rigidity may be considered a release phenomenon due to the drive of the facilitatory tract when inhibition is lost.

2. The cerebellum has both facilitatory and inhibitory roles. Damage to the posterior lobe causes hypotonia in the affected muscles. The anterior lobe exerts a reciprocal influence on extensors and flexors, being predominatly inhibitory to the extensors. Consequently, removal of the anterior lobe causes an increase of tone in the extensor muscles.

3. The basal ganglia exert their effects on the extrapyramidal system in two ways. One is by means of a circuit through the thalamus and back to the motor cortex; the other is through the substantia nigra and reticular formation. Dopamine transmission is involved in both systems. Lesions of the basal ganglia due to depletion of dopamine cause a state of imbalance resulting in rigidity and muscular tremor.

4. The premotor cortex exerts an inhibitory influence on spinal motoneurons; the effects are mediated by the reticular formation. In addition, the cortex contains a number of suppressor bands, which keep it under continuous inhibitory control. Lesions of the extrapyramidal cortex or its descending pathways result in the development of spasticity, exaggerated reflexes, and clonus.

5. The limbic system exercises bilateral control of organized defense reactions. It normally exerts a modulatory influence on behavior, and damage can lead to the appearance of rage and other emotional disturbances.

SELECTED BIBLIOGRAPHY

Eccles, J.C. The cerebellum as a computer: patterns in space and time. *J. Physiol,* 229:1-32, 1973.

Eccles, J.C., Ito, M., and Szentagothai, J. *The Cerebellum as a Neuronal Machine.* Berlin: Springer-Verlag, 1967.

Granit, R., and Phillips, C.G. Excitatory and inhibitory processes acting upon individual Purkinje cells of the cerebellum in cats. *J. Physiol.* 133:520-547, 1956.

Hines, M. The anterior border of the monkey's motor cortex and the production of spasticity. *Amer. J. Physiol.* 116:76, 1936.

Jansen, J., and Brodal, A. *Aspects of Cerebellar Anatomy.* Oslo: Johan Grundt, 1954.

Kennard, M.A. Experimental analysis of the functions of the basal ganglia in monkeys and chimpanzees. *J. Neurophysiol.* 7:127-148, 1944.

Magoun, H.W., and Rhines, R. An inhibitory mechanism in the bulbar reticular formation. *J. Neurophysiol.* 9:165-171, 1946.

Magoun, H.W., and Rhines, R. *Spasticity: The Stretch Reflex and Extrapyramidal Systems.* Springfield, Ill.: Charles C. Thomas, 1947.

Marr, D. A theory of cerebellar cortex. *J. Physiol.* 202:437-470, 1969.

McDowell, F.H., and Markham, C.H. (Eds.). *Recent Advances in Parkinson's Disease.* Philadelphia: Davis, 1971.

Thack, W.T. Cerebellar output: properties, synthesis and uses. *Brain Res.* 40:89-97, 1972.

CHAPTER XIII

Voluntary Movement

In 1870, Hughlings Jackson, in his analysis of epileptic disorders, was the first to offer the concept of an organized control of movement in the cerebral cortex. Jackson postulated the existence of localized motor areas in the brain from studies of clinical cases, and soon afterward, two other investigators, Fritsch and Hitzig, discovered that electrical stimulation of the cortex caused movements in the opposite limbs. In 1875, Ferrier demonstrated the existence of discrete motor points and described the deficits resulting from localized cortical ablations. By the end of the nineteenth century Sherrington had begun his classical research on the nervous system, which helped to elucidate many of the problems connected with movement patterns initiated from the motor cortex. This was followed by detailed studies of the excitable areas, the construction of representative maps, and the tracing of pathways between the cerebral cortex, the cerebellum, and the spinal motoneurons.

In more recent years, attention has been directed to the role of the pyramidal and extrapyramidal systems, studied by making lesions in animals and comparing the performances left intact with their clinical correlates. Important deductions have been made to suggest that pyramidal function makes possible the discrete usage of the musculature, especially of the digits, while extrapyramidal influences provide the adjustments requisite for postural movements. Finally, it is accepted that all movements initiated by the motor cortex are guided by instructions from the sense organs. These may take the form of a learning process, whereby speedier and more efficient actions come with increasing practice. The movements are then performed more or less automatically and entirely outside consciousness, but they are still dependent upon information from the sense organs if they are to be executed smoothly and correctly. Interruption in the flow of information may lead to gross impairment of voluntary movement when the motor systems and the muscles themselves are apparently undamaged. For example, the cerebellum is constantly processing information from the sense organs. The intention tremor and ataxia, which are commonly seen in cerebellar disease, are largely due to this factor.

Mechanics of voluntary movement

Voluntary movement is based upon two important properties–coordination and strength.

1. Coordination between different groups of muscles involves protagonists, antagonists, and synergists:

- The protagonist is the prime mover of a joint. Thus contraction of the biceps causes flexion of the elbow joint; contraction of the triceps causes extension of the elbow joint.
- The antagonist is the opposing muscle at a joint. Thus in order to bend the elbow, the extensor muscle must relax. This is the principle of reciprocal innervation.
- The synergists are the associated muscles around a joint. They cooperate to make the movement more effective and facilitate its execution.

2. The strength of contraction depends upon the character of motoneuron discharge. This in turn is determined by the rate of discharge of individual motoneurons, the number of motor units in action, and the phase of activity. When a large number of motoneurons discharge in phase, the action is powerful and synchronous as is the case in lifting a heavy weight. When the motoneurons discharge out of phase, the action is less powerful, but steadier and asynchronous; such action is required for delicate movements.

EXCITO-MOTOR CORTEX

Functional anatomy

The motor areas of the cerebral cortex lie mainly in the frontal lobe of each hemisphere, rostral to the central sulcus or fissure of Rolando (Fig. 107). They comprise:

- The precentral motor cortex (areas 4, 6, and 8).
- The motor speech center (Broca's area 44).
- The supplementary motor area (on the medial surface).
- The second motor area (on the Sylvian fissure).

The precentral motor cortex is regarded as the principal site through which the brain expresses its control over the skeletal muscles. It is the part of the cortex that receives afferent projections from the thalamus and gives rise to the cortico-spinal tracts. It is the source of initiated movements,

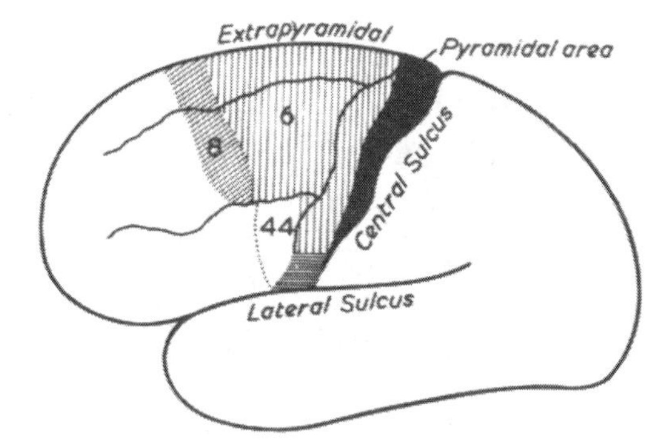

Fig. 107. Map of excito-motor areas of human cortex. The pyramidal area occupies the precentral gyrus anterior to the central sulcus (fissure of Rolando). The extrapyramidal area lies between it and the frontal eye fields (area 8). The second motor area is located above the lateral sulcus (Sylvian fissure). The motor speech center (area 44) is at the posterior end of the inferior frontal gyrus.

stimulation producing movement patterns on the opposite side of the body, while ablation results in paralysis. These facts have a firm basis in both experimental work on animals and in clinical observations. The precentral motor cortex has three functional parts.

1. Primary motor area (area 4). In primates, area 4 occupies the precentral gyrus, extending from the superior border of the hemisphere down to the Sylvian fissure. The region nearest to the central sulcus is known as the area giganto-pyramidalis, which contains the giant cells of Betz. Its functions are entirely motor and it exercises discrete control of the fine movements used in voluntary acts of skill and precision.

2. Premotor area (area 6). The boundaries of the premotor area are not so accurately defined by anatomical markings. It lies immediately rostral to the primary motor area, extending from the superior border of the hemisphere down to Broca's area at the posterior end of the inferior frontal gyrus. It is bounded on each side by the supressor strips 4s and 8s. It is a motor area within which there is a crude localization of function, but it does not contain any Betz cells. It is also involved in the control of posture, as mentioned in the previous chapter.

3. Frontal eye fields (area 8). In the human being, area 8 is restricted to the middle frontal gyrus just rostral to the premotor area, but this area appears to be more extensive in monkeys. It is a motor area concerned with ocular movements and other reactions associated with the eyes.

Structural organization

1. General features. The cerebral cortex in man is about 4 mm thick. Six layers of cells can be recognized (Fig. 108):

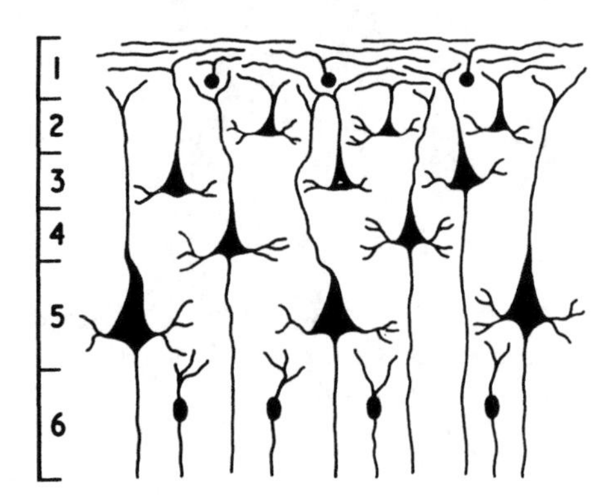

Fig. 108. Diagrammatic representation of the layers of cells in the motor cortex. The pyramidal cells increase in size in the deeper layers; the giant cells of Betz are found in layer 5. Apical dendrites run toward the pial surface, while the basal dendrites spread laterally. The axons emerge from the base of the pyramids and descend into the white matter.

Layer I is a superficial or molecular layer of short-axon cells, together with a plexus of fibers running parallel to the surface and the apical dendrites from neurons lying deeper in the cortex.

Layer II, the external granular layer, contains densely packed granule cells and small pyramids, each with the apex directed toward the surface.

Layer III, the external pyramidal layer, contains medium-sized pyramidal cells whose axons emerge from the base.

Layer IV is the internal granular layer, which is composed of many small granule cells with short-branching axons.

Layer V, the internal pyramidal layer, contains large pyramidal cells whose axons pass into the white matter.

Layer VI is a fusiform layer of spindle-shaped cells whose dendrites pass toward the surface and whose axons descend into the white matter.

2. Special features. The precentral motor cortex has developed certain features of architecture that distinguish it from other regions:

- The granule cells have largely been replaced by pyramidal cells, which tend to increase in size with increasing depth. They have both apical and basal dendrites. The former ascend to the superficial layers of the cortex, while the latter form complex synaptic arborizations with the terminals of thalamo-cortical afferents.
- The giant cells of Betz are found in layer V of the primary motor area

along the lip of the central sulcus. According to Lassek, there are about 34,000 Betz cells in each motor area, accounting for not more than 3 percent of fibers in the pyramidal tract.

• The cellular arrangements in the premotor area resemble those of the primary motor cortex except for the absence of Betz cells. All the excitomotor regions are characterized by an agranular type of cortex.

Pyramidal tract

Origin. The cortical origin of the pyramidal tract was first investigated by the method of retrograde degeneration following a discrete lesion. The limitations of the methods are well known since lesions may be incomplete or else they may interrupt fibers of passage which have no connection with the cells examined. It is also difficult to draw conclusions from the method of recording antidromic potentials from the surface of the cortex because spread of stimulus current cannot be strictly controlled.

On stimulating the motor cortex and recording from the pyramidal tract, Patton and Amassian obtained early and late responses, the first deflections being due to direct excitation of pyramidal axons and the later ones accounted for by excitation of intracortical neurons. It would seem, therefore, that the origin of the pyramidal tract is far more extensive than was originally believed. The largest fibers with the fastest conduction velocities arise from area 4 which includes the giant cells of Betz. These, however, form only a minority of the fibers passing into the spinal cord through the medullary pyramids. The greatest contribution to the pyramidal tract consists of small myelinated fibers and some unmyelinated ones originating from the motor areas and to some extent from the adjacent parietal lobe.

Course. In its descent through the white matter of the cerebral hemisphere, the pyramidal tract is intermingled with other pathways from the cortex bound for synaptic relays in the brain stem. They converge on the internal capsule, occupying the genu and posterior limb between the thalamus and globus pallidus. From the internal capsule, the tract descends through the basal part of the cerebral peduncle to the pons, where many collaterals are given off to the pontine nuclei and reticular formation. In the upper part of the medulla, the tract forms a compact bundle recognized as a prominence on the ventral surface, but in the lower part of the medulla, the tract breaks up into three distinct bundles:

1. About 75 percent of the fibers cross over to the opposite side to form the crossed pyramidal tract in the lateral column of the spinal cord.

2. A proportion of the remaining fibers do not decussate, but enter the lateral column of the cord on their own side.

3. Other uncrossed fibers pass into the ventral part of the cord but do not proceed beyond the thoracic segmental level.

Termination. In the spinal cord, many uncrossed fibers ultimately terminate on the opposite side, with the result that the motor cortex of one cerebral hemisphere exerts its influence on the musculature on the opposite side of the body. The terminal synapses of the pyramidal fibers are found at the base of the dorsal horn and around short internuncial neurons which connect with the anterior horn cells. Some pyramidal fibers terminate directly on the alpha motoneurons, particularly on those supplying the upper limb. Such monosynaptic pathways are seen only in primates and could be important in allowing the motor cortex to initiate sudden movements like those involved in defense reactions.

STIMULATION EXPERIMENTS

Primary motor cortex

Type of stimulation

Faradic or repetitive stimulation of the cortical surface in the anesthetized monkey results in coordinated movements of the opposite limbs. The responses are generally organized contractions of extensors or flexors in which the antagonistic muscles are reflexively inhibited. The movements are brought about by recruitment of pyramidal neurons near the electrode site causing a succession of impulse volleys in the pyramidal tracts. There is a build-up of spinal interneuronal activity and rapid depolarization of related motoneurons. Increasing the frequency of stimulation increases synaptic transmitting potency, with a consequent increase in size of the motoneuron discharge. The excitability of the motor cortex is dependent upon many other factors, including the effects of previous stimulation, level of anesthesia, pressure of the circulating blood, and hydrogen ion concentration. Spreading depression caused by dehydration may also affect the electrical activity of the cortex.

Single shock stimulation at threshold intensity levels usually initiates more limited motor responses. Different proportions of excitatory and inhibitory effects are obtained from adjacent points on the cortical surface; when only a small population of cells is excited the contractions become discrete and limited to a few muscle groups. Under natural conditions, the movements of individual joints are brought about by the combined action of several overlapping motor points.

Map of motor area

The body is represented upside down in the motor cortex with the leg area occupying the medial surface and upper part of the lateral surface, the face area occupying the lower part of the lateral surface and the arm area in between (Fig. 109). Sherrington's observations on the ape disclosed a considerable degree of functional overlapping within these subdivisions as well as differences in the amount of cortex devoted to the various regions of the body. Thus the cortical representation of the fingers, lips, and tongue is much greater than the amount devoted to other parts, while the representation of the trunk muscles is relatively small. The arrangement suggests that a greater number of neurons are required for finely graded movements and acts of skill or precision. The motor points eliciting discrete movements are located in the central sulcus in the position of the Betz cells. Movements become less discrete when they emanate from points in a more rostral direction.

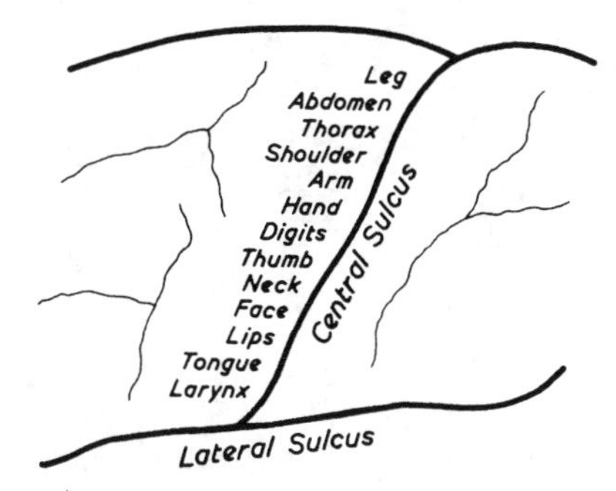

Fig. 109. Map of representative areas in the primary motor cortex. The body, represented inversely, is roughly divided into leg, arm, and face areas. Note relatively small representation of the whole of the trunk in contrast to the large amount of cortex devoted to the hand, lips, and tongue.

Movements closely resembling those seen in animal experiments were observed by Penfield, who stimulated the motor cortex in conscious patients following craniotomy under local anesthesia. Using minimal current strength, fine movements of the thumb, lips, and tongue could be readily elicited. The results confirmed the evidence of localization in the motor cortex obtained from mapping experiments in animals.

Cortical epilepsy

When the stimulus strength is above threshold or when the excitability of the cortex is raised, for example, owing to the presence of an irritable lesion, the movements evoked by the stimulus may spread to adjoining regions and affect the entire limb or even the whole of one side of the body. The spread of an irritable focus in an epileptic fit was first described by Hughlings Jackson, who postulated the existence of cortical representation from the sequence of convulsion patterns. The epileptic attack can begin at any point in the motor cortex, but its spread occurs in an orderly sequence. Thus contraction of the thumb muscles is followed by contractions in the hand, wrist, forearm, and shoulder of the same limb.

Premotor cortex

Intact animal

At threshold intensities, electrical stimulation of the premotor area results in complex movement patterns in the opposite limb. The proximal joints are principally affected. The whole range of joint movements may be involved, including flexion, extension, and rotation. As the strength of the stimulus is increased, the movements become rhythmic in character, spreading eventually to all the joints of the limb. The pathways mediating these responses pass through area 4 to join the cortico-spinal tracts.

Section between area 6 and 4

Stimulation of the premotor area after making a section between it and the primary motor area fails to elicit simple movements of flexion and extension, but the lesion does not abolish the responses altogether. Torsion movements of the trunk and pelvis may be present as well as complex movement patterns in the leg. These responses appear to be independent of area 4 and are lost if the premotor cortex itself is undercut.

Functional differences

The premotor cortex is provided with two outlet pathways—one set of fibers passing through area 4, and the other descending in the cortico-spinal tract to the brain stem motor centers. The threshold for stimulation of the premotor cortex is higher than that of the primary motor cortex. There is some localization of function but nothing as discrete as that of the primary area. The movements evoked are complex in character, acting mainly on the proximal joints with little influence at all on the fine movements of the digits. These facts underline essential differences observed by electrical excitation of the cortex and emphasize the difficulty of interpreting how even the simplest voluntary movement is accomplished under natural conditions.

ABLATION EXPERIMENTS

The purpose of these experiments was an attempt to reproduce in monkeys and in the higher apes some of the more commonly observed neurological deficits in patients, with a view to establishing their physiological basis.

Lesions in the central sulcus

In primates, a lesion restricted to the posterior part of area 4, near the central sulcus, results in a flaccid paralysis on the opposite side of the body. The loss of tone in the affected muscles suggests that the pyramidal tract normally exerts a facilitatory influence. Immediately after the operation, there is complete suppression of reflex activity, but about a week later, stroking the plantar surface of the foot evokes extension of the big toe. This is known as the sign of Babinski. The loss of motor power in the muscles is always greatest in the digits.

Lesions in area 4

If the arm area alone is removed, the paralysis is limited to the upper limb. There is usually weakness of the shoulder and elbow and absence of movements in the hand. While recovery may occur after a few weeks, the skillful use of the digits remains impaired. Total ablation of area 4 results in a flaccid paralysis on the opposite side. During the course of recovery, the muscles in the distal part of the extremities may become moderately spastic. This was attributed by Fulton to the fact that area 4 receives some fibers from the premotor cortex.

Sign of Babinski. This is one of the most valuable clinical signs in neurology. It occurs normally in infants, but otherwise, it indicates damage to the pyramidal tract in any part of its course. Stimulation of the sole of the foot usually results in plantar flexion of the big toe. After a pyramidal lesion, stimulation causes extension of the big toe and sometimes fanning of all the toes. Fanning is most conspicuous when the lesion encroaches upon the premotor area.

Lesions of the pyramidal tract

Above the medullary decussation, the signs appear on the opposite side; in lesions of the spinal cord the signs are on the same side. The following deficits are usually encountered:

1. Impairment of voluntary movement. There is marked weakness of the muscles, especially for discrete movements of the digits. Simple movements like kicking and scratching are still possible, indicating that there is no true paralysis. On the other hand, skillful manipulations are performed with great difficulty, and the muscles are readily fatigued.

2. Loss of tone. There is a decrease in muscle tone in the affected limb; the leg tends to be more flaccid than the arm.

3. Diminished reflexes. The superficial reflexes (abdominal and cremasteric) are raised in threshold or are lost. The planter reflex is typically extensor. The deep reflexes may be unchanged, but the grasp reflex may be altered by an inability to open the hand or foot after grasping an object.

4. Vasomotor disturbances. The skin temperature on the affected side is usually lower than that on the normal side. This may be due to failure of vasodilator mechanisms in the cord and overactivity of tonic constrictor reflexes.

Lesions in area 6

1. If care is taken to avoid damage to area 4, the limbs can be used with the same precision as they were previously, and gross movements, such as those involved in climbing, are either unaffected or rapidly regained. However, movements that are acquired by training may be impaired. For example, manipulations performed on a problem box may become disorganized, although re-training is possible.

2. Isolated lesions of the premotor cortex produce *dystonia,* an increased resistance to passive movement due to heightened muscle spindle activity in the opposite limbs. There is no true spasticity unless area 4s is included in the lesion.

3. The positive grasp reflex is seen in the contralateral extremities. It involves a slow flexion of the digits in response to contact. The Babinski sign is not present, but fanning of the toes is usually observed.

Combined lesions

The motor disturbances resulting from a combined lesion of area 4 and area 6 resemble those commonly observed in hemiplegic patients. The principal signs are as follows:

1. Voluntary movement. There is at first complete paralysis on the opposite side of the body. Movements that involve both sides of the body, for example, those of respiration, are not affected. Recovery occurs after some weeks, with return of power to the arm and leg. There is, however, little improvement in the use of the hand and foot. Loss of discrete movements in the digits may be a permanent feature.
2. Posture. Muscle tone is increased on the affected side. The usual picture is that of spasticity, with the upper limb held adducted at the shoulder, and the lower limb firmly extended at the knee and ankle. If an attempt is made to flex the knee, resistance is encountered. On applying greater force, a lengthening reaction is elicited and the limb suddenly flexes as in the clasp-knife rigidity of the hemiplegic patient.
3. Reflex action. The deep reflexes are increased on the affected side. Knee and ankle clonus can often be elicited owing to the heightened muscle tone. The superficial reflexes are lost. The Babinski sign is present and fanning of the toes occurs with stronger stimulation.

In general, the results of a combined cortical lesion are due to disturbances of both pyramidal and extrapyramidal influences on the lower motor centers. The selective control of movements is impaired according to the extent of pyramidal damage and appears to be more debilitating in man than in other primates. Restoration of function is also more prolonged in man and less complete especially in the distal joints. The presence of spasticity indicates involvement of extrapyramidal pathways, since the excitability of the motoneurons is enhanced on withdrawal of inhibition.

Significance of the extrapyramidal system

An appraisal of the results of the various experiments described above will give insight into the possible functional role of the extrapyramidal system.

1. After removal of the primary motor area, the remaining corticofugal pathways are capable of carrying impulses to the motor nuclei of the brain stem and to the spinal cord. Voluntary movement is therefore possible and, with the passage of time, a fair degree of compensation takes place. Animals learn to use their limbs as before, and only the discrete control of movement is lost. In man, on the other hand, owing to the greater use of

precision movements, the limitations imposed on the cerebral cortex after a pyramidal lesion are substantially increased. The extrapyramidal system does not possess the spatial or temporal organization necessary for the finer control of movement and cannot compensate for the loss of this function.

2. After removal of the premotor cortex, the ability to exercise selective control of the muscles is retained, and rapid initiation and cessation of movements are possible. This suggests that the extrapyramidal system is less concerned with the function of individual muscles than with the task of preparing the musculature as a whole for a smooth, progressive performance. The hypertonia that ensues must surely interfere with the execution of even the simplest movement, and the addition of spasticity is sufficient to emphasize the importance of inhibitory influences in the organization of the motor systems.

SERVO-CONTROL OF MOVEMENT

Afferent impulses play an important part in the control of voluntary movement. Information about the state of the muscles, the distribution of tone and the position of the joints is derived from proprioceptive structures. Other afferent sources come from the labyrinths, the sense organs in the skin, and from the special senses of hearing and vision. Before dispatching its commands to the muscles, the cerebral cortex makes use of all the available information, for it is clear that if a limb is completely de-afferented, gross ataxia is likely to ensue. In carrying out its functions, the cortex must be continuously informed about the progress of a movement so that it can apply a correcting action and make adjustments, as necessary, to produce the desired effect. In addition, the cortex uses information for learning; the sequence and patterns of a particular act are recorded to improve performance at the next attempt.

The central nervous system has developed three separate mechanisms of servo-control—in the spinal cord, cerebellum, and basal ganglia.

1. Spinal cord. Information conveyed by muscle afferents contributes to the reflex excitability of the anterior horn cells. Although reflex contraction of the muscle tends to resist any changes in muscle length, the sensitivity of the muscle spindles is largely determined by impulses from higher centers, particularly those impinging on the gamma motoneurons. Hence, a movement initiated by the pyramidal cortex is always accompanied by a corresponding gamma discharge, which serves to correct errors of loading and to dampen oscillations by appropriate feedback.

2. Cerebellum. Information conveyed to the cerebellum from the various sense organs in the body is subjected to continuous processing before its

influence is communicated to other parts of the central nervous system. The output is particularly concerned with the smooth and effective control of movement. The deep cerebellar nuclei exert their effects directly on the spinal cord or else project upward to the motor areas of the opposite cerebral cortex. In either case, each cerebellar hemisphere controls the musculature on its own side of the body. It is of interest that descending influences from the motor cortex complete a dynamic loop through mossy fiber and climbing fiber pathways back to the cerebellum (Fig. 110). As a movement is being carried out, the motor cortex can signal its requirements to the cerebellum so that a revised output pattern is immediately available to accelerate or check the actions or to prevent sudden jerks and overshooting.

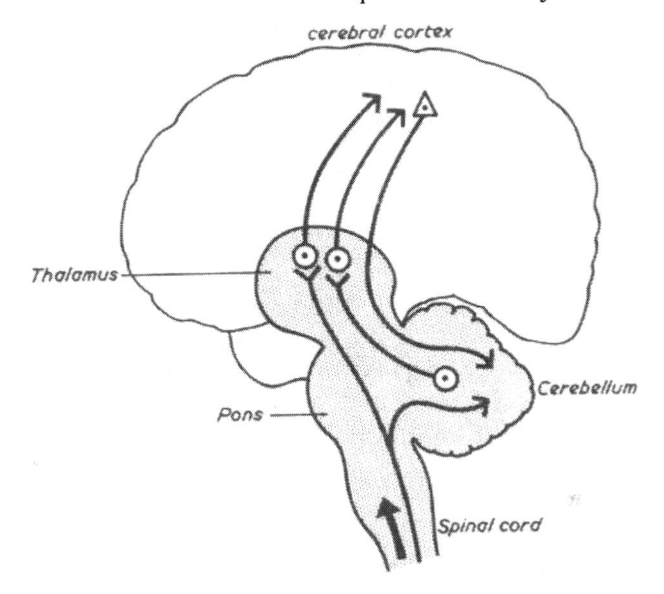

Fig. 110. Plan of cerebellar control system with direction of the pathways indicated by arrows. Impulses from the sense organs are conducted through the spinal cord to the cerebellum and via thalamic relays to the cerebral cortex. Impulses from the deep cerebellar nuclei ascend via the thalamus to the cerebral cortex, which in turn projects back to the cerebellum.

The significance of these controls is only too obvious when they are disrupted by injury or disease. All voluntary movements may then be disturbed. The commonest signs are errors in the direction, duration, force, range, and sequence of the movements performed. Patients walk with their feet widely apart; they may exhibit a coarse, rhythmic tremor of the limbs, and there are also typical disturbances in speaking and writing. The muscles are not paralyzed, but the motor cortex is considerably handicapped in planning movements correctly in the absence of a controlling network.

3. Basal ganglia. The influence of the basal ganglia on voluntary movement has been determined largely from clinical studies, since abnormal spontaneous movements result from lesions of these ganglia in man. The abnormalities tend to emphasize the inhibitory functions of dopaminergic transmission systems, for there is no doubt that the administration of levadopa, a metabolic precursor of dopamine, has beneficial effects in treatment. The clinical features of basal ganglia disease represent a state of imbalance in which excitatory transmitter systems are left unopposed. That is why the use of certain anti-cholinergic drugs, such as belladonna or benzhexol, is advocated for the relief of tremor in mild cases. Hence, it can be envisaged that in the normal functioning of the basal ganglia, there is an effective control of cortical excitability through a circulation of impulses flowing between the motor cortex, globus pallidus, thalamus, and motor cortex. It appears that movements initiated by voluntary effort require a harmonious action between the pyramidal and the extrapyramidal components of the motor system at its highest level.

ORGANIZATION OF VOLUNTARY MOVEMENT

Movements are brought about by impulse discharges in the motor nerves supplying the muscles. Movements initiated reflexively are integrated at lower levels outside the realm of consciousness; but those carried out by voluntary effort, even apparently simple movemements, require careful planning and detailed instructions if a desired result is to be attained. The participation of the cerebral cortex is essential.

The cortico-spinal tracts contribute both excitatory and inhibitory influences on the motoneurons responsible for muscular contraction. The pyramidal component is important for starting and stopping the movements and for the accomplishment of skills with the fingers like playing on a musical instrument. The discrete use of the muscles is permanently lost if the pyramidal tract is damaged. On the other hand, ordinary activities like walking or moving the arms are maintained by the extrapyramidal system. Both components act on internuncial mechanisms operating the alpha and gamma routes to the muscles, but sudden movements might take a direct route through the alpha motoneurons.

The flow of information into the nervous system is perhaps just as important as the descending motor tracts for smooth and effective performance. The way in which afferent impulses regulate motor functions is illustrated by a number of different servo-mechanisms. In the spinal cord, the steady flow of spindle discharges ensures that the resting tone of the muscles is appropriate for existing postures. During voluntary effort, activation of the gamma efferent loop causes an increase of spindle firing which, in turn, contributes to the maintenance of a smooth contraction. De-afferentation of a limb leads to ataxia, but not to paralysis.

In the cerebellum, there is continuous processing of information from the sense organs through which commands from the motor cortex can be revised. The cerebellum reads the situation at every stage during the progress of a movement and evolves corrective output patterns to nullify errors and dampen oscillations. The extent of cerebellar control is seen in the deficiencies that ensue after injury or disease. Again, there is no true paralysis of the muscles, but their contractions are poorly coordinated and movements fundamentally deranged.

In the basal ganglia, the integrity of regulatory mechanisms appears to be essential for maintaining normal levels of cortical excitability. Impulses fed back by a return loop through the thalamus modulate the discharges of the motor area. The tremor and abnormal movements of basal ganglia disease are suggestive of an unbalanced state in which loss of inhibition is the dominant feature. Different types of clinical disorders are frequently encountered, but depletion of dopamine stores within the nerve cells is the most constant pathological finding.

Finally, it has been suggested that programming of movement patterns is laid down in the memory and that these memories are recalled before any movement is repeated. Once a program has been learned, the most complicated movements can be carried out efficiently, without the aid of vision or even conscious control. The value of programming is enormous. It permits elaborate and precise use of the musculature, seen perhaps at its best in the functions of speech. But its efficiency is undoubtedly linked to the various afferent mechanisms and regulatory networks that serve it.

SELECTED BIBLIOGRAPHY

Bernhard, C.G., and Bohm, E. Cortical representation and functional significance of the corticomotoneural system. *Arch. Neurol. Psychiat.* 72:473-502, 1954.

Brooks, V.B., and Stoney, S.D. Jr. Motor mechanisms: the role of the pyramidal system in motor control. *Ann. Rev. Physiol.* 33:337-392, 1971.

Bucy, P.C. *The Precentral Motor Cortex.* Urbana, Ill.: Illinois Press, 1949.

Carpenter, M.B., and McMasters, R.E. Lesions of the substantia nigra in the Rhesus monkey. *Amer. J. Anat.* 114:293-319, 1964.

Eccles, J.C. Cerebral synaptic mechanisms. In *Brain and Conscious Experience.* West Berlin: Springer-Verlag, 1966.

Evarts, E.V. Relation of pyramidal tract activity to force exerted during voluntary movement. *J. Neurophysiol.* 31:14-27, 1968.

Granit, R. *The Basis of Motor Control.* London: Academic Press, 1970.

Penfield, W., and Rasmussen, T. *The Cerebral Cortex of Man.* New York: Macmillan, 1950.

Penfield, W., and Welch, K. The supplementary motor area of the cerebral cortex. *Arch. Neurol. Psychiat.* 66:289-317.

Phillips, C.G., and Porter, R. *Corticospinal Neurones.* New York: Academic Press, 1977.

Rushworth, G. Spasticity and rigidity—an experimental study and review. *J. Neurol. Neurosurg. Psychiat.* 23:99-118, 1960.

Stefanis, C., and Jasper, H. Recurrent collateral inhibition in pyramidal tract neurons. *J. Neurophysiol.* 27:855-877, 1964.

Stern, G. The effects of lesions in the substantia nigra. *Brain* 89:449-478, 1966.

Tower, S.S. Pyramidal lesion in the monkey. *Brain* 63:36-90, 1940.

CHAPTER XIV

Disturbances of Motor Functions

The disturbances resulting from interruption of motor pathways are more easily examined in patients than in experimental animals. An attempt is first made to localize the site and to assess the extent of the damage, and then certain tests are conducted to help confirm the diagnosis. If the disturbance is confined to the anterior horn cells or to the motor nerves and their muscles, it is described as a lower motor neuron lesion; if the disturbance arises anywhere above this level, it is called an upper motor neuron lesion. Motor disabilities may be due to a primary neurological cause. Alternatively, they may develop in association with other diseases, for example, diabetes or pernicious anemia. Symptoms generally complained of by the patient are loss of control, weakness, stiffness, fatigue, and pain, which may be due to muscular spasm or cramp. On examination, the principal physical signs are loss of power, wasting or atrophy of muscles, and paralysis. Muscle tone may be altered with resulting abnormalities of reflexes, posture, and gait. Fasciculations of fine muscle groups, coarse tremors, or involuntary movements may also be present.

Examination of the motor system

Clinical tests

1. Muscle size. A tape measure is used to compare corresponding parts of the limbs for size; muscle atrophy or hypertrophy is noted. The contours and outlines are inspected for symmetry; drooping of the mouth is nearly always apparent in facial weakness. Light tapping of a muscle may reveal unusual irritability.
2. Muscle tone. The resistance to passive movements is compared on the two sides. A marked increase in resistance is exhibited in a spastic limb; a reduction of tone suggests a lower motor neuron lesion.
3. Muscle power. All the major joints are tested by voluntary movements

and then with the examiner creating resistance. Weakness of the pelvic muscles or the presence of a foot-drop may affect the gait and suggest a pyramidal lesion, although many ataxias result from sensory disorders. Weakness may be caused by a disturbance in the peripheral nerves, at the neuromuscular junction or in the muscles themselves.

4. Reflexes. The superficial reflexes are tested by stroking the skin and the deep reflexes by tapping a tendon. The responses on corresponding sides are compared; they may be increased, diminished, or absent. Knee and ankle clonus and the Babinski response are important signs of upper motor neuron disease.

Electromyography

The technique of recording the electrical activity of voluntary muscle was described in Chapter IV. The normal electromyogram reveals characteristics features of motor unit discharges on voluntary contraction. Abnormal findings are as follows:

1. In primary muscle disease, in which there is a reduction of the contractile elements, the action potentials are small in amplitude and polyphasic in outline.
2. In partial denervation, the muscle fibers contract spontaneously, giving brief, fibrillation potentials of small amplitude.
3. In motor neuron disease, there is a reduction in the number of motor units compared with the normal trace. Surviving units are often larger than normal and can be distinguished as discrete potentials.

Nerve conduction tests

The clinical application of nerve conduction measurements is helpful in the diagnosis of peripheral nerve lesions and for assessing the progress of recovery. The simplest method is to measure the threshold intensity required to excite the motor nerve and to compare this intensity with the threshold on the opposite side. Delayed conduction between the wrist and thenar muscles is found in carpal tunnel syndrome due to compression of the median nerve.

Muscle biopsy

In this technique, a sample of muscle is removed under local anesthesia and then immediately frozen in liquid nitrogen. Subsequent analysis may reveal pathological changes in the individual muscle fibers. These may include abnormalities of the myofilaments and banding patterns, or there may be alterations of enzyme processes utilized in muscle metabolism.

Primary muscle disease

In the following conditions, the primary disturbance is in the muscle itself, often due to some inherited or genetic defect.

1. Myotonia congenita. This condition is usually present at birth. Muscle biopsy reveals the presence of rod-like structures in continuity with the Z lines. The affected muscles contract promptly but fail to relax normally. There is usually no wasting, but excitability is increased and relaxation prolonged. As a result, the muscles are extremely sensitive to mechanical stimulation, producing a long-lasting electrical discharge or tetanus. The patient has difficulty in relaxing the grip and complains of stiffness after a brief exertion.

2. Muscular dystrophy. This is a group of disorders in which the affected muscles show progressive degeneration of individual muscle fibers, accompanied by weakness, loss of tone, and diminished tendon reflexes. The dystrophies are divided into various clinical types on the basis of their physical signs and genetic patterns:

Duchenne muscular dystrophy is the commonest type, affecting the proximal limb muscles especially in the lower limb. The patient has a characteristic waddling gait. The main histological changes are degeneration and loss of muscle fibers, variation in fiber size and their replacement by connective tissue and fat.

Pseudo-hypertrophic muscular dystrophy usually appears in childhood. The condition affects the calf, quadriceps, and deltoid muscles where the accumulation of fat accounts for the main bulk of the muscle. The patient has difficulty in raising himself up from the ground and attains the erect position by climbing up his legs and knees with his hands. The degeneration changes probably arise from a deficiency in the function of myoglobin, resulting in a chronic state of hypoxia, which leads to arrest of the developing muscle fibers.

3. Periodic muscular paralysis occurs in young adults with attacks of weakness or paralysis at irregular intervals. There is an abnormality of the muscle membrane in relation to the movements of potassium ions. Symptoms develop whenever there is a fall of potassium level in the blood as occurs, for example, after a heavy meal or severe exercise.

4. Poliomyositis. This disease may occur at any age. Weakness of the muscles is accompanied by tenderness and aching pains as part of a general inflamatory illness. Muscle biopsy may show abnormalities of fiber size, but

no hypertrophy. Electromyographic studies may reveal fibrillation potentials and small polyphasic spikes like those seen in denervation.

5. Contracture. This is a state of shortening of a muscle not caused by voluntary contraction. It develops in many chronic muscular diseases and can be overcome in the early stages by stretching. Later, however, fibrotic changes ensue and may require surgical intervention.

Neuromuscular junction

The disorder commonly associated with the neuromuscular junction is known as myasthenia gravis. In this condition, the patient complains of weakness and fatigue of certain muscle groups especially those of the face. Inability to raise the upper eyelid is a classical symptom called ptosis. There may also be difficulty in chewing and swallowing due to weakness of the masseter muscles. The defect in myasthenia gravis is ascribed to a post-synaptic block of transmitter action. A reduction in the number of end-plate receptors prevents normal functioning of acetylcholine. It has been suggested that the receptors combine with substances released from the thymus gland, since in many cases of myasthenia gravis the thymus is enlarged. Certainly there is no evidence of a deficiency of acetylcholine production, and the muscle fibers themselves appear to be normal. Since acetylcholine is hydrolized by the enzyme cholinesterase, the administration of anticholinesterase drugs, e.g., neostigmin, increases the available concentration of the transmitter to produce an immediate clinical improvement.

Peripheral nerve

A nerve trunk contains efferent and afferent fibers. Consequently, the clinical picture may be a mixture of motor and sensory deficits. Lesions of a peripheral nerve may be incomplete or complete.

1. Incomplete lesions resulting from minor injuries and compressions or from the effects of circulating toxins cause weakness and wasting of the muscles, diminished force of contraction, and early onset of fatigue.

2. Complete lesions resulting from long-term compression or from section of the nerve cause paralysis of the affected muscles, absence of tone, and loss of reflexes.

Recovery of function occurs by a process of regeneration. Recovery after nerve section occurs more rapidly if the divided ends of the nerve are sutured soon after the injury. The degree of damage and the progress of recovery can be assessed by measuring the conduction velocity. The median nerve may be taken as an example. The nerve is stimulated at the elbow and the motor nerve action potential is recorded at the wrist. The latency of the action potential and the distance between elbow and wrist are noted. Thus, in a normal adult with a latency of 4 msec and a conduction distance of 24 cm, the conduction velocity is 60 m/sec.

Motor neuron

There are a number of clinical disorders resulting from damage to anterior horn cells. They have in common a severe reduction of voluntary power and wasting of the muscles.

1. Progressive muscular atrophy arises from degenerative changes of the cell bodies in the anterior horn and may extend into the motor nuclei of the brain stem. The muscles of the limbs and trunk are affected.
2. Peroneal muscular atrophy is an inherited condition manifested by extreme wasting of the peroneal muscles in the legs, while the thigh muscles remain relatively normal.
3. Poliomyelitis is attributed to a viral infection of anterior horn cells and internuncial neurons. The result is a lower motor neuron type of paralysis and wasting of the muscles, although some neurons may escape destruction and their muscle fibers do not atrophy.

Motor neuron disease leads to fairly specific electromyographic abnormalities. These include fasciculation, fibrillation, and giant motor unit spikes.

Fasciculation is due to the spontaneous activity of complete motor units which is often visible through the skin. The potentials differ in amplitude and waveform and they fire irregularly.

Fibrillation is due to the spontaneous activity of single muscle fibers whose motor end-plates show increased sensitivity to acetylcholine. As only a very few muscle fibers are involved, the potentials are of low amplitude and simple waveform, and they fire regularly. They usually appear about two weeks after denervation.

Giant motor unit spikes up to 10 mv or more in amplitude are found in conditions where there is incomplete denervation. They are due to the sprouting of intact terminals and enlargement of the fields of innervation of surviving anterior horn cells. On voluntary contraction, the trace shows a reduction in the number of units firing compared with that of the normal, but the high amplitude spikes are easily distinguished.

Pyramidal tract

Interruption of this tract causes impairment of voluntary movement on the same side of the body and loss of discrete use of the digits. Muscle tone is reduced and the tendon jerks sluggish. The superficial reflexes are usually absent on the side of the lesion, and the Babinski sign is readily elicited. The clinical picture is therefore that of a flaccid paralysis resembling the deficits produced by an experimental lesion. Injuries and diseases of the spinal cord, however, are rarely restricted in this way so that the clinical signs are often more complicated and at first tend to be more extensive because of the effects of edema and contusion.

Motor cortex

The commonest disturbances of the motor cortex are epilepsy and hemiplegia.

1. Epilepsy. This is an abnormal condition of excessive neuronal discharges occurring in any region of the brain and resulting in attacks that range from a simple seizure of short duration to major convulsions with loss of consciousness. The condition may be idiopathic, arising from a developmental abnormality, or symptomatic, caused by the presence of a cerebral lesion. There are usually characteristic electroencephalographic findings (Chapter XXIV).

Motor epilepsy produces involuntary contractions of the musculature on both sides of the body leading to a generalized convulsion. Following the attack, the patient may appear confused and behave irresponsibly.

Jacksonian epilepsy begins as a local muscle contraction and then spreads in an orderly fashion from one muscle group to the next. The spread of excitation conforms to the idea of a wave of activity moving across the motor area of the cortex.

2. Hemiplegia. This condition is usually caused by cerebral vascular disease. Mild forms are due to arterial spasm occurring at any age or they are due to cerebral thrombosis in the elderly. In more serious cases, hemiplegia results from a cerebral hemorrhage, often in patients suffering from high blood pressure. The onset is sudden, with immediate loss of consciousness. At this stage, examination reveals paralysis on the side of the body opposite from the hemorrhage, flaccid muscles, and absent reflexes. Later, the classical signs of a spastic paralysis make their appearance. Voluntary movement of the arm and leg is lost, but facial movements of expression are retained. Tone is increased in the affected muscles, and the limbs assume abnormal

postures. Resistance to passive movements of the joints is increased; clasp-knife rigidity and knee and ankle clonus can be elicited. The superficial reflexes are absent on the contralateral side, the tendon jerks are brisk, and a typical Babinski response can be demonstrated. All these signs in the hemiplegic patient point to a disturbance of both pyramidal and extra-pyramidal functions.

In most cases a fair degree of recovery can be anticipated. The patient learns to walk again with little more than a slight limp, and a good deal of power returns to the arm. The individual is generally left, however, with some weakness in the hand and foot, and the skillful use of the fingers may be permanently impaired. Recovery is attributed partly to absorption of blood clot and edema in the surrounding tissues and partly to compensation by the use of pathways left undamaged.

Basal ganglia

The principal syndromes resulting from basal ganglia disturbances are Parkinson's disease and chorea.

1. Parkinson's disease or paralysis agitans is a slowly progressive disorder of the extrapyramidal system occurring in late adult life and associated with degenerative changes in the substantia nigra and other nuclear masses. The clinical signs are as follows:

• Abnormal movements. Typically, there is a rhythmical tremor occurring at rest and varying from simple movements of the thumb and wrist, known as "pill-rolling," to the coarse shaking of the whole limb. The tremor is absent in sleep, suppressed by voluntary movement, but increased by emotion. Many patients have a disturbed balance, walk in short steps, and tend to fall backward.

• Rigidity. Although muscle tone is increased all over the body, there is a general attitude of flexion, with the head bent forward; the gait becomes shuffling in character. The rigidity produces feelings of heaviness and stiffness which interfere with the smooth execution of voluntary movements. Because rigidity also occurs in the muscles of the face and tongue, the lack of expression, the monotonous voice, the mask-like features, and the absence of gesture add to the patient's disability.

2. Chorea. This disorder is caused by scattered lesions in the basal ganglia which may develop in childhood after an attack of rheumatic fever (Sydenham's chorea) or may appear as an inherited condition in adults (Huntington's chorea). Sometimes it affects elderly persons following a stroke, in which case the abnormal movements are limited to one side and

described as *hemiballism.* The use of the limbs is often interrupted by muscle jerks, and the gait is complicated by sudden lurching of the leg. Involuntary movements of the mouth, tongue, and eyes distort the features.

Cerebellum

Clinical signs of cerebellar dysfunction are manifested in a variety of lesions affecting the cerebellum itself–e.g., tumor or abscess–or any of the ascending and descending pathways that connect it to the brain stem, spinal cord, and cerebral hemispheres. In unilateral lesions, the disturbances are confined to one side of the body; otherwise, the signs are generally bilateral, as is the case in disseminated sclerosis.

1. Posture. Cerebellar rigidity is seen only when lesions involve the anterior lobe. This condition is the result of loss of tonic influences on the inhibitory portion of the reticular formation. More commonly, the muscles become hypotonic, and the limbs sway about loosely. The knee jerk is pendular in type. If the arms are held forward and the eyes are closed, the limb on the affected side tends to droop and swing outward. Nystagmus is invariably present.

2. Movement. Incoordination is a prominent feature. The patient has difficulty in running his heel down the front of his leg and is unable to carry out rapid turning movements of the wrist (adiadochokinesia). The finger-nose test is performed clumsily or with excessive force. Intention tremor is conspicuous when the patient attempts to lift an object. Speech is slow and imperfect, and there is weakness of conjugate ocular movements. In walking, the patient tends to sway toward the affected side, the feet are put down with irregular force, and the arms do not swing with a natural rhythm.

3. Balance. Vomiting, giddiness, and nausea are symptoms of flocculonodular involvement. The patient may have difficulty in standing without support and tends to deviate from a straight path, walking as in a drunken gait.

SPEECH

Many physiological processes are involved in the faculty of speech. To recognize the meaning of sounds, speech must be intelligible, a language has to be learned, and memories of auditory and visual impressions must be stored. To communicate by sounds, we make use of the respiratory apparatus by which expired air throws the vocal cords into vibration to produce the voice. To convert sounds into articulate words, the voice is modified by the mouth, lips, and tongue in a harmonious action. Finally, to give full expression to the purpose and emotional aspects of speech, the tone of the voice and the quality imparted to speech can be altered at will. Disorders of speech may arise from interference with any one of these processes. Clinically, it is useful to classify the disorders into two groups—dysarthria and aphasia. Dysarthria results from defects of the mechanisms controlling the muscles used in speech. Aphasia is a neurological syndrome caused by damage to one or more of the speech centers in the brain.

Normal speech development

At birth, the infant is able to cry and soon learns to recognize sounds. Articulation develops by the end of the first year, and vocabulary increases to express simple needs. At the age of two, a child can use word sequences and short phrases, and thereafter fluency develops rapidly. The development of speech depends primarily on the power to hear others speak and, therefore, an intact auditory system is essential. Once speech has become an established faculty, loss of hearing does not necessarily involve the loss of speech, although its clarity may be impaired.

Acquisition of language increases with the development of general intelligence and the ability to read. This in turn requires storage in the brain of the visual symbols of speech. Reading and writing are learned at school as part of the educational attainments of the child. Both functions require the close coordination of sensory and motor mechanisms with those areas of the cerebral cortex involved in translating the comprehension of language into articulate expression.

It is generally believed that the cortical areas for speech are developed in most persons in the left cerebral hemisphere. This has traditionally been taken to imply the probability of cerebral dominance. While infants show little preference for either hand in grasping an object, the tendency to use the right hand is noticed generally before the child is two years old. Left-handedness is found in only about 10 percent of the population and usually continues throughout life. It remains an open question as to whether speech development in left-handed persons occurs in the right cerebral hemisphere. Certainly a good deal of clinical evidence from studies of aphasia in both right-handed and left-handed patients supports the idea of left cerebral dominance in the control of speech.

Localization of cortical speech areas

The existence of separate anatomical speech areas has been suggested on the basis of a localized lesion underlying a specific language defect. Since the faculty of speech is an integrated function of the brain as a whole, it is unlikely that anatomical correlates can give more than a crude evaluation of a clinical syndrome. Nevertheless, it is useful to consider the cortical localization of speech areas in terms of their sensory and motor aspects:

Sensory aspects

- the auditory speech area is in the transverse temporal gyrus.
- the visual speech area is in the medial occipital lobe adjacent to the striate area.
- the language comprehension area is in the supramarginal and angular gyri.

Motor aspects

- Broca's area is in the inferior frontal gyrus.
- Excito-motor area is in the precentral gyrus.

Disorders of speech

Dysarthria

This term is applied to a disturbance of the neuromuscular functions necessary for articulation and phonation. These may be either sensory (e.g., cerebellar deficiency) or motor (e.g., a lesion of the cranial nerves supplying the palate, larynx, and tongue).

1. In diseases of the pyramidal system, speech may be slurred because of weakness of the muscles of the lips and tongue.
2. In diseases of the extrapyramidal system, speech may be slow and monotonous because of tremor and stiffness of the muscles, which make enunciation difficult.
3. In cerebellar disease speech may be thickened or explosive in character because of a loss of afferent control.

Aphasia

This term is applied to an impairment in the use of language, both spoken and written, and generally denotes a breakdown in comprehension of auditory and visual symbols or in the ability to express ideas in the form of speech. In children, who are not othewise retarded, aphasia may result from a failure in the development of the speech areas or of the transcortical conduction pathways which link them. Later in life, aphasia commonly occurs after injuries or in diseases of the dominant hemisphere which may affect either the sensory or motor aspects of the speech mechanism.

Types of sensory aphasia. These are generally classified as receptive disorders of speech in which the patient is not dumb but, owing to loss of comprehension, cannot recognize his own errors; speech is frequently void of meaning. The individual may be unable to name objects correctly or to answer simple questions. Such gross disturbances are seen after hemorrhage of the middle cerebral artery causing damage in the temporo-parietal region. Aphasia also occurs with head injuries, tumors, and degenerative lesions.

1. Pure word deafness. When the lesion is restricted to the auditory speech area, hearing ability may not be affected, but the patient has difficulty in recognizing the meaning of sounds. Consequently, he cannot appreciate music or conduct intelligent conversation.

2. Pure word blindness. A lesion in the occipital lobe may isolate the striate cortex from the visual speech area. As a result the patient cannot recognize the printed word or patterns even though he can copy them. The condition is known as alexia and is present in children of normal intelligence who show great slowness in learning to read. If the lesion involves the angular gyrus, the disability may include agraphia, or the difficulty of copying print into writing.

3. Nominal aphasia. Lesions of the posterior parietal and adjacent temporal lobes may cause loss of some auditory memories without much defect of comprehension. The main disturbance is an inability to find the correct words or to name familiar objects.

4. Jargon aphasia. More severe lesions of the posterior parietal region in the dominant hemisphere produce unintelligible language defects called jargon. The patient uses word forms that do not exist or substitutes new words without appreciating their significance and talks volubly when excited.

Motor aphasia. The efferent aspect of language is a function of the frontal lobe, where speech production and fluency of articulation are formulated. Lesions in this area represent a serious handicap to the patient, although early initiation of therapy sometimes produces improvements.

2. Developmental aphasia is seen in children who are usually silent and who prefer to express themselves through actions rather than words. The muscles of articulation function normally but speech output is delayed, possibly because of inadequate storage of motor patterns.

2. Broca's aphasia. This is usually seen following a vascular lesion in the posterior part of the inferior frontal gyrus. The patient knows what he wants to say but is unable to express more than an occasional word. In mild cases there are loss of fluency and errors of grammar as well as hesitancy in responding to a question, although comprehension is perfectly normal. In severe cases the patient is completely dumb; the intellectual faculties, however, remain unimpaired.

In summary, it is possible to characterize an aphasic syndrome by noting the ability to comprehend the spoken and written language and by observing the fluency of the speech output.

CHAPTER XV

Somatic Sensory System

The purpose of this chapter is to follow the course of sensory impulses from the skin and deeper tissues to higher levels of the brain where sensations are aroused. The process begins at the receptors that provide information on a great variety of environmental changes. The brain must not only distinguish one sensation from another but must be capable of detecting very small differences in detail and quality. The source of the environmental change must also be detected as accurately as possible. Of the vast amount of information available, only a proportion reaches through to consciousness, for the brain attaches importance to incoming signals on the basis of previous experience and rejects those of no immediate value. In order to fulfill these diverse functions, the sensory impulses flow into two distinct functional channels—lemniscal and anterolateral—which operate in quite different ways.

The *lemniscal system* is a fast conducting route for impulses from one side of the body to the opposite sensory cortex. A striking feature of its organization is that its anatomical arrangements are maintained through successive synaptic relays to provide a faithful representation of its peripheral fields; at the cortical level a point-to-point projection of the body surface is preserved. The lemniscal system is therefore used to convey impressions from the receptors about the site, intensity, and specific qualities of the stimuli that excite them.

The *anterolateral system* is a slower conducting route within a multisynaptic framework that lacks precise topographical patterns. Many of its neural components converge on common targets which project bilaterally on diffuse cortical fields in both cerebral hemispheres. Impulses transmitted by the anterolateral system are not directly linked to the stimulus parameters. They serve as a major filter in sensory processing and no doubt play an important role in determining cortical excitability at a psychological level.

In recent years, a good deal of information on how impulses are handled in the central nervous system has been derived from studies of single unit discharges. Activity is evoked equally well by exciting tactile or joint receptors or by stimulating their afferent fibers. The unit responds to an incoming signal by discharging one or more spikes, indicating its excitable state and topographical position with respect to the region stimulated. Altering the strength or position of the stimulus causes corresponding changes in the firing pattern of the unit. If a large population of cells is sampled, various forms of excitatory and inhibitory reactions may be observed. Microelectrode studies permit analysis of events that take place between the onset of the stimulus and the time of arrival of the impulses but, unfortunately, do not explain the production of a conscious sensation.

Finally, no account of the somatic sensory systems is complete without reference to the existence of corticofugal pathways operating on the ascending relays. It appears that the sensory cortex can exercise a modulatory influence on impulse transmission at all important synaptic levels.

SPINAL CORD

Impulses entering the spinal cord may take part in local reflex actions or may be conducted to higher levels. A proportion of the traffic terminates in the cerebellum, reticular formation, and other structures in the brain stem. Impulses destined for the cerebral cortex are carried in the posterior columns and anterolateral tracts, which together comprise two functional systems:

The lemniscal system

Functional anatomy. The spinal component of this system is formed by the central processes of the spinal ganglia entering the cord through the medial bundle of a dorsal root and ascending in one of the posterior columns on the same side. The tract lying nearest to the mid-line carries fibers from the lumbar region and lower limb and terminates in the gracile nucleus of the medulla. The other tract, carrying fibers from the thoracic region and upper limb, terminates in the cuneate nucleus. The majority of fibers in both tracts are first order neurons, although some fibers may arise from cells within the spinal cord itself. The cuneate tract only commences in the mid-thoracic region and, in consequence, is placed on the outer side. Thus a topographical arrangement of the dermatomes is maintained.

Modalities. The types of sensation conveyed by the posterior columns are touch, pressure, vibration, position sense, and joint movements.

Touch and pressure are cutaneous sensations. They give information about dermatomal localization and about the nature of the object that comes into contact with the skin, for example, the size and shape of the object.

Vibration is a form of mechanoreception derived from movement detectors in the skin and periosteum. The sense of vibration is appreciated over a wide frequency range, with peak frequencies about 250 Hz.

Position sense, or kinesthesia, depends upon afferent impulses from the joints. It provides information on the disposition of the limbs and the ability to detect angular displacements as well as the direction and speed of a movement.

Functions. The primary afferent neurons of the posterior columns form an orderly system of fast-conducting pathways terminating in the gracile and cuneate nuclei. They represent the first stage of central transmission for impulses from the periphery and therefore convey the immediate patterns of receptor discharge without the modifications imposed by later stages. The topographic arrangement of the primary neurons ensures that a precise and detailed pattern of the receptive fields is preserved at this level. Axons emerging from the gracile and cuneate nuclei cross the midline to join the contralateral medial lemniscus. Their main function subserves position sense, joint movement, and vibration.

Within the posterior columns there occurs a cluster of small cells with radiating dendrites which subserves the sensory modality of light touch. These cells can be activated by natural stimulation of the skin or by gentle movement of hairs. They do not show a topographic representation of the body surface, and some of them have large and occasionally bilateral receptive fields. In certain species like the cat, they form a distinct group of cells known as the lateral cervical nucleus, whose properties add a new feature to the sensory input, since interactions occur between them and the primary afferent neurons.

The anterolateral system

Functional anatomy. The spinal component of this system is formed by the lateral bundles of rootlets entering the cord through the dorsal roots. They consist of small myelinated A fibers and unmyelinated C fibers, terminating almost immediately on cells in the posterior horn. Second order neurons and numerous collaterals are relayed upward through the cord in a network of tracts which is made increasingly complicated by the contributions of successive dermatomes.

1. Some fibers ascend only a few segments before terminating in the substantia gelatinosa at the tip of the posterior horn. The emerging axons

then cross the mid-line anterior to the central canal and ascend in the contralateral spino-thalamic tract to reach the ventrobasal complex of the thalamus.

2. The majority of fibers relayed from the posterior horn ascend in the anterolateral regions of the cord; some pass to the opposite side while others remain uncrossed. A spino-bulbar component contributes to the reticular formation of the brain stem.

Modalities. The types of sensation conveyed by the anterolateral system are pain, temperature, touch, and light pressure.

Pain and temperature fibers are conducted in the spino-thalamic tracts and, for the most part, remain topographically discrete. Pain pathways are also found in the anterolateral columns and may involve unmyelinated as well as myelinated fibers. They ascend by multiple relays with crossing and recrossing of the cord.

Touch and light pressure are functionally equivalent to the posterior column pathway but lack the discrimination attributes of the lemniscal system.

Functions. The diffuse character of the anterolateral system, the slower-conducting fibers, and bilateral pathways suggest that localization and fine discrimination are not particularly relevant. On the other hand, the occurrence of multiple relays and opportunitites for convergence allow limited possibilities for modifying the sensory input before it gains access to higher synaptic levels. This property has very important implications for impulses carrying pain.

BRAIN STEM

Ascending pathways in the brain stem provide a series of intervening relays through which impulses from the spinal cord reach the thalamus. They include the lemniscal, trigeminal, and reticular systems.

Lemniscal system

The medial lemniscus emerges from the sensory decussation in the medulla, ascends through the pons and mid-brain (where it lies dorsal to the substantia nigra), and terminates in the ventrobasal complex of the thalamus. The majority of its fibers are second order neurons arising from the dorsal column nuclei–gracile and cuneate. In experimental recordings, access to these nuclei is gained by removing the arch of the atlas and the bone overlying the caudal part of the cerebellum.

The *nucleus gracilis* lies on either side of the mid-line, extending from the obex to the level of the second cervical segment. Within this region the cellular arrangements exhibit marked functional differences. The cells in the rostral part of the nucleus have the largest receptive fields, indicating a high degree of synaptic convergence. Many are spontaneously active and respond to mechanical or electrical stimulation by discharging a burst of spikes. They are not modality specific and play little part in spatial discrimination. The cells in the middle part of the nucleus have the smallest receptive fields. Their spontaneous activity can be suppressed by stimuli applied to the surrounding areas of skin. They may be fired antidromically by stimulating the contralateral medial lemniscus, disclosing that most of the axons of the middle section project to the thalamus. The caudal part of the nucelus reveals mixed facilitatory and inhibitory responses to peripheral stimulation and little spontaneous activity. Some of the cells give rise to the external arcuate fibers passing into the cerebellum.

The *nucleus cuneatus* is found in the lower part of the medulla, lateral to the nucleus gracilis. Its cells respond to both cutaneous and proprioceptive stimuli. The former are usually silent in the absence of stimulation, but a tactile stimulus results in a sustained discharge with short inter-spike intervals. In contrast, proprioceptive cells fire repetitively with the limbs in a position of rest. When the limbs are displaced, the frequency of firing is increased. Thus the initial patterns of receptor discharge are transformed into new patterns at the first synaptic relay.

Trigeminal system

The sensory root of the fifth cranial nerve represents the peripheral axons of cells in the trigeminal ganglion. It lies near the apex of the petrous part of the temporal bone. The root divides into three branches–ophthalamic, maxillary, and mandibular–which supply the skin of the face, the teeth, and the nasal cavity. The central processes of the ganglion cells enter the pons, where they divide into short ascending and long descending branches. The ascending fibers terminate in the main sensory nucleus within the pons, while the descending fibers enter the nucleus of the spinal tract in the upper cervical region of the cord. Second order neurons from the main sensory nucleus and nucleus oralis of the spinal tract cross the mid-line and ascend with the medial lemniscus to the contralateral ventrobasal complex of the thalamus. The majority of the axons from the spinal nucleus join the anterolateral system through which they reach the intralaminar and posterior groups of thalamic neurons.

Impulses conveyed through the main sensory nucleus serve discriminative functions for the whole trigeminal area and a crude form of topographical representation is preserved. Impulses conveyed through the spinal nucleus serve the modalities of pain and temperature; their discriminative functions are not so marked. The separation of the pathways for transmission of sensory impulses is related to differences in the types of sensation carried by the lemniscal and anterolateral systems.

Reticular system

In their course through the brain stem, the ascending sensory fibers contribute many collaterals to the reticular formation, which also receives inputs from the anterolateral columns, the spino-thalamic tract, and the cerebellum. The result is that a complex network of interneurons is built up in the central portion of the medulla and pons and extending upward to end in the intralaminar nuclei of the thalamus. The reticular formation has been called a third neural system. Its functions will be described in Chapter XXIV, and only the sensory influences will be mentioned here.

Reticular neurons respond to stimulation of a wide variety of sense organs. Their receptive fields are large and often bilateral; some are excited, others inhibited, while a third group shows mixed effects. Because of overlapping receptive fields, there is little evidence of topographical organization. Afferents from many different sources may converge on the same neuron with consequent possibilities for interaction.

The properties of reticular neurons fit many of the characteristics of the anterolateral system. As they are not modality specific, they transmit few of the attributes of the receptor and have poor discriminative qualities. On the other hand, they respond readily to nociceptive stimuli, exerting a driving force or "energizer" on thalamic relays to the cerebral cortex. For this reason, the reticular formation has an important role in heightening the excitability of all afferent channels and maintaining a state of alertness. In addition, inhibitory interactions in the network of interneurons may be effective in delaying the onward transmission of impulses. This is comparable to a screening device or filter for weak and unwanted signals in the sensory pathway.

THALAMUS

The thalamus is the main gateway for the transmission of sensory impulses to the cerebral cortex. It contains several discrete collections of nuclei and, in addition, there are numerous cells scattered over the surface and in between the nuclear masses. The projections of the thalamus are arranged in two functional systems–*specific* and *diffuse.* The specific thalamocortical projections are third order neurons conveying lemniscal afferents to the postcentral region of the parietal lobe. The diffuse thalamocortical projections are continuous with the upper end of the reticular formation and supply all parts of the cortex in both cerebral hemispheres. Functionally, the thalamus and the cerebral cortex are inseparable. Activation of the cortex by thalamic relays excites specific sensory areas relating to the intensity, modality, and spatial aspects of the sensation and also determines the general level of attentiveness. In turn, the cerebral cortex transmits impulses back to the thalamus, whereby sensory information from all over the body is effectively controlled.

Functional anatomy

The thalamus is situated deep in the substance of the brain, lying between the third ventricle and the internal capsule. Internally, it is divided into three parts known as the anterior, medial, and lateral nuclei.

The anterior nuclei

Afferents from the mammillary bodies relay through the anterior nuclei to the anterior cingulate gyrus. They form part of the limbic organization for expressing emotion, as explained in Chapter XXII.

The medial nuclei

These include the following three nuclei:

1. The dorsomedial nucleus. This nuclear mass occupies the center of the thalamus. It receives large projections from the prefrontal areas, including the orbital surface of the frontal lobe. Its role in connection with the higher functions of the nervous system will be discussed in Chapter XXIII.

2. The centrum medianum. This thalamic nucleus is associated with the ansa lenticularis. Many of the fibers originate from the globus pallidus and take part in the cortico-strial-pallidal circuit, as described in Chapter XII.

3. The intralaminar nuclei. These form a diffuse collection of cells lying between the specific nuclear masses. They belong to the anterolateral sensory system and project bilaterally on the cerebral cortex.

The lateral nuclei

The lateral nuclear mass, which is the largest of the three divisions, is subdivided into a number of discrete groups:

1. The ventrobasal complex. This is the most important of all the relay stations for sensory transmission and will be discussed in more detail below. The medial portion, or arcuate nucleus, represents the sensory innervation of the trigeminal nerve. The lateral portion serves the dorsal column and spino-thalamic systems.

2. The posterior nuclei. This group of nuclei lies between the ventrobasal complex and the lateral geniculate body. The afferents belong to the antero-lateral system and include some non-specific components of the spino-thalamic tract. Fibers from the posterior group project on the second sensory area and are involved in the transmission of nocuous signals.

3. The antero-ventral nucleus. These cells receive fibers from the globus pallidus, and their axons complete the cortico-strial circuit, as previously described.

4. The ventro-lateral nucleus. This is an important relay station for impulses from the cerebellum. Spatial orientation of the projections to the excito-motor cortex is maintained.

The various groupings of thalamic nuclei and their respective functions are represented in the scheme of Figure 111.

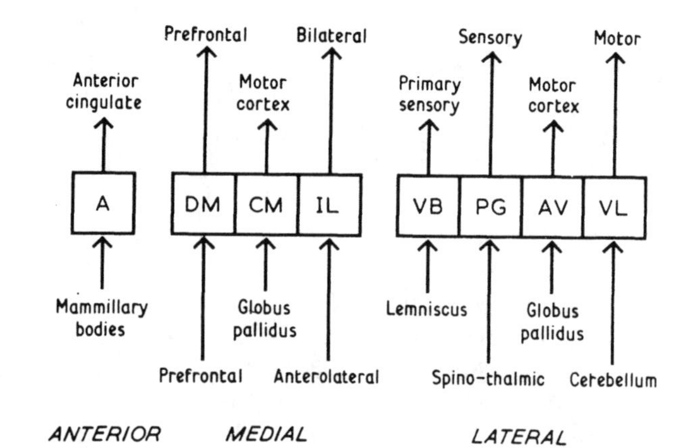

Fig. 111. Scheme of spino-thalamic nuclei showing principal afferent and efferent projections. (A, anterior nucleus; DM, dorsomedial; CM, centrum medianum; IL, intralaminar; VB, ventrobasal; PG, posterior group; AV, anteroventral; VL, ventrolateral.)

Ventrobasal complex

This is the principal relay station of the lemniscal system and of the specific components of the spino-thalamic tract. The neurons within this complex are highly specific for sensory modality, tactile elements being distributed posteriorly and joint afferents anteriorly. Topographical localization is also strictly maintained: The head is represented in the medial portion, the trunk is intermediate, and the leg is in the lateral portion.

Effects of peripheral stimulation

1. The receptive fields for neurons responding to light mechanical stimulation of the skin are small and discrete. This reflects the fine degree of resolution to be found in the lemniscal system.
2. The latencies are relatively short, indicating the presence of a direct pathway between the dorsal column nuclei and the thalamus.
3. The typical response of a VB neuron to a peripheral stimulus is a high frequency repetitive discharge (Fig. 112). Adjacent neurons may also respond, but at a longer latency and at a lower frequency. This suggests that neurons in the center of a discharge zone receive the largest number of terminal afferents, while those surrounding it are excited to a lesser extent.

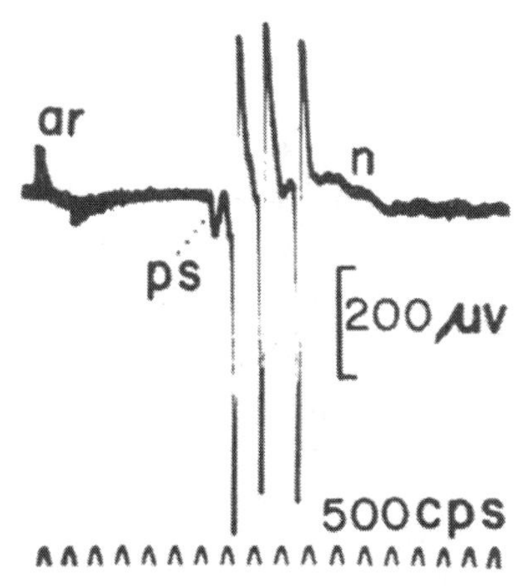

Fig. 112. Evoked response of a VB neuron. The stimulus (ar) was the displacement of a few hairs by a brush applied to the contralateral forepaw of the cat. Note presynaptic potential (ps) and train of spikes superimposed on a primary wave (n). (After Rose & Mountcastle.)

4. VB neurons follow increased frequency of stimulation with increased frequency of discharge up to about 70/sec.

5. Impulses from joint receptors elicit a high frequency discharge when the position of a joint is changed. The rate of discharge then falls to a steady value which remains a function of the joint angle. Some thalamic neurons are maximally stimulated when a joint is fully extended; others are maximally stimulated when the joint is fully flexed. Thus, as a joint is moved from one position to another, there is a change in the pattern of frequencies in a whole population of VB neurons. This provides information not only on the particular joint that is being moved, but also on the degree of joint displacement.

Sensory interaction

When two stimuli are applied in quick succession, the response to the second stimulus may be increased or diminished. The increased response is known as facilitation. It may be due to an increase of excitation of the thalamic relay cells or to a decrease of corticothalamic inhibition. Facilitatory effects are recognized by an increase in size of the evoked primary wave, a larger number of spikes discharged, and a reduction in the latency of the second response. Inhibitory effects are recognized by a reversal of these properties. With single unit recordings, there is a progressive reduction in the probability of firing until the unit ceases to discharge altogether. The depression of excitability in thalamic relay cells may persist for as long as 150 msec.

Thalamic inhibition

The problem of explaining the repetitive discharges of thalamic relay cells as well as the predominance of inhibitory interactions can be attributed to the role played by axon collaterals.

1. When a VB neuron is excited by an afferent nerve volley, impulses are generated in axon collaterals which cause excitation of neighboring relay cells. Extensive convergence of collaterals from lemniscal fibers contributes to the excitatory action and repetitive firing of the VB neuron.

2. Some axon collaterals of thalamic relay cells terminate on short inhibitory interneurons. Such cells can operate in one of two ways: In *presynaptic inhibition,* the inhibitory interneuron is distributed to the synaptic knobs of the lemniscal fibers (Fig. 113, A). In *postsynaptic inhibition,* the inhibitory interneuron is distributed to the relay cells (Fig. 113, B).

3. Thalamic inhibition plays an important role in modifying the information carried by the sensory pathways to the cerebral cortex. The lemniscal

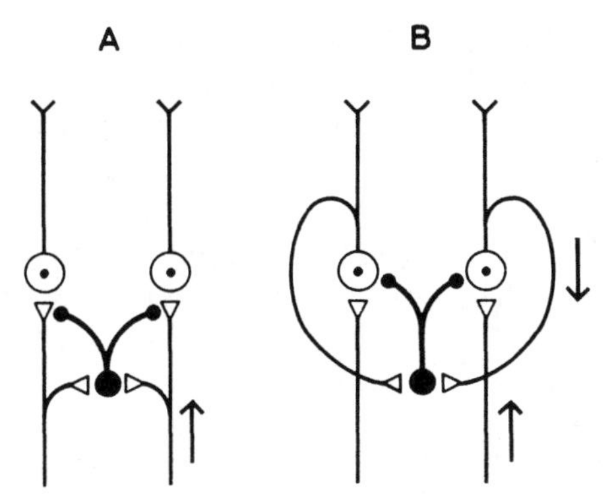

Fig. 113. Action of thalamic relay cells. A, presynaptic inhibition: Afferent collaterals excite the inhibitory interneuron, which acts on the terminals of the lemniscal fibers. B, postsynaptic inhibition: Thalamo-cortical collaterals excite the inhibitory interneuron, which in turn acts on the relay cell. (Modified from Eccles.)

volley has both an excitatory and an inhibitory action on the relay cells, thus accounting for the way in which rhythmic, repetitive firing is generated.

In summary, ventrobasal neurons provide a faithful representation of the activity transmitted by receptors serving the modalities of touch, pressure, vibration, and joint sense. They transmit accurate information on the position and intensity of the stimulus as required for the specific localizing and discriminative functions of the sensory cortex. They transform the individual discharges of lemniscal fibers into new impulse patterns by mechanisms of sensory interaction and afferent inhibition.

Posterior nuclei

This area of the thalamus receives an inflow from the anterolateral system, including the non-specific elements of the spino-thalamic tracts. There is no topographic representation of the body and no particular sensory modality can be identified. Indeed, the majority of posterior neurons respond to afferents from many different sources, especially those elicited by noxious stimuli. The effects of peripheral stimulation are as follows:

1. The receptive fields are relatively large and may be contralateral or ipsilateral.

2. Posterior group neurons are non-specific—i.e., the same neuron can be excited by cutaneous, visceral, auditory, and visual stimuli. Impulses from joint receptors are an important exception.

3. The threshold of stimulation covers a wide range of the fiber spectrum.

Some neurons are activated by low threshold A beta fibers, being responsive to light mechanical stimuli; others respond only to high threshold noxious stimuli, suggesting their implication in the transmission of pain.

4. The latencies of the evoked responses are relatively long, indicating the polysynaptic nature of the anterolateral pathway.

5. The high degree of afferent convergence enhances the possibilities of interaction and the regulation of cortical excitability.

In summary, posterior group neurons have none of the discriminative features of the lemniscal system. They respond to a variety of stimuli of different modalities, especially the high threshold afferents subserving pain. In many ways their properties are suited to the more generalized form of sensory drives transmitted by the diffuse thalamic projections.

SENSORY CORTEX

The processing of information from the sense organs reaches its highest synaptic level in the cerebral cortex. The main sensory projections through the dorsal columns and ventrobasal complex terminate in the contralateral sensory areas of the cortex. Within this projection system the specific modality of the sense organ is preserved. It is precisely organized to serve the localizing and discriminative functions of cutaneous and kinesthetic sensibilities.

Functional anatomy

There are two distinct sensory areas in the cerebral cortex, primary and secondary. The human primary sensory area is found in the postcentral gyrus posterior to the fissure of Rolando. Here the contralateral parts of the body are represented in a regular topographical pattern. The second sensory area lies on the superior lip of the Sylvian fissure; both sides of the body are represented in a sequence of overlapping dermatomes. In the cat, the primary sensory area is known as somato-sensory area I. Easily identified after craniotomy, it is anterior to the ansate sulcus. The corresponding second sensory area is known as somato-sensory area II; it is found in the ectosylvian gyrus.

Primary sensory area

Distribution of excitable points

Woolsey used the evoked potential technique to map out points on the cortical surface that responded to a peripheral electrical stimulus. Diagrams were constructed of the sensory representations in different animal species; they showed the relative expansion of the primary area in the more highly developed species. Electrical stimulation of the brain in the conscious human subject has been described by Penfield; the sensory experiences induced were, of course, referred to the body surface. All the evidence revealed a point-to-point linkage between the periphery and the cortex, the greatest representation being for the most sensitive regions of the body.

1. The primary sensory area can be divided roughly into leg, arm, and face areas by moving the recording electrode in a rostral-caudal direction. The leg is represented on the medial surface and upper part of the postcentral gyrus; the arm is intermediate, and the face is represented on the lower part of the gyrus. As in the motor cortex, the different areas of the body are inversely represented (Fig. 114).

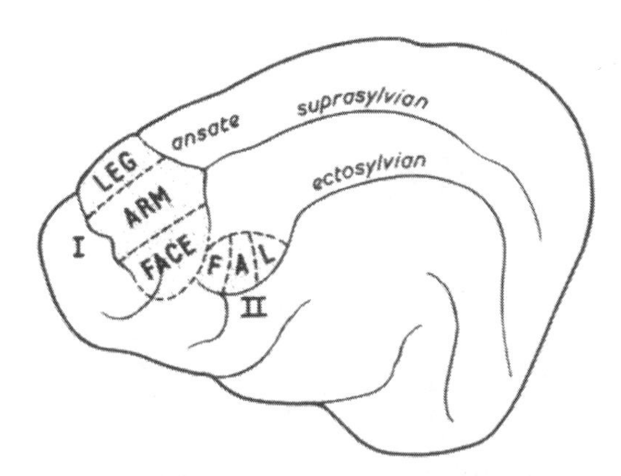

Fig. 114. Primary and secondary sensory areas of cat cerebral cortex. Sensory area I lies anterior to the ansate sulcus and is seen to be roughly divided into leg, arm, and face areas. Sensory area II lies immediately below in the rostral part of the anterior ectosylvian gyrus. It is subdivided into Face (F), Arm (A), and Leg (L) areas.

2. The most sensitive regions of the body have the widest cortical bands, and the least sensitive regions have the narrowest. Thus large cortical areas are devoted to the toes and fingers, while the amount for the leg, trunk, and shoulder is relatively small. Individual points can be mapped for the digits of the hand and for the mouth, lips, and tongue (Fig. 115).

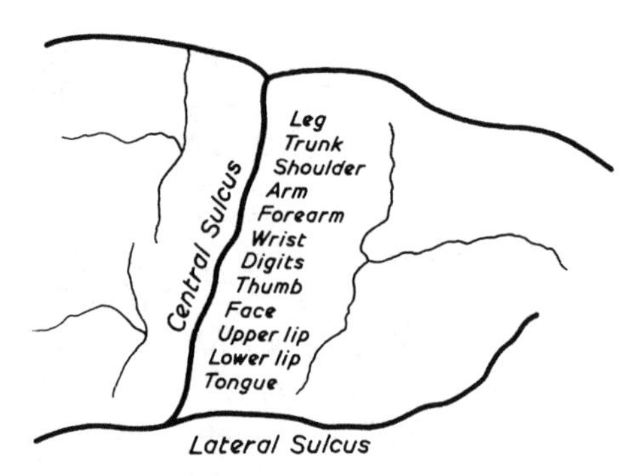

Fig. 115. Map of the human postcentral gyrus showing the distribution of sensory projections from the body surface. The widest cortical bands are devoted to the skin areas of the thumb and fingers and to the lips and tongue.

3. Cortical neurons responding to sensory stimulation are not wholly confined to the postcentral gyrus; some neurons are located in the adjacent parietal and motor regions. However, for each cortical field there is a central discharge zone, or area of maximal activity, and a surrounding zone where the evoked potentials are smaller in amplitude. The postcentral gyrus is the most important region for somatic sensation and has the lowest excitability threshold.

Organization of sensory modalities

Mountcastle discovered that neurons sampled in a vertical penetration of the cortex were all of the same modality type and had identical receptive fields. It seems that the individual neurons of the sensory cortex are arranged in vertical columns and that each column is activated by a different mode of stimulation. Thus, neurons responding to the tactile stimuli are separated from those responding to kinesthetic stimuli. Interaction may occur between adjacent columns, resulting in a zone of excitation surrounded by a zone of inhibition. The advantage of such a system is that it allows the cortex to identify the quality and intensity of the stimulus and provides a means of discrimination.

1. Tactile sensibility. The tactile sense depends upon information about the site, intensity, and form of a mechanical stimulus delivered to the skin. The site of the stimulus is determined by the spatial distributions of the thalamo-cortical projections representing a point-to-point linkage between the cortex and the body surface.

The intensity of the stimulus is determined by the strength of the incoming sensory signals and consequent size of the evoked discharge. The form of the stimulus takes into account the differential excitatory-inhibitory effects between the vertical columns. This type of organization also provides a means for two-point discrimination.

2. Kinesthetic sensibility. Sensory inflow from the muscles, joints, and ligaments gives information about limb movements and position. Each movement elicits a high frequency discharge that subsides to a steady discharge rate, as determined by the angular position of the joint. The cortical neurons signal the degree of joint displacement by variations in discharge frequency.

Second sensory area

In 1941 Adrian described dual receiving areas in the cerebral cortex of the cat. In the same year Marshall, Woolsey, and Bard located a representative area in the rostral strip of the anterior ectosylvian gyrus which seemed to be independent of the primary sensory area. The term somato-sensory area II was later proposed by Woolsey, and his nomenclature is now used to define primary and secondary areas for the auditory, visual, and motor systems as well. Duality in the cerebral cortex must, therefore, be regarded as the rule rather than the exception, yet the precise role of the second sensory area is not clear. Indeed, it is interesting to note that after complete ablation of an animal's second sensory area, its sensory functions remain essentially unimpaired.

Microelectrode studies of single cortical units reveal a number of distinctive properties:

1. Many units give short-latency responses following peripheral nerve stimulation. The pathways mediating the responses belong to the lemniscal system, which has a low excitatory threshold and serves discriminating functions.

2. Many units are particularly sensitive to impulses of visceral origin, but not to impulses from joints. Both ipsilateral and contralateral stimuli appear to be equally effective.

3. Other units have discharge properties characteristic of the anterolateral system. The pathways mediating the responses are derived from relay cells in the posterior nuclei of the thalamus. Such units are activated by high threshold afferents of different modalities; they have large receptive fields and a bilateral distribution. They may be involved in the sense of pain.

Summary of lemniscal and anterolateral systems

The lemniscal system serves the functions of accurate localization and discrimination. It conveys information from the sense organs in a precise, detailed manner, and its ascending pathways retain their specific character from the periphery to their destination in the cerebral cortex. The principal features of the lemniscal system are as follows:

1. Restricted receptive fields. Each sensory unit subtends only a small area of the periphery.
2. Orderly topographic pattern. Peripheral space is transmitted faithfully at each synaptic level.
3. Point-to-point representation. A spatial orientation of the fibers from individual parts of the body is maintained.
4. Specific modality types. Each vertical column of cortical neurons has a preferred sensitivity for only one kind of stimulus.
5. Rapid transmissions. The large myelinated fibers of the lemniscal system ensure rapid transmission of signals with few intervening synapses.
6. Linear stimulus-response relationship. The intensity functions of the stimulus are related to the frequency and number of spikes discharged.
7. High discriminative capability. Information on the form and quality of the stimulus is coded centrally by interacting links between the cortical columns. Two-point sensibility is dependent upon surrounding inhibition.

The anterolateral system is concerned partly with mechanisms underlying pain and temperature and partly with controlling the general level of cortical excitability. It has a less precise role than the lemniscal system, operating over a broad spectrum of sensory inputs, but is capable of exerting a major driving force. The principal features of the anterolateral system are as follows:

1. Large receptive fields. Dermatomal overlap is very pronounced; there is a lack of strict topographical organization.
2. Non-specific modalities. Anterolateral neurons show little preference for one stimulus modality and most respond to noxious stimuli.
3. Bilateral character. Impulses are not confined to one side of the cord or brain stem and spread diffusely on both cerebral hemispheres.
4. Slow transmission. The fibers belong to the A delta and C groups of afferents, some of which are not myelinated. Conduction velocity is relatively slow, with many synaptic relays on course.
5. Distorted stimulus-response relationship. Interaction occurs at sites of convergence with a consequent modification of impulse patterns and a loss of fidelity for components of first-order afferents.
6. Low discriminative capability. The anterolateral system cannot detect fine gradations of intensity, contour, and form; localizing ability is also

deficient. On the other hand, its widespread influence on central neural processing suggests that more likely it has a role in the elaboration of sensory perceptions.

SINGLE UNIT ACTIVITY

While the primary responses recorded from the pial surface of the sensory cortex are the classic methods for mapping the representation of the periphery, a lot more information has been deduced from analysis of the discharge properties of the individual units making up the population. Extracellular recordings of single cortical neurons are displayed on the oscilloscope as diphasic positive-negative spikes when microelectrodes with small tips are used in sampling. When evoked by a tactile stimulus, the characteristic response is a short, high frequency burst of spikes. On the other hand, an electrical stimulus is commonly followed by a single spike discharge. Electrical stimulation is more suitable for precise timing of the events and for controlling stimulus intensity. The response of a cortical unit to sensory stimulation may be as follows:

1. The silent unit may become active usually during the period of the primary wave.
2. The spontaneously discharging unit may become more active and then briefly silent, before resuming its original firing rate.
3. A unit discharging spontaneously at high frequency may show a temporary reduction of firing rate or may cease to discharge for a variable period of time.

Alterations in the intensity of peripheral stimulation may lead to changes in the latency of the evoked response—in the number of spikes discharged and in the probability that the unit will respond on any given trial. Such variations in the activity of cortical units suggest the means by which information on the nature of the stimulus is coded in the sensory cortex.

Relation between stimulus intensity and latency

The time between the application of the stimulus and the first spike of a unit discharge varies considerably at low stimulus intensities. For example, in the cat under light barbiturate anesthesia, a cortical unit responding to stimulation of the contralateral forepaw has a latency range between 10 msec and 25 msec. Increasing the intensity of stimulation tends to reduce the mean latency of the response. It can be seen in the histograms of Figure 116 that a strong peripheral stimulus results in a greater incidence of short-latency responses and in a tendency for the long latencies to drop out. The decrease of synaptic delay with increasing stimulus strength is due in part to recruitment of fast conduction pathways to the cortex.

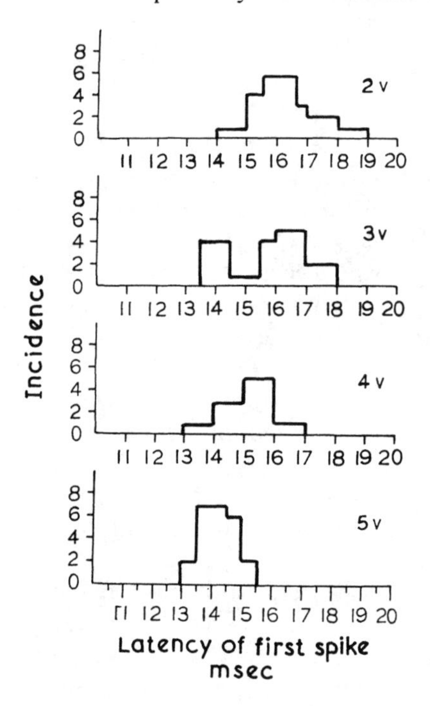

Fig. 116. Histograms of latency ranges with increasing intensities of stimulation. Latency of first spike in each evoked discharge is plotted against incidence. Note descrease of latency and greater incidence of shorter latencies with increasing stimulus intensities.

Relation between stimulus intensity and number of spikes

In many units, an increase of stimulus intensity may cause an increase in rate of firing and in the number of spikes per discharge. Thus a strong stimulus to a sensory nerve may convert a single spike discharge into a burst of six or seven spikes (Fig. 117).

Relation between stimulus intensity and probability

Increasing the intensity of stimulation increases the probability that a cortical unit will discharge at least one spike on any given trial. The graph in Figure 118 shows that at low intensities of stimulation, the number of

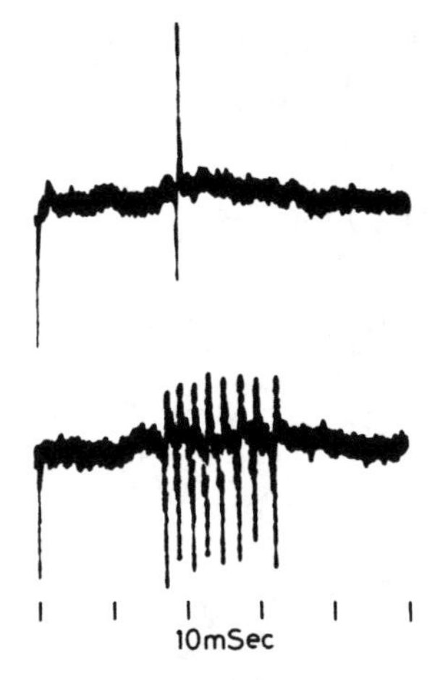

Fig. 117. Recordings of evoked unit discharges from cat sensory cortex. Upper trace shows a single spike discharge at threshold stimulation. Lower trace depicts a multiple spike discharge from the same unit following a stronger stimulus.

responses is small compared with the number observed at higher intensities. The general form of the graph illustrates a gradual change in response probability over the initial range of intensities followed by a rapid rise to a high level of probability with maximal stimulation. The stronger stimulus

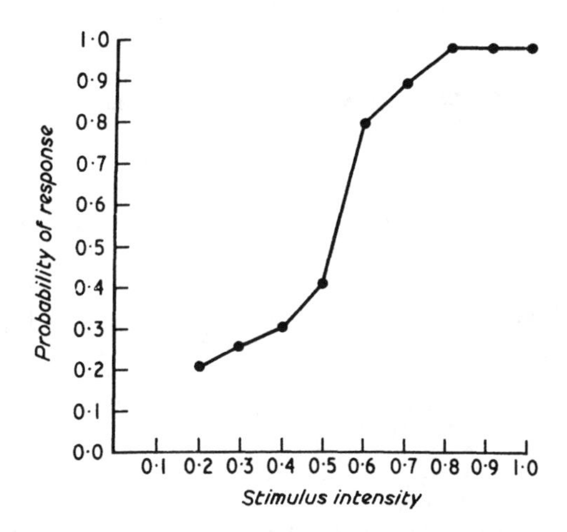

Fig. 118. Relation between stimulus intensity and probability of response. The fraction of stimulus intensity below maximum is shown on the abscissa. The probability factor represented on the ordinate is calculated from the number of times the unit fires in successive trials and the number of times it does not. Note gradual change in probability over the initial range of intensities followed by a rapid rise with higher intensities.

not only excites more afferent fibers but opens up many alternative pathways to the cortex that were not previously involved.

Statistical methods of analysis

Microelectrode recordings of the discharge patterns of single cortical neurons show that a majority fire continually at a low rate, with occasional bursts of activity, and that stimulation changes such patterns in a variety of ways. An example is given in Figure 119 in which the discharge pattern is time-locked with the onset of the stimulus. In order to show that the time-locked response is consistent and not due to chance, a histogram may be constructed to verify that a number of responses do occur at a preferred time following the stimulus. Three useful applications of the histogram technique will be considered here. They provide a set of predictions about the temporal aspects of a unit responding to an incoming signal, but they are, of course, inadequate for multi-unit studies of cell populations and interaction behavior, which require special statistical treatment.

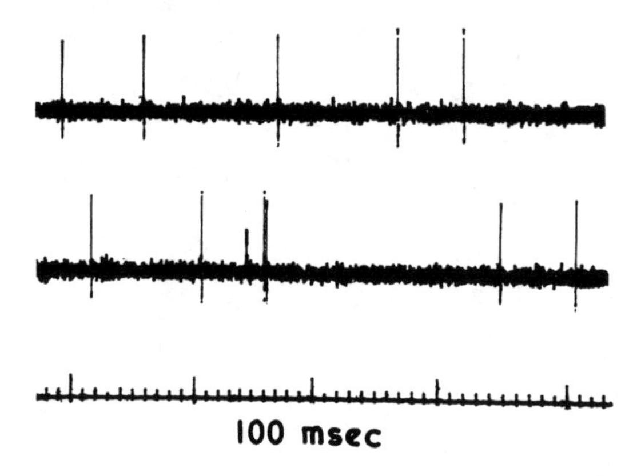

Fig. 119. Effect of peripheral nerve stimulation on a spontaneously discharging cortical unit. Upper trace shows the unit firing irregularly in the absence of stimulation. Lower trace shows how the unit becomes time-locked to the stimulus, followed by a period of post-excitatory depression.

1. Post stimulus time histogram. This is essentially a relationship between the presentation of a stimulus and the time of occurrence of all the spikes influenced by it. Spikes that exceed a certain amplitude level are processed by the computer and displayed as a number of peaks, each peak indicating a high probability of firing at a particular time after stimulation (Fig. 120).

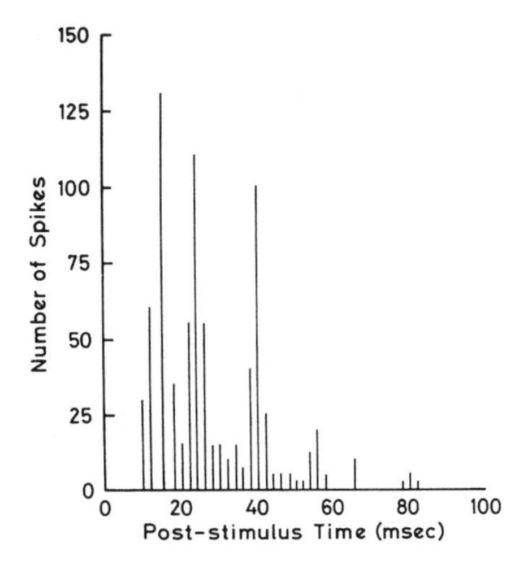

Fig. 120. Post-stimulus time histogram of a cortical unit influenced by sensory nerve stimulation. Time of unit firing after the stimulus is computed from 128 successive sweeps.

The post stimulus time histogram represents the response pattern of the unit sampled and allows comparisons to be made between the behavior of different units responding to identical stimuli. It also registers the behavior of different units in response to changes in the rate of stimulation. Most units in the sensory area have the ability to follow repetitive stimulation up to about 200/sec. At these rates, the times of the peaks do not alter significantly, but at higher rates of stimulation, there are usually latency shifts and a tendency for more peaks at the shorter latencies. Finally, it is believed that the post stimulus histogram may reflect differences in stimulus modality, since the duration and pattern of the evoked discharge often depend upon the type of stimulus employed.

2. Interval histogram. In a long train of repetitive discharges the intervals between adjacent spikes may vary considerably. Apart from local changes in cortical excitability, several factors may be responsible for the fluctuations observed, including sensory adaptation to a steady stimulus and, of course, any changes occurring in the parameters of the stimulus. Short or long intervals may appear at any position in the train. The interspike interval histogram is essentially a plot of the number of intervals of a given width against a time axis (Fig. 121). The shape of the histogram shows the overall distribution of intervals with the occurrence of one or more peaks along its extent.

By taking successive time slices during sustained activity of the unit, an estimate can be made of the probability of the unit firing during a certain time interval. The time between two adjacent spikes is called a first-order

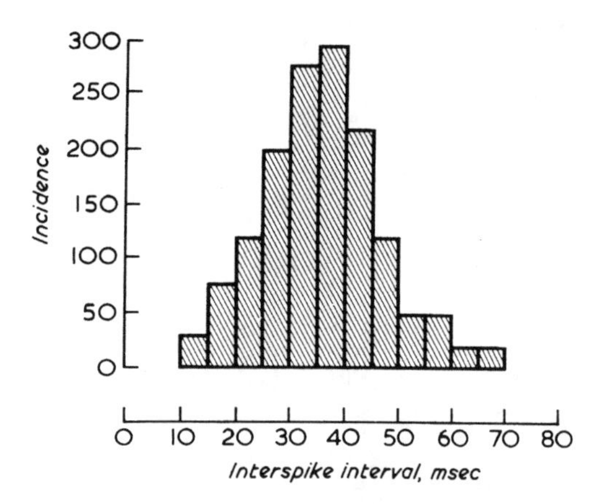

Fig. 121. Example of an interspike interval histogram. The distribution of spikes during sustained activity is plotted on a time axis. The histogram shows the occurrence of peaks between 30 msec and 40 msec with decaying distributions on either side. Other types of distribution are characterized by peaks at very brief intervals or by a second peak at a later interval.

interval; the time between two non-adjacent spikes is called a second- or third-order interval. These higher order intervals may be used to determine the probability of the unit firing a spike after a given interval of time irrespective of the number of intervening spikes. In spike train analysis, the probability of encountering an event as a function of time is specified a renewal process—i.e., the length of each interval is quite independent of the previous interspike interval. In contrast, the joint interval distribution of pairs of intervals gives the interval distribution conditional upon the previous interval being a particular value. This may be useful in analyzing the nature of any correlations existing between successive interspike intervals.

3. Cross-correlation histogram. This technique is of value in the interpretation of two or more spike trains recorded simultaneously. It is essentially a relationship between the interval distributions of one unit discharge and the interval distributions of another. If the two spike trains are firing independently, the spikes of the first train will occur at moments of time completely unrelated to the occurrence of spikes in the second train. No peaks will be present in the histogram to suggest overlap of time intervals except by chance. On the other hand, if the two units happen to be related—for example, when they are driven by a common presynaptic input—then peaks will appear in the histogram at time intervals that partially or completely coincide (Fig. 122). The cross-correlation histogram provides statistical evidence for dependent firing between two spike trains. It implies that the discharge of a spike in one unit is associated with the occurrence of a

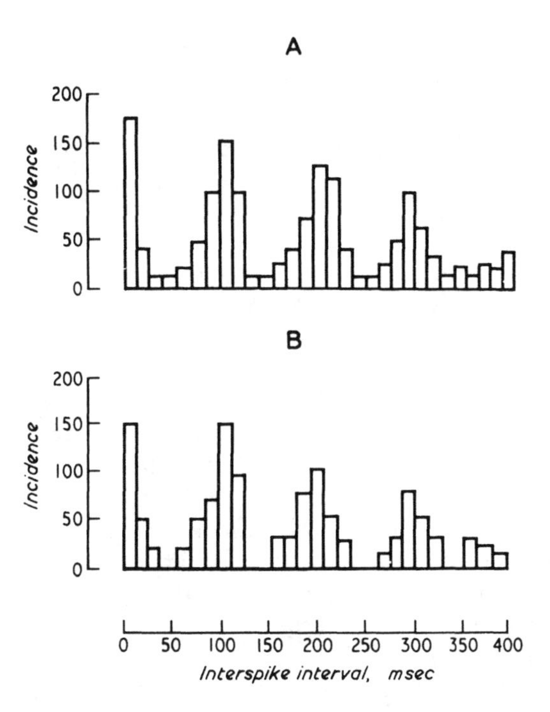

Fig. 122. Cross-correlation histogram of the spike activities of two cortical units, A and B, simultaneously recorded. Coincidences in time between the main peaks in A and B suggest that they are driven by the same presynaptic input.

spike in the other unit. Accordingly, if there is a change in firing rate in the first train, this will be reflected by an acceleration or slowing in the second train. Acceleration is evidenced in the histogram by peaks, and slowing by dips. Such correlations in the temporal patterns of cortical unit discharges constitute a form of dependence or time-locked relationship between different elements in a neuronal network.

DESCENDING INFLUENCES ON SENSORY TRANSMISSION

Anatomical studies show that there are descending projections from the sensory cortex to lower levels of the brain and spinal cord. One of the methods used entails placing lesions at selected points in the cortex and allowing the corticofugal fibers to degenerate. The pathways can then be traced to their terminations and identified by various staining reactions. Another method, found especially useful for mapping cortico-thalamic projections, is that of local strychninization combined with electrical recording.

It is now established that fibers emanating from the primary and secondary areas of the sensory cortex descend to the thalamus, hypothalamus, and to the trigeminal and pontine nuclei of the brain stem, and reach the first synaptic relays in the dorsal columns. Similar corticofugal pathways terminate on the posterior horn relays of the spinothalamic system. Thus on anatomical evidence alone, we know there is a feedback loop for each synaptic level of the sensory inflow.

Cortical influence on the ascending sensory systems has been amply confirmed by electrophysiological experiment. For example, the potentials evoked at the synaptic relays following peripheral nerve stimulation can often be modified by stimulating the sensory cortex. The amplitude of the evoked responses is sometimes increased, but more usually reduced or completely suppressed. Studies of single unit discharges in the dorsal column nuclei show that the firing rates are altered by corticofugal activation. Thus, the response of a cuneate unit to a peripheral stimulus is typically a high frequency, sustained discharge of spikes. If the sensory cortex is stimulated just prior to the cutaneous stimulus, the discharge of the unit may be depressed. The cause of the depression is due partly to internuncial blocking and partly to presynaptic effects on the primary afferent terminals. The net result is sensory inhibition.

Cortical influences on the thalamic relay cells may be facilitatory or inhibitory. Facilitation occurs when the cortical stimulus produces an increase in firing frequency of the relay cell. The neuronal mechanism responsible for this effect is not a simple one. It may be due to several factors–(1) direct synaptic action of cortico-thalamic fibers; (2) activation of the ascending reticular system; and (3) antidromic invasion of thalamic relay cells, causing the generation of impulses in their axon collaterals, which in turn excite adjacent thalamic relay cells.

Inhibitory interaction may be demonstrated as follows:

1. If a cortical stimulus is used for conditioning and a cutaneous stimulus for testing, presentation of the paired stimuli may result in complete inhibition of the test response lasting up to 100 msec or more.

2. If the sensory cortex is ablated or made unresponsive by deep anesthesia, the response of a thalamic unit is greatly increased. This suggests that the cortex normally exerts a tonic inhibitory influence on thalamic transmission. The mechanism of thalamic inhibition has already been discussed. It seems likely that the effects are brought about through short interneurons acting postsynaptically on the thalamic relay cells or presynaptically on the terminals of the ascending afferents.

All of these findings confirm the general view that the transmission of impulses from the periphery to the sensory cortex is subject to modification by the cortex through corticofugal processes of excitation and inhibition.

The patterns of activity observed in the cortical discharges must therefore represent the end product of a series of transformations occurring at the intervening synaptic relays. Obviously, corticofugal control of the sensory input allows the cortex to alter excitability thresholds and regulate the flow of different sensory signals. It might also help to focus attention on specific types of information from the receptors by excluding those outside its immediate requirements.

SELECTED BIBLIOGRAPHY

Adrian, E.D. Afferent discharges to the cerebral cortex from peripheral sense organs. *J. Physiol.* 100:159-191, 1941.

Amassian, V.E. Evoked single cortical unit activity in the somatic sensory areas. *Electroen. Clin. Neurophysiol.* 5:415-438, 1953.

Amassian, V.E., Waller, H.J., and Macy, J. Jr. Neural mechanism of the primary somatosensory evoked potential. *Ann. N.Y. Acad. Sc.* 112:5-32, 1964.

Anderson, P., Eccles, J.C., and Sears, T.A. The ventro-basal complex of the thalamus. *J. Physiol.* 174:370-399, 1964.

Brooks, V.B., Rudomin, P., and Slayman, C.L. Sensory activation of neurons in the cat's cerebral cortex. *J. Neurophysiol.* 24:286-301, 1961.

Kuypers, H.G.J.M., and Tuerk, J.D. The distribution of the cortical fibres within the nuclei cuneatus and gracilis in the cat. *J. Anat.* 98:143-162, 1964.

Marshall, W.H., Woolsey, C.N., and Bard, P. Observations on cortical somatic sensory mechanisms of cat and monkey. *J. Neurophysiol.* 4:1-24, 1941.

Mountcastle, V.B. Modality and topographic properties of single neurons of cat's somatic sensory cortex. *J. Neurophysiol.* 20:408-434, 1957.

Poggio, G.F., and Viernstein, L.J. Time series analysis of impulse sequences of thalamic somatic sensory neurons. *J. Neurophysiol.* 27:517-545, 1964.

Rose, J.E., and Mountcastle, V.B. Activity of single neurons in the tactile thalamic region of the cat in response to a transient peripheral stimulus. *Johns Hopkins Hos. Bull.* 94:238-282, 1954.

Sugitani, M. Electrophysiological and sensory properties of the thalamic reticular neurones related to somatic sensation in rats. *J. Physiol.* 290:79-95, 1979.

Towe, A.L., and Amassian, V.E. Patterns of activity in single cortical units following stimulation of the digits in monkeys. *J. Neurophysiol.* 21:292-311, 1958.

CHAPTER XVI

Physiology of Pain

Pain is a sensation that involves a relationship between the nervous system and the mind. It can result from any form of stimulation carried to excess and it can arouse all kinds of bodily reactions depending on the site, intensity, and duration of the stimulus and on the tolerance of the individual. Pain is difficult to study in animals for they cannot describe their sensations, but animals that are hurt do show obvious signs of distress and react to a painful stimulus in a predictable way—e.g., avoidance, attack, and accompanying reflex actions. Experimental work on anesthetised animals has only a limited value and very often deductions have to be made when nociceptive stimuli are employed.

In human subjects, sensations that are described as painful are also somewhat confusing and may bear little relationship to the cause or the presence of disease. Many diseases produce no pain at all. Nevertheless, pain is an important warning sign with regard to injury; its persistence is helpful not only in making a diagnosis, but in selecting the correct form of treatment. In this respect the clinician has to consider both the neurological and psychological aspects of his patient's condition before appropriate measures for the relief of pain can be undertaken.

PERIPHERAL MECHANISM

Since the nervous system is involved in all pain sensations, it is necessary to examine the neural elements that transform a painful stimulus into a sensory experience. The mechanisms subserving pain are based on inputs from receptors. Pain is evoked by nerve impulses transmitted to the central nervous system by peripheral nerves and carried through lemniscal and anterolateral pathways to the cerebral cortex. There is therefore a parallel anatomical basis for pain and other sensory modalities. The main distinction appears to be the predominance of the anterolateral system for coding the impulse patterns and for imposing the modifications that determine the special qualities and character of a painful sensation.

Receptors

There are no encapsulated end organs. The only form of nerve termination distributed widely enough to serve sensitivity to pain is the free nerve ending. In some parts of the body, such as the cornea, it is the only form of innervation. Free nerve endings are found in the superficial layers of the skin, in the periosteum, falx and tentorium and around joint capsules, and also in the arterial walls and deeper tissues of the body, although their distribution in the viscera is not so extensive. The terminals consist of a network of fine unmyelinated axons interweaving between the cells on which they sometimes end within folds of Schwann cell cytoplasm. The free endings are not considered specific for painful stimuli, since they have a range of sensitivity for other modalities.

Adequate stimulus

Noxious stimuli of any type possess the property of evoking pain.

1. In the skin, injury is the most effective stimulus. Injury liberates pain-producing substances from the damaged cells. Likely substances include histamine, bradykinin, and prostaglandin; these all arouse intense pain when injected beneath the normal skin. Mechanical stimulation such as pricking or pinching and the extremes of hot and cold may act by lowering the threshold of the nerve endings. Local reactions to injury produce an area of vasodilatation and wheal formation, resulting in a heightened sensibility. The surrounding area becomes tender, and stimuli previously ignored now become painful. This condition is termed clinically "hyperalgesia."

2. Nerve pain arises from irritation of the sensory fibers—for example, the pain of a growing tumor. Or the nerve trunk may be subjected to compression as it passes through a bony canal.

3. Muscular pain is often caused by spasm, irritation, or ischemia. Spasm or inadequate relaxation occurs with sudden stretching; prolonged spasm compresses the intramuscular blood vessels and cuts off the blood flow. Irritation is usually due to an underlying inflammatory condition. Ischemia allows the accumulation of polypeptides and other pain-producing substances. It commonly affects the legs of patients with narrowed arteries.

Insufficient oxygen to the limb muscles produces pain on walking, which disappears during periods of rest.

4. Cardiac pain is produced by deficiency of the coronary circulation and release of metabolic end-products, which stimulate the pain nerve endings. Narrowing of the coronary arteries prevents effective vasodilatation in exercise; the blood supply of the heart muscle is insufficient for its increased needs. The patient suffers from angina of effort.

5. Visceral pain may be caused by overdistension of a hollow organ—for example, in acute retention of the urinary bladder. More commonly, pain occurs in the form of cramps resulting from rhythmic contractions of the smooth muscles when it is irritated or attempting to overcome an obstruction.

Measurment of pain threshold

Although a wide variety of stimuli may give rise to pain, they may all be classified as mechanical, electrical, chemical, or thermal. Any of these types of stimuli can be used for measuring the pain threshold. This may be defined as the lowest intensity of a stimulus that will evoke the first perceptible trace of pain.

1. Mechanical method. A simple method for measuring the pain threshold is that of exerting pressure on the skin and soft tissues. The apparatus consists of a metal syringe within which is mounted a steel spring. The plastic tip of the plunger is placed on the forehead, and the pressure increased until the subject reports pain. The force in grams exerted on the tissues of the tip indicates the degree of compression of the spring and is read off from a scale.

2. Electrical method. A pair of silver-saline electrodes is wrapped around a finger, and single shocks are delivered from a stimulator at progressively increasing intensities. Measurements are made of the current flow required to produce a sensation of pricking pain.

3. Thermal radiation method. The purpose of this method is to determine the skin temperature at which a pricking pain is evoked by thermal radiation applied for three seconds to the body surface. The area of skin to be tested is first blackened with India ink to prevent penetration of heat to the subcutaneous tissue. Then heat from a lamp is focused by a condensing lens through a fixed aperture, while a shutter regulates the duration of exposure. Pain threshold is measured in terms of the number of millicalories per second per square centimeter of thermal radiation. Most subjects feel a pricking sensation at the level of about 250 mc/sec. The threshold for burn-

ing pain is lower than that for pricking pain, but if the skin temperature is allowed to reach 45° C, pain is aroused immediately in all subjects.

Subjective assessment of pain

The intensity of a stimulus necessary to cause pain is relatively constant, whichever method of measurement is employed. In other words, the majority of persons are neither hypersensitive nor insensitive as far as their thresholds for pain are concerned. On the other hand, individuals do vary considerably in their reactions to a painful stimulus, since tolerance is conditioned largely by upbringing and ethnic influences. To determine how much a person is likely to react to pain is therefore extremely difficult.

Beginning with the threshold for pain, a scale can be constructed up to the point where the most intense pain can be distinguished. This is known as the *dol* scale. Each point on the scale represents the smallest change in stimulus intensity that can be recognized as a change in pain intensity. The procedure is to increase the stimulus until the subject detects a difference in the degree of pain evoked. The determination of this just noticeable difference (JND) is plotted on the scale on the basis that 1 dol = 2 JND. The total number of dols that can be detected by the average person is 11, when the maximal level is reached. This represents the limit of discrimination of the sensory system for pain, for any increase of stimulus intensity above maximal does not evoke any greater appreciation of the pain. In clinical practice, descriptive terms for the assessment of pain are commonly used. Thus a patient may describe his pain as "mild," "moderate," "severe," "intense," or "unbearable."

Adaptation

Pain nerve endings are slowly adapting and sometimes do not adapt at all to persistent noxious stimuli. This failure to adapt is most important if pain is to serve its purpose as a warning of injury and destruction. The pain threshold, however, may be changed in a variety of ways—for example, by the local application of cold or heat or by the excitation of other afferents in the nervous system. It is also a common experience that pain is often ignored when the attention is distracted. Or pain may be caused to disappear by voluntary effort.

Types of afferent fibers

Pain is served by two sets of afferent fibers differing in diameter, conduction velocity, and threshold:

1. A delta fibers have the lowest threshold to pain. They are small myelinated fibers conducting at velocities between 3 m/sec and 20 m/sec. They represent a *fast* route for transmitting a sharp, pricking type of sensation.

2. C fibers have the highest threshold to pain. They are all unmyelinated and conduct at velocities between 0.5 m/sec and 2 m/sec. They represent a *slow* route for transmitting a prolonged, burning type of sensation.

The two sets of fibers are responsible for the double quality of pain sensation. If the A delta fibers are blocked, only the second kind of pain is elicited by a strong stimulus, and this tends to become more painful over a period of time. However, all C fibers are not nociceptive, while other types of fibers increase their discharge frequency when the stimulus reaches noxious levels and may contribute to the total experience of pain.

CENTRAL MECHANISM

The pathways conducting impulses from the periphery to the higher realms of consciousness are of great interest to a surgeon seeking to relieve pain without damaging other vital structures. The problem is not a simple one because pain pathways ascend by multiple relays with crossing and recrossing in the cord and diffuse distribution of the fibers at higher levels. The first order neurons terminate on cells in the posterior horn. Thereafter, the pain fibers take a dual course to the thalamus, many crossing over to the opposite spino-thalamic tract, and others passing through the anterolateral columns to the reticular formation of the medulla and pons. All pathways eventually reach the cerebral cortex, but once again, the problem of tracing afferent fibers to their terminals is extremely difficult since many of them are unmyelinated and appear to establish extensive bilateral connections within the cortical networks.

Spinal cord

Pain fibers enter the spinal cord in the lateral division of the dorsal roots and immediately give off collaterals which pass to the cells of the substantia gelatinosa in the posterior horn. Axons from these cells take one of several possible routes; they may form local circuits for reflex actions, they may give rise to the spino-thalamic tract, or they may enter the anterolateral columns.

1. Local circuits. The axons of cells in the substantia gelatinosa connect with cells in other layers of the posterior horn through which the motoneurons can be influenced. One or more segments of the cord may be involved. The functions of these local circuits are obviously important:

• They provide opportunities for modification of the primary impulse patterns.

• They provide a tonic internuncial background which may affect the discharge of pain impulses to higher levels.

• They provide a means for descending controls originating in the brain.

• They play a part in the withdrawal reflex and in other reactions to painful stimuli.

2. Spino-thalamic tract. The "fast" pain fibers are mostly second-order neurons from posterior horn cells, which ascend a few segments of the cord before crossing in the anterior commissure to the opposite side. The tract extends the full length of the cord and is joined in the brain stem by the trigeminal lemniscus; its terminal portion ends in the posterior group of thalamic nuclei.

3. Anterolateral columns. The "slow" pain fibers form a network ascending bilaterally through the anterolateral columns of the cord to the reticular formation of the brain stem. They are mostly high threshold afferents ending in the intralaminar nuclei of the thalamus. They are capable of exciting the reticular activating system to promote defense against injury.

Thalamus

The separate pathways for pain impulses are maintained at the thalamic level, where spino-thalamic components in the posterior nuclei serve localizing and discriminative functions for the contralateral body, while the anterolateral components of the reticular and intralaminar nuclei serve the more general functions of painful reactions. The evidence is as follows:

1. Stimulation of the thalamus in conscious animals with chronic implanted electrodes elicits behavioral responses suggesting pain.

2. Recordings from single thalamic units show that a proportion respond only to activation of C fibers when a strong cutaneous stimulus is applied.

3. In man, vascular lesions in the reticular nuclei of the thalamus may be associated with severe and unpleasant pain. On the other hand, a discrete electrolytic lesion in the caudal part of the ventrobasal complex may cause the disappearance of pain on the opposite side.

The thalamus may be regarded as a higher center for the recognition of painful stimuli. Under normal conditions, its connections with the cerebral cortex are no doubt responsible for the discriminative and emotional aspects of painful sensations. When cut off from cortical control, discrimination is lost and the reactions aroused become unusually excessive and persist when the stimulus is removed.

Sensory cortex

The ascending fibers for touch and kinesthesia mix with those for pain and temperature, as they all terminate at the cortical level. Pain fibers from the body wall and viscera are distributed to both sensory areas.

1. Primary sensory area. Localization of pain probably results from simultaneous stimulation of tactile receptors along with pain nerve endings. Thus a pinprick in the skin can be located with considerable precision since the topographic representation of the body surface is preserved in the thalamo-cortical projections on the postcentral gyrus. Stimulation of the sensory area in conscious human subjects elicits painful sensations referred to the opposite side of the body, except for the face area which elicits bilateral responses. On the other hand, removal of a specific part of the primary cortex very often leaves pain sensibility intact. This can only mean that the primary sensory area does not play a dominant role in pain sensation.

2. Secondary sensory area. There is reason to believe that high threshold afferents from the skin and abdominal viscera are associated with the conduction of pain nerve impulses. A delta and C fibers ascend through the anterolateral columns to the posteroventral, intralaminar, and reticular nuclei of the thalamus, from which projections converge on cells in the second sensory area of the cerebral cortex. Activation of these fibers produces bilateral surface responses and also evokes the discharge of single cortical units. A splanchnic receiving area is found as a narrow strip wedged between arm and leg areas, but overlapping tactile and proprioceptive projections. The receptive fields are larger in size than those in the primary sensory area and topographical organization is less precise. Furthermore, as many different afferent pathways converge on the same cortical unit, discrete information is not available. All this evidence suggests that the second sensory area is responsive to noxious stimuli and may have a participant role in the mechanism of visceral pain. It is well known, clinically, that visceral pain is seldom localized with any degree of accuracy.

Prefrontal cortex

The subjective aspects of pain sometimes cause suffering that the patient describes as intolerable. Such patients generally have a low tolerance for pain, react excessively to a mild stimulus, and appear to be wrapped up in their own fears and morbid emotions. The prefrontal lobes have long been considered the domain of emotional experience and, despite many uncertainties, their involvement in awareness and reactivity to pain is now generally accepted. Impulses of somatic and visceral origin are relayed to the prefrontal cortex along projections from the diffuse thalamic nuclei. If these pathways are cut or the nuclei destroyed, the reactions commonly associated with pain may not be present. The condition of congenital indifference to pain may also be explained in terms of a thalamo-cortical defect. After the operation known as prefrontal leucotomy, the threshold for noxious stimuli remains unchanged, and the patient feels the pain as before. His behavior, however, is altered to the extent that he no longer complains of the pain or demands frequent medication to relieve it.

CHARACTER OF PAIN

Information on the quality, intensity, and duration of pain is helpful in clinical diagnosis and afterward in confirming the success of any treatment. A distinction must be made between pain from the body wall and pain from the viscera. In addition, it must be determined whether the site of the pain is simply a reflection of an injury or of disease at a different site. Finally, related sensations such as itch and tickle and local tenderness are all important for building a useful history.

Somatic pain

Pain arising from the body wall may be superficial or deep. *Superficial* pain arises from the skin. It is either pricking or burning in character. Pricking pain is always accurately localized and subsides quickly when the stimulus is removed; it is mediated by fast A delta fibers. Burning pain is more persistent and generally evokes a withdrawal response as part of a defense reaction; activity is transmitted to the brain by unmyelinated C fibers. Both types of pain are elicited by noxious stimuli that damage the skin. Stimuli that do not normally arouse pain may do so if the tissue is already inflamed, thus lowering the pain threshold.

Deep pain arises from the muscles, joints, fascia, tendons, and periosteum. It is usually aching in character, more diffusely spread than superficial pain, although it can be localized. Often it is sustained over long periods of time. It is generally accompanied by reflex muscular spasms, which are themselves a source of pain and local tenderness.

Headache is the commonest example of deep somatic pain brought on by muscular tension, eye strain, or general body fatigue. Other causes of headache are increased intracranial pressure or mechanical distension of the cerebral vessels. Migraine is a special kind of headache due to sudden vasodilatation in the branches of the external carotid artery followed by extravasation of fluid in the perivascular spaces where pain-producing chemical substances accumulate.

Visceral pain

Pain arising from the interior of the body may come from the viscus itself or from its parietal coverings. *True visceral* pain originates from organs in the thoracic and abdominal cavities. In mild cases, it may merely represent a feeling of discomfort, aching in character and poorly localized; it may be due to distension of a hollow viscus or to local irritation, as occurs in intestinal colic. In more serious cases, pain is severe, gripping in character, accompanied by sweating and anxiety. However, it is not usually associated with superficial tenderness or skeletal muscle spasms. This kind of severe pain is often due to strong, rhythmic contractions of the gall bladder or ureter in passing a stone. It also arises from ischemia and from any site of inflammation.

Visceral pain resembles deep somatic pain in its diffuse character, but differs from superficial pain in many important respects:

1. Adequate stimulus. The viscera are normally insensitive to stimuli that produce pain in the skin. A surgeon can cut or apply diathermy to the intestine in a conscious patient without causing pain.
2. Quality. True visceral pain of low intensity is described as a dull ache in contrast to the sharp character of parietal and cutaneous pain.
3. Localization. Visceral pain is vague and poorly localized. This may be because of the poverty of afferent fibers supplying the viscera when compared with the rich innervation of the skin.
4. Reflex action. If the disease has not spread to the parietal coverings, there will be no superficial tenderness or reflex muscular rigidity.
5. Nerve block. True visceral pain cannot be relieved by injecting local anesthetics into the skin, a procedure that gives temporary relief for most somatic pains.

Parietal pain originates from the serous coverings of the viscera and from nerve endings in the peritoneum, pleura, and pericardium. These structures are all supplied by the fast A delta fibers of somatic nerves. The character of parietal pain is sharp and fairly well localized. It is accompanied by superficial tenderness and reflex muscular rigidity. These facts are important in clinical diagnosis. Capps and Coleman inserted a probe through the chest wall of patients with pleural effusion; they found that the pain was accurately localized to the point of contact with the pleural membrane. Patients with pericardial effusion complain of tightness in the chest beneath the sternum. In the early stages of acute appendicitis, before the inflammation has spread to the parietal peritoneum, pain is felt in the region of the umbilicus, since the afferent pathways enter the spinal cord at approximately the level of T10. The pain is diffuse and aching in quality but the abdominal muscles remain relaxed. Later, as the disease progresses, pain is accurately localized to the site of the appendix in the right iliac fossa. It now becomes sharp and severe in character, associated with extreme tenderness and muscular rigidity. These signs are the result of impulses passing from the parietal peritoneum through spinal nerves and entering the cord at approximately the level of L1.

Referred pain

Pain may be experienced in a part of the body at a distance from the actual site of disease. This kind of referred pain may be due to the anatomical arrangement of dermatomes or, as is more frequently the case, to the convergence of cutaneous and visceral afferents within the spinal cord.

Dermatomal mechanism

The experiments of Capps and Coleman on the human diaphragm demonstrate how pain is referred to the dermatomes supplied by spinal nerves. The margins of the diaphragm are supplied by the lower six intercostal nerves; pain arising from stimulation of this region is referred to the anterior abdominal wall, which is innervated by the same intercostal nerves. The central portion of the diaphragm is supplied by afferents in the phrenic nerves which enter the spinal cord through the posterior roots of the third and fourth cervical segments. Pain arising from stimulation of this region is referred to the shoulder and neck, in the distribution of the same posterior nerve roots. When infection spreads to the diaphragm from an empyema or liver disease, the impulses ascending the phrenic nerve are referred to the dermatomes of these roots.

Convergence mechanism

Pain arising from the viscera and other deep structures is often referred to the body wall. The impulses are conducted into the spinal cord where they synapse with spino-thalamic neurons that receive pain fibers from the skin. The sensations aroused in the cerbral cortex are then interpreted as coming from the skin itself. These reference patterns produced by convergence of visceral and cutaneous pathways are obviously important in making a correct diagnosis. The pain of cardiac disease is a classic example. Pain originates in the ischemic muscle of the heart, but is referred to the chest wall with extension down the medial side of the left arm and up the neck to the angle of the jaw. Sometimes the pain is referred to the epigastrium, where it can simulate many of the signs of an acute abdominal condition. Pain impulses from the heart pass into the spinal cord via the upper thoracic posterior nerve roots. Excitation spreads diffusely through adjacent segments producing hyperalgesia, tenderness, and painful muscular contractions over a very wide area.

Itch and tickle

These sensations have much in common with that of pain and are served by similar neural mechanisms. Their purpose is to call attention to mild surface stimuli and to elicit reactions leading to removal of the irritant. Itch is usually caused by chemical irritation and can sometimes be very intense, though rarely painful. Tickle is evoked by light mechanical stimulation, such as stroking the skin, and is not always considered unpleasant. Both kinds of stimuli promote a desire to scratch.

It is generally assumed that itch is a separate modality distinct from pain and served by sensitive free nerve endings in the superficial layers of the skin. The evidence is as follows:

1. By means of the blister technique (e.g., the application of cathardin), the epidermis can be separated from the dermis. If histamine is added to the unbroken blister, only itching results and no pain, whereas histamine injected into the base of the blister, where it can penetrate more deeply, gives rise to pain. Itch is not evoked from any deep tissue of the body.
2. Itch and tickle sensations are mediated by unmyelinated C fibers similar to those that transmit the burning type of pain. On the other hand, pain also has a large A fiber component.
3. Itch and tickle provoke scratching, while pain elicits withdrawal or evasive reflexes.
4. Itch can be inhibited by painful stimuli by a process of interaction

between the large and small fibers in the cord. This process will be described later in this chapter under the heading "gate control theory."

5. The administration of morphine for the relief of pain can make itching worse.

Despite all the evidence that itch and pain are separate modalities, there is no doubt that, clinically, they are closely related. The intense itching of pruritus, urticaria, and allergic dermatitis is derived from the release of histamine and proteolytic enzymes, which in higher concentrations act as pain-producing agents. Furthermore, the pathways responsible for itch and cutaneous pain appear to converge centrally so that any apparent dissociation is lost. Many measures directed toward the relief of pain also cause the disappearance of itch.

THEORIES OF PAIN

The task of explaining the central mechanism of pain has been attempted by several investigators in the past, and the literature is now full of their discarded theories. A satisfactory theory must account for variations in the quality and intensity of pain, differences in character and localization, the problem of reference, the influence of descending controls, and the mechanisms that call forth a host of physiological reactions. The theories considered here are particularly relevant to the management of patients with chronic or intractable pain, since the theories offer a rational basis for present therapies and treatment.

Specific theory

The idea that pain is a specific sensory modality with its own nerve endings and fiber tracts has long been advocated on clinical grounds. There is also a good deal of experimental work to support this contention. Single fiber studies show that high threshold nociceptive afferents are present in the skin of most mammals and that they are distinct from receptors for touch and temperature. Earlier, Adrian, and Zotterman had recorded high frequency discharges in small myelinated fibers following strong mechanical stimulation of the skin. More recently, Perl identified mechanically excitable spots sensitive only to injurious stimuli and showing little or no sensitivity to innocuous stimuli or to extremes of heat. He concluded that the first sensation of pain arose from specific nerve endings in the superficial layers of the skin.

The studies of Iggo on unmyelinated C fibers reveal that there is a considerable range of sensitivity to various forms of energy change. Most of the C fibers are excited by low intensity mechanical stimulation. But a proportion have high mechanical thresholds, are slowly-adapting, and respond to damaging or noxious heat stimulation. These findings are of particular significance in the context of pain. They suggest the idea of "selective sensitivity" of the nerve endings for specific nociceptive changes.

While the idea that identifiable nociceptors exist in the skin continues to gain strong support, the central mechanism for pain transmission is far from settled. Why pain coming from deep structures is commonly referred to the body surface is the subject of much speculation. It was argued that cutaneous and visceral pain shared a common segmental innervation and created an "irritable focus" in the spinal cord. This was the basis of Mackenzie's "viscero-sensory reflex" and Ruch's "convergence" theory, in which it was suggested that a common pool of neurons shared the same spino-thalamic tract. Neurons having a specific nociceptive input have been found in the marginal zone of the posterior horn. Other posterior horn neurons are known to be excited by noxious heating of the skin. It is necessary, however, to follow the impulses to higher synaptic levels before a complete analysis can be made. Furthermore, the concept that pain is a specific modality is difficult to establish in view of the considerable psychological overtones that pain arouses. For this reason, surgical lesions of the central nervous system are rarely successful in abolishing pain. The conclusion to be drawn is that there are many other variables in the pain mechanism besides that of a specific sensory input.

Pattern theory

According to this theory, pain is reported when a characteristic pattern of nerve impulses is evoked by intense non-specific changes in the environment. Supporters of the theory like Weddell and Sinclair assume that all types of nerve endings can participate in cutaneous pain when stimulated excessively. The impulses converge centrally at sites where summation and interaction take place and the resultant discharge is carried to higher synaptic levels where further modification occurs. The cerebral cortex receives an impulse pattern which is interpreted as pain, although it is subject to additional influences from the cortex itself. Painful stimuli evoke reproducible patterns of activity that may be long-lasting or permanent and, thus, leave a memory trace of the experience.

The pattern theory ignores the discovery of high-threshold receptor units in the skin and the fact that pain is not produced until the smallest myelinated fibers are activated and becomes unbearable only when the unmyelinated C fibers are involved. On the other hand, the theory can be used more readily to explain the pain of a phantom limb. Phantoms appear in patients as positive pain long after the limb has been amputated and, of course, the pain cannot be due to stimulation of non-existent nerve endings. At one time it was thought that pain was caused by irritation of the cut end of the nerves in the stump, but phantoms have been reported in patients with total lesions of the spinal cord. According to the pattern theory, a phantom can be explained as a memory trace of a painful experience, recalled by the patient in the same way as the brain recalls other modalities of sensations.

Gate control theory

In an attempt to reconcile the main features of the specific and pattern theories, Melzack and Wall introduced their gate control theory of pain. It was based on the assumption that impulses entering the spinal cord passed through a central control system before reaching the brain. In other words, information about the presence of injury is received at the cortical level as impulse patterns modified by a series of interacting processes. The theory rejects the classical concept that the pain pathway from the periphery is a straight-through transmission system to the cerebral cortex.

In its original form, the theory proposes that the cells transmitting pain through the spinal cord (T cells) are under the control of the substantia gelatinosa neurons in the posterior horn. This is the site for convergence of (1) large sensory fibers from cutaneous mechanoreceptors, (2) small diameter fibers–A delta and C fibers–from the peripheral structures including the viscera, and (3) descending fibers from the brain. The mechanism of gate control is illustrated in the diagram of Figure 123.

It was supposed that the substantia gelatinosa neurons exerted an inhibitory effect on the presynaptic afferent terminals supplying the T cells. When a stimulus is applied to the large fibers, the T cells are excited directly, but their output is soon diminished because of feedback inhibition through collaterals of the large fibers terminating in the substantia gelatinosa. The net effect, therefore, is that the gate is closed. When a stimulus is applied to the small fibers, the T cells are again directly excited, while the inhibitory effects of the substantia gelatinosa are decreased. The net effect is that the gate is opened for pain transmission.

The gate control theory has been criticized on the grounds that it ignores the specific sensitivity of free nerve endings for noxious stimuli. The convergent excitatory and inhibitory influences of A and C fibers through feedback mechanisms have not been confirmed, while the existence of

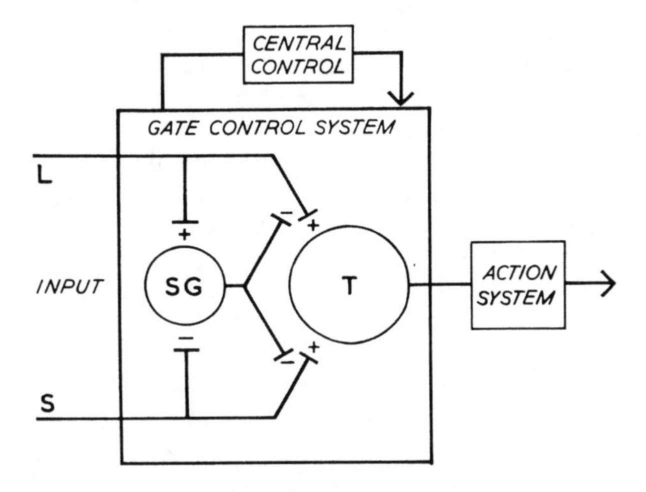

Fig. 123. Diagram of the gate control theory of pain. The large-diameter fibers (L) excite the first central transmission cells (T), which project to the cells of the action system. The output is diminished by feedback exerted on the cells of the substantia gelatinosa (SG). The small-diameter fibers (S) excite the transmission cells and decrease the inhibitory effect of the substantia gelatinosa, thus opening the gate for pain transmission. The diagram also indicates a central control by descending impulses from the brain. (After Melzack and Wall.)

presynaptic control of the T cells remains in debate. Nevertheless, the idea of gate control for pain mechanisms is an attractive one. Extensive convergence of all types of afferents does take place in the posterior horn. Opportunities occur for excitatory and inhibitory interactions that can grade the level of impulse activity in the ascending sensory systems. The level of impulse activity is the result of a fine balance between the large and small fiber components which have mutually antagonistic effects on the pain pathway. Pain can arise when the small fiber component is strong enough to overcome the effects of inhibition. Conversely, pain is suppressed when either the large fiber component or corticofugal control predominates. While speculation continues on the details of the gate mechanism, there seems no reason to doubt the existence of control by convergence and interaction. Indeed, this may be the basis for the relief of certain types of pain by mild stimulation of the large sensory fibers in the peripheral nerves and dorsal columns.

RELIEF OF PAIN

Measures that are used for the prevention or relief of pain must take into account the psychological needs of the patient as well as the signs of a physical illness. The obvious remedy is to remove the cause. Unfortunately, this is not always practicable, especially when emotions and anxiety play an important part. When the cause is known, pain of physical origin may be treated by attempting to control the impulse traffic at any appropriate point between the sensory nerve endings and the cerebral cortex. Thus, impulses can be blocked at their source or in the peripheral nerves or at selected sites within the central nervous system. When the cause is unknown, or when the illness is accompanied by acute distress and depression, pain can be alleviated only by attending to the general needs of the patient in an attempt to restore his confidence.

Physical remedies

1. Removal of the cause. If the nature of the trouble causing pain can be diagnosed, treatment is directed toward correcting the damage: A foreign body may be extracted, a fractured bone set, or an acute appendix removed.

2. Local anesthesia. Analgesic drugs (e.g., procaine, lignocaine) infiltrating the painful area or injected into the region of the nerve trunk will block the conduction of nerve impulses. This method is commonly used in dental practice and for minor surgical operations. Long-lasting local anesthetics, such as benzocaine, are employed for the relief of chronic pain. For instance, they may be used for blocking the Gasserian ganglion for trigeminal neuraliga or the celiac plexus for upper abdominal pain.

3. Physiotherapy. Massage and the local application of heat give a sensation of comfort and help to reduce pain by counter-irritation and hyperemia. Diapulse therapy in the form of penetrating electromagnetic energy aids recovery by reducing swelling, especially in deep-seated injuries. The application of ice packs or an ethyl chloride spray is sometimes helpful as a temporary measure.

4. Electrical methods. A source of electrical energy from certain fish was used in classical times to produce numbness until pain ceased. In the seventeenth and eighteenth centuries various electrostatic machines were invented, and their popularity increased as knowledge progressed and more effective instruments became available. The use of local electrical stimulation for pain relief has been frequently reported in the literature, but it is only in recent years that its clinical application has taken advantage of modern electronic techniques. At the present time several forms of electrical stimulation of afferent nerves and of the central nervous system have been introduced with favorable results:

- Cutaneous stimulation. The types of stimulator used deliver a square wave pulse of variable width, frequency, and voltage. Carbon rubber electrodes are applied to the painful area, and a combination of stimulus parameters chosen that prove to be the most effective. Patients with chronic pain of muscular origin appear to benefit from this form of therapy.

· Peripheral nerve stimulation. The application of a weak constant current to produce blocking of nerve impulses is based on the membrane theory of action potentials. Anodal currents oppose the local changes in the nerve by restoring the polarized state or by producing hyperpolarization of the membrane. For dental analgesia the positive lead from a dry battery is connected to the drill, which must be insulated at the handle, and the negative lead is usually attached by a metal clip to the patient's ear. The strength of current required is regulated by a potentiometer.

· Dorsal column stimulation. In this method an attempt is made to block pain pathways in the spinal cord by means of sensory interaction. A stimulating electrode is implanted beneath the dura and connected by a subcutaneous wire to a contact button below the clavicle. The dorsal column is stimulated in the T5 segment for pain below the thoracic region and in the C2 segment for pain above. It is supposed that excitation of the large A fibers produces inhibition of impulses carrying pain in the small A delta and C fibers. The patient uses the stimulator as often as necessary; relief is obtained in a high proportion of chronically affected patients who have resisted other forms of treatment.

5. Acupuncture. This method originated in China where it is practiced extensively on patients undergoing major surgery. The full cooperation of the patient is essential. He remains fully conscious throughout the procedures and participates as a member of a team, obeying instructions, yet obviously insensitive to pain. Acupuncture needles are inserted into the skin at a number of selected sites. They are either twisted by hand or connected to an electronic vibrator. The frequency of stimulation can be increased when it is necessary to perform a painful manipulation. The emphasis on prolonged mechanical stimulation of the cutaneous nerves suggests that control of pain is achieved by interaction effects of central mechanisms between low and high threshold afferents.

Neurosurgical techniques

Several surgical procedures are available for the relief of chronic pain. They are indicated either for specific painful conditions like causalgia, peripheral arterial disease, and trigeminal neuralgia, or for intractable pain arising from malignant disease. Operations are worth considering for severe pain that is not alleviated by drugs. The sites of operation may be at any level of the pain pathways from their entry into the spinal cord to their terminations in the cerebral cortex.

1. Sympathectomy. In this operation the sympathetic chain is divided between the third and fourth thoracic ganglia for the upper limb and the second and third lumbar ganglia for the lower limb. It is the operation of choice for causalgia, an unpleasant condition associated with burning pain and profound emotional disturbances. In pain resulting from peripheral vascular disease, sympathectomy produces vasodilatation and an increased blood flow to the muscles of the leg, thus relieving ischemia during walking.

2. Posterior root section. The roots to be divided are determined from the dermatomal situation of the pain when this is unilateral and localized to a few segments. One disadvantage is that all modalities of sensation are lost, leaving the denervated limb subject to trophic changes and locomotor disorders. There is, however, considerable overlap between adjacent dermatomes and if one or two roots can be left intact, some function can be preserved. Another disadvantage is that pain may recur owing to postoperative fibrosis.

3. Cordotomy. Section of the spino-thalamic tract is suitable for unilateral pain in the lower limbs. The section, which must be made on the side opposite to the patient's pain, can be performed by open operation under general anesthesia or by electrolytic lesions guided by X-ray control under local anesthesia. Cervical cordotomy for pain in the arm and trunk is usually carried out at C2, but at this segmental level, the problem is to avoid damage to the descending respiratory tracts. A successful cordotomy produces complete loss of pain and temperature sensations in the opposite half of the body below the section. Light touch is impaired slightly, but joint position and vibration remain intact.

Anterolateral cordotomy is suitable for pain felt centrally in the abdomen or chest. Bilateral section, however, tends to produce retention of urine, and after a time many patients experience a return of their pain.

4. Thalamotomy. In brain operations for the relief of chronic pain, the surgeon attempts to interrupt pathways in the thalamus itself or in the thalamo-cortical projections. The technique employs stereotaxic procedures for accurate placement of lesions by electrocoagulation and is sometimes used for the relief of pain unresponsive to other methods.

5. Leucotomy. Operations involving the prefrontal lobes tend to cause changes that accompany the reactions to pain. They may be very successful in relieving anxiety and tension and in increasing the patient's tolerance of pain to an acceptable level.

Drug therapy

There is no ideal drug for the relief of pain since so much depends upon individual tolerance and side effects. The treatment of patients with chronic pain introduces additional problems of toxicity and addiction. The best guideline for dosage is to use the smallest amount of drug that works.

• Analgesic drugs such as aspirin, salicylate, and paracetamol have the advantage of being non-habit-forming and are useful for relieving mild forms of pain from any cause. More powerful drugs of an almost unlimited variety are also available.

• Narcotic durgs such as pethidine and morphine are valuable in controlling restlessness associated with severe pain as they tend to produce a soporific effect. Unfortunately, narcotics given at regular intervals lead to early tolerance so that doses may have to be increased to some extent. The action of narcotics on the central nervous system is that of depression, particularly on the reticular activating system, while the possibility of respiratory depression is a positive danger. For this reason narcotics should be given at fixed intervals or combined with a non-addictive drug.

Psychological therapy

Helping patients overcome their anxieties and fears can be a great help in relieving their pain:

1. Medication. Apart from the use of specific drugs, various forms of medication, either alone or in combination with a mild analgesic, are frequently administered to make the pain more bearable. They are popularly known as tranquilizers. Many patients obtain relief from placebo medication (usually saline or distilled water) when anxiety is the dominant factor. In more agitated states, chlorpromazine in small doses or the tricyclic compounds (e.g., amitrytpyline) have a useful sedative effect. The object is to increase the confidence of the patient to the point where he feels that his pain is under control.

2. Transcendental meditation. This method, also known as the "science of creative intelligence," is based on the regular practice of complete rest with the eyes closed and the mind concentrated on a sound, word, or part of the body. The technique promotes bodily relaxation in which there is a significant lowering of the basal metabolic rate, decrease of respiratory rate and blood pressure, and stabilizing of all physiological systems. At the same time the brain becomes more alert and capable of greater production and inventiveness. Advocates of the method claim that psychological stability develops automatically when the mind and emotions are subject to relaxa-

tion. Obviously, this state of restful alertness will tend to reduce inner tensions and anxieties in most individuals and help to alleviate painful experiences that are primarily emotional in origin.

3. Hypnosis. In selected cases, where again the pain is primarily emotional in origin, hypnosis can be very useful. Lack of confidence plays an important part in making the pain of physical lesions worse, and some patients with chronic pain appear to live in a state of continuing emotional stress. Under hypnosis, the patient's attitude toward his personal problems can be changed so as to narrow his attention or exclude the experience of pain. (Further details under this heading will be described in Chapter XXIV.)

4. Enkephalins. It is believed that many aspects of pain perception may involve the release of naturally occurring peptide substance called endorphins. The best known of these biochemical compounds are the enkephalins, which are pentapeptides containing methionine or leucine side chains. Enkephalins imitate the action of morphine by diminishing pain and causing respiratory depression. They are distributed mainly in the limbic system, hypothalamus, and pituitary gland. Possibly they serve as neurotransmitters in the modulation of central inhibitory mechanisms.

SELECTED BIBLIOGRAPHY

Abram, W.P., Allen, J.A., and Roddie, I.C. The effect of pain on human sweating. *J. Physiol.* 235:741-747, 1973.

Bessou, P., and Perl, E.P. Response of cutaneous sensory units with unmyelinated fibers to noxious stimuli. *J. Neurophysiol.* 32:1025-1043, 1969.

Bonica, J.J. (ed). *International Symposium on Pain.* New York: Raven Press, 1979.

Hughes, J. Isolation of an endogenous compound from the brain with pharmacological properties similar to morphine. *Brain Res.* 88:295-308, 1975.

Keele, C.A., and Smith, R. (eds). *The Assessment of Pain in Man and Animals.* Edinburgh and London: Livingstone, 1961.

Lim, R.K.S. Pain. *Ann Rev. Physiol.* 32:269-288, 1970.

Melzack, R., and Wall, P.D. Gate control theory of pain. *Science* 150:971-979, 1965.

Poggio, G.F., and Mountcastle, V.B. A study of the functional contributions of the lemniscal and spinothalamic systems to somatic sensibility. Central nervous mechanisms in pain. *Bull. Johns Hopkins Hosp.* 106:266-316, 1960.

Rexed, B. The cytoarchitectonic organizaton of the spinal cord of the cat. *J. Comp. Neurol.* 96:415-466, 1952.

Wall, P.D., and Sweet, W.H. Temporary abolition of pain in man. *Science* 155:108-109, 1967.

Willis, W.D., Trevino, D.L., and Coulter, J.D. Responses of primate spinothalamic tract neurons to natural stimulation of hindlimb. *J. Neurophysiol.* 37:358-372, 1974.

CHAPTER XVII

Central Processes of Vision

The functions of the eye as a special kind of receptor with optical properties and a photochemical mechanism were described in Chapter VIII. It was shown that the eye possesses two kinds of photosensitive elements, one for night vision and one for daylight vision, thus allowing it to function over a wide range of light intensities. The breakdown of visual pigments initiates a complex electrical response beginning as a receptor potential in the rods and cones and spreading through second-order neurons to activate the retinal ganglion cells. The next stage is the conversion of the response into a train of action potentials for relay along the optic nerve fibers to the brain. The optic nerves, therefore, represent third-order neurons, in contrast to most other sensory mechanisms which have their first neural relays in the central nervous system itself. The purpose of this chapter is to describe the later stages of visual processing and to explain how the brain is able to detect the characteristics of the image in both white light and color.

CENTRAL VISUAL PATHWAYS

Most of the fibers in the optic nerve originate from the ganglion cells of the retina. After piercing the eyeball, fibers from the nasal half of each retina cross in the optic chiasm to enter the optic tract of the opposite side; fibers from the temporal half of each retina continue through the optic chiasm, without crossing, to enter the optic tract on the same side. Each optic tract runs backwards to the lateral geniculate body, from which fourth-order neurons emerge as the optic radiations. Some of the fibers, however, proceed to the superior colliculus, which is a center for the pupillary light reflex. The optic radiations lie close to the posterior horn of the lateral ventricle on their way to the striate cortex where they terminate on the medial surface of the occipital lobe. The striate cortex is the primary receiving area of the visual system. It extends over the lips of the calcarine fissure, and here the fibers regroup to maintain a high degree of topographical representation for the various retinal regions. Beyond the striate area, the visual cortex extends into the parietal and temporal lobes, frequently called the prestriate or association areas, where all tasks based on visual impressions are correctly interpreted.

Optic nerve

The axons of the ganglion cells, which constitute the optic nerve, transmit impulses continuously at low rates even in the dark. This spontaneous activity or dark discharge results from the combined effects of excitatory and inhibitory neural components in the retinal network. When light enters the eye the visual signal is superimposed on this background activity. The majority of optic nerve fibers have conduction velocities between 16 m/sec and 70 m/sec. The latencies are obtained by stimulating the optic fibers or tract and recording the antidromic response at the optic disc.

1. Response to diffuse illumination. If the retina is illuminated by diffuse light, all the photoreceptors are stimulated equally. The background discharge is then modified by an increase of frequency as the brightness of the light increases. Other factors involved are the duration of exposure and the state of adaptation of the retinal neurons. A fall of frequency therefore occurs if a steady light intensity is maintained.

2. Response to a spot of light. If the retina is illuminated by a spot of light, a contrast border is formed between light and dark areas. The photoreceptors on the light side of the border are excited, while those on the dark side of the border are inhibited. Corresponding changes occur in the ganglion cells. When impulses are recorded from the optic nerve fibers, it is possible to observe the effects of contrast borders on the discharge patterns. Excited fibers exhibit both phasic and tonic patterns. The phasic response consists of an initial burst of high frequency spikes, and the tonic response is a maintained discharge at a frequency somewhat higher than the background level. Inhibited fibers exhibit a transient discharge when the spot is shifted or the light turned off. The retina is particularly sensitive to contrasting light, as opposed to diffuse light, because of the mechanism of surrounding inhibition. Thus any change in the contour of an image, such as occurs around borders and edges, causes a change of impulse pattern in the same way as does a spot of light.

Lateral geniculate body

1. Functional anatomy. This nucleus receives the terminal fibers of the optic tract and gives rise to the optic radiations. The optic tract is a compact

bundle carrying fibers from the temporal half of the ipsilateral retina and from the nasal half of the contralateral retina. The destination of the fibers from each retinal quadrant has been determined by placing discrete lesions in the retina and noting the sites of degeneration produced in the cells of the nucleus. The results are illustrated in Figure 124. There is a lateral area for the lower quadrant of the retina, a medial area for the upper quadrant and an extensive central area for the fibers of the macula.

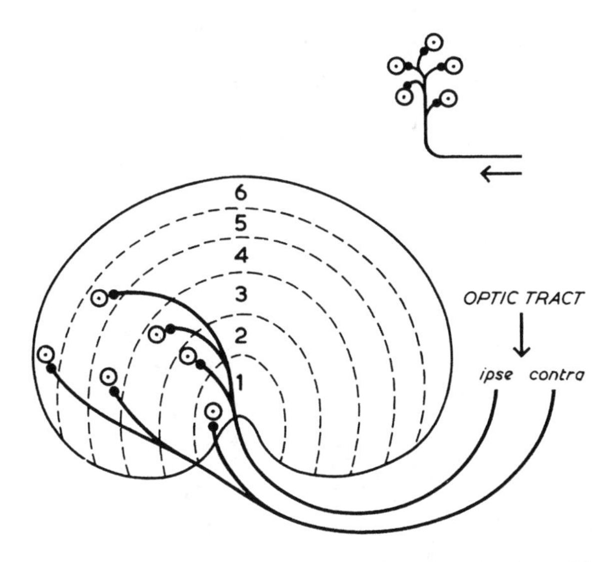

Fig. 124. Section through the lateral geniculate body showing termination of fibers in the optic tract. Fibers from the temporal half of the ipsilateral retina terminate in layers 2, 3, and 5 and those from the nasal half of the contralateral retina terminate in layers 1, 4, and 6. The layers are numbered from the hilum outward. The inset shows how an individual fiber ends in a spray of five to six branches, each making synaptic contact with the body of one geniculate cell.

The cells of the lateral geniculate body are arranged in layers numbered 1 to 6 from within outward. Fibers from the ispilateral hemiretina terminate in layers 2, 3, and 5; those from the contralateral hemiretina terminate in layers 1, 4, and 6. On entering its appropriate layers, each fiber breaks up into about five or six branches, one branch making synaptic contact with one cell. There is no convergence, so each geniculate neuron is stimulated by a single synapse. If one optic tract is cut, all the cells of the corresponding lateral geniculate body are found to show atrophic changes.

2. Properties of geniculate neurons. The discharge properties of single geniculate neurons can be studied by observing reactions to electrical stimulation of the optic nerve or to flashes of light directed on the retina. In addition, antidromic stimulation of the striate cortex gives information on

the post-synaptic events in the optic radiations. There is a striking similarity between the response of lateral geniculate neurons and the responses of retinal ganglion cells. In both cases, the receptive fields are concentric, with an excitatory "on" center and an inhibitory "off" center, or the reverse. Also, many of the units are particularly responsive to contrast borders or to movements of objects. Such similarities of recordings provide evidence for a direct relay of visual impulses through the lateral geniculate nucleus to the striate cortex. However, there are some properties of geniculate neurons which can only be explained as the result of interactions between the cellular layers of the nucleus. One of them is the more pronounced suppression of firing at the geniculate level compared with that in the retina. It appears that some modification of the visual input is very likely at this level.

Electrical stimulation of the optic nerve

A large number of units in the lateral geniculate body are spontaneously active. The frequency of this discharge varies considerably, even in complete darkness. When a single shock is applied to the optic nerve, the spontaneous activity disappears and the unit responds by discharging one or more spikes. As the microelectrode advances through the various layers of the nucleus, it is possible to demonstrate the distribution of the nerve fibers from the two eyes. The cells in the superficial layer are called A units and are driven exclusively by the contralateral eye; the cells in the middle layer are called A_1 units and are driven by fibers from the ipsilateral eye; cells encountered in the deepest layer are called B units. The latter have receptive fields larger than those of the other cells and originate from crossed fibers only. These arrangements of the optic nerve terminations suggest an orderly projection of the visual fields on the cellular layers of the nucleus without intermingling of the units: two layers of units respond to contralateral optic nerve stimulation separated by an intermediate layer of units responding to ipsilateral stimulation.

Over the past few years, there has been an accumulation of evidence that seriously challenges the above concept of topographical representation in the lateral geniculate body. It appears that a number of units found in transition areas between the anatomical layers respond to stimuli from both eyes. Indeed, the response of a unit to stimulation of one side can be considerably modified by stimulating the opposite side. This can mean only that there is overlap of converging fibers from the two eyes as well as some binocular interaction.

Photic stimulation of the retina

The projection of the visual fields on the lateral geniculate body may be studied by using an optical stimulator such as a spot of light shone on a screen or by moving objects of various sizes and shapes in front of the animal's eyes.

Stimulation by spots of light. The great majority of geniculate neurons have small, well defined receptive fields. In their general arrangement the fields consist of a center and a surround. Figure 125 shows a typical "on" response for the center and a reduction in unit firing for the surround. When the surround alone is stimulated the firing of the unit is suppressed. A typical "off" response follows the stimulus in the lower trace.

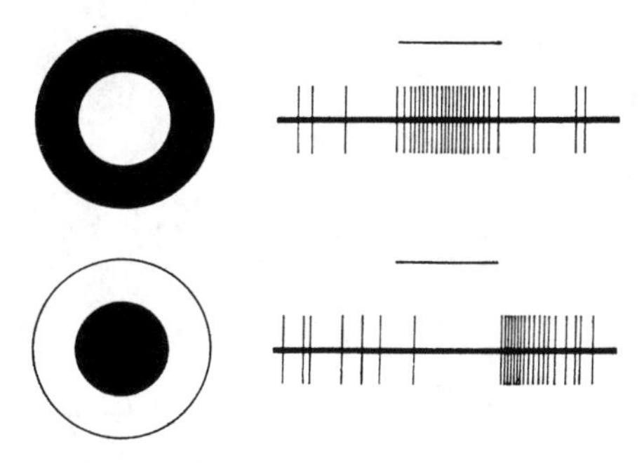

Fig. 125. Recordings from lateral geniculate body showing arrangement of receptive fields. The upper trace reflects a typical "on" response from the center of the field. The lower trace shows suppression of firing from the surround followed by an "off" response. The line above each record indicates when the light is on.

The discharges of geniculate neurons fall into two main categories, phasic and tonic. Phasic units are generally silent in the absence of stimulation and respond to light after a short latency by a burst of spikes. Their influence on surround inhibition is minimal. Tonic units are always spontaneously active. They give a transient burst of spikes when stimulated, followed by a sustained discharge. Activity in the surround of the field is totally suppressed (Fig. 126).

Stimulation by moving light. Nearly all geniculate neurons respond to light and dark borders moving over a wide range of velocities. Phasic units have a clear-cut preference for fast moving objects, while tonic units respond best to slow movements. Both types have a center-surround organization of their receptive fields—either an excitatory center with an inhibitory surround or an inhibitory center with an excitatory surround.

In summary, the cells of the lateral geniculate body are directly activated by the axons of retinal ganglion cells, and their discharges, in turn, form the

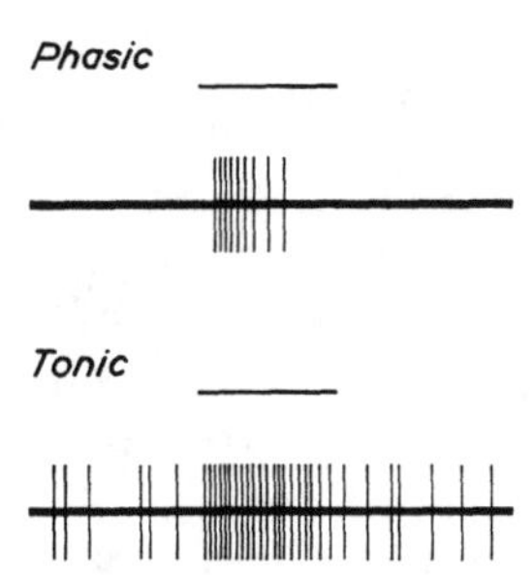

Fig. 126. Comparison between phasic and tonic discharges of lateral geniculate neurons. Upper trace shows response of a phasic unit otherwise silent. The lower trace shows response of a spontaneously active tonic unit. Line above each trace indicates light stimulus to center of field.

principal visual input to the striate cortex. Incoming fibers from the two eyes terminate on different cell layers within the nucleus, where corresponding retinal fields are represented and probably interact. The responses evoked from both stationary and moving light indicate that geniculate neurons have a center-surround organization, similar in many respects to that of the ganglion cells, and that they are particularly sensitive to the contrasting stimuli of light and dark borders.

Striate cortex

Functional anatomy

The primary visual cortex (area 17) occupies the medial surface of the occipital lobe, above and below the calcarine fissure. The right half of each retina connects with the right visual cortex, and the left halves connect with the left visual cortex. The upper quadrant of each retina is represented on the upper lip of the calcarine fissure, extending into the cuneus; the lower quadrants are represented on the lower lips of the fissure, extending into the lingual gyrus. The macula is represented at the occipital pole of the visual cortex and may extend more anteriorly since macular vision is spared when the occipital lobe is damaged. The high degree of topographical organization of the visual system is consistent with the ability to identify specific points in the visual field. It is clear, however, that other aspects of cortical function are fundamental for obtaining a complete interpretation of visual impressions. These aspects include spatial orientation, movement direction, depth, and color, which are served by the distinctive properties of striate neurons.

Types of cells

The striate cortex is made up of densely packed cells arranged in horizontal layers extending from the pial surface to the white matter. Altogether there are six layers containing a large number of functionally different cell types. The principal types are designated "simple," "complex," and "hypercomplex."

1. Simple cells. Hubel and Wiesel classified cells as simple because their receptive fields could be mapped into "on" and "off" areas. Simple cells are widely distributed in the various cortical layers but are most numerous in layer 4. They have little or no spontaneous activity and respond with a brief burst of spikes when light falls on the retina or when a slit of light having a particular angle of orientation is moved slowly across the receptive field. The discharge is presumably the result of the excitatory center component in the receptive field of a geniculate neuron. When light falls on the inhibitory part of the receptive field, the surround zone of the geniculate neuron is activated.This results in suppression of firing of the cortical cell (Fig. 127). Thus the receptive field of a simple cell has both excitatory and inhibitory regions.

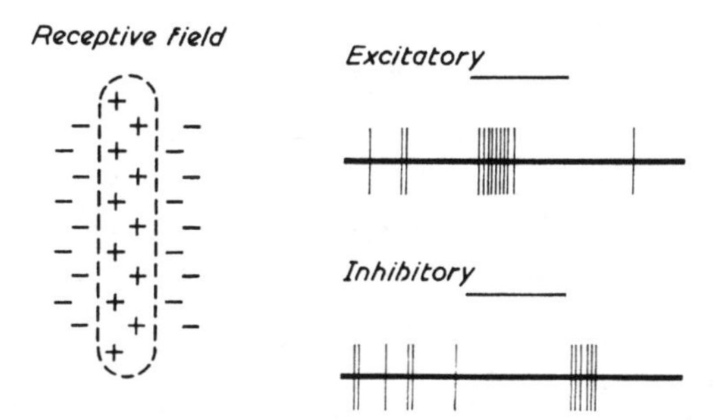

Fig. 127. Receptive fields of a simple striate cell showing a narrow vertical region from which responses are evoked by circular spots of light on the retina (+). This region is flanked on either side by areas giving decreased responses (–). The traces on the right of the figure show corresponding excitatory and inhibitory effects on the firing rate of the unit. The line above each trace indicates when the light is on.

Moving a light stimulus in the visual field is the most effective way of identifying simple cells. If a narrow slit of light is moved across the eye, a brief discharge is evoked each time an excitatory region is stimulated. As the boundaries between the excitatory and inhibitory regions are not concentric as in the retina, the direction of movement is critical. By moving the slit of light in various directions–vertical, horizontal, or oblique–the shape of the receptive fields can be mapped. For example, in the recordings of

Figure 128A, the movement of a vertical slit evokes a discharge while a horizontal movement does not. Some simple cells respond to one direction of movement and are inhibited by movements in the reverse direction (Fig. 128B). It can be concluded that the majority of simple cells possess a sensitivity for a particular orientation of the stimulus which corresponds to the axis of orientation of their receptive fields; also many cells respond to movements in a preferred direction.

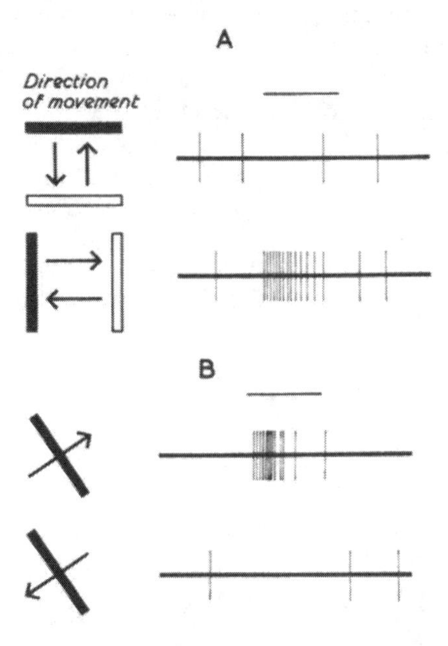

Fig. 128. Responses of a simple striate cell to movements of a slit of light across the receptive field. A, up and down movements of a horizontal slit do not evoke a reponse. Side to side movements of a vertical slit of light give a burst of impulses in each direction. B, obliquely oriented slit evokes a discharge in one direction but not in the opposite direction. Lines above tracing indicate when slit of light crosses the receptive field. (Modified from Hubel and Wiesel.)

2. Complex cells. These cells are the commonest of all the types found in the visual cortex, although they are generally absent in layer 4. They differ from simple cells in that their receptive fields are not rigidly separated into excitatory and inhibitory regions and that they tend to give mixed "on"–"off" responses. Most complex cells exhibit a high resting discharge and respond well to moving stimuli by continuous firing. Diffuse light has little influence on the firing, but the presentation of slits or borders is very effective. There are many different categories of complex cells in regard to their discharge characteristics, firing patterns, and receptive field organization.

However, as simple cells do, they show preferences for stimulus orientation and direction of movement. One category is sensitive to a restricted band of wavelengths and may therefore function in the processing of color vision.

3. Hypercomplex cells. These cells represent a higher order of visual processing and are not directly influenced by the geniculate projections; they may serve as a link between the striate cortex and the visual patterns beyond it. Many hypercomplex cells are located superficially in layers 2 and 3 of the cortex. They are all sensitive to stimulus orientation and, in addition, exhibit selectivity to the length of the stimulus. In other words, when plotting the receptive field of a hypercomplex cell by moving an appropriately oriented slit, there is a region of central excitation bounded on either side by an inhibitory region. The central excitatory region is related to an optimal length of the slit, and any extension beyond this length merely causes suppression of the response. This would suggest that the hypercomplex cell derives an inhibitory synaptic input, possibly from a complex cell in its immediate neighborhood. The view is supported by the action of bicuculline, an inhibitory antagonist, applied iontophoretically on a hypercomplex cell. Bicucullin causes a marked extension of the excitatory region at the expense of the inhibitory region and thus tends to eliminate the property of length detection. Under normal conditions, it would seem that the discharge evoked by a stimulus of given length differs somewhat from the discharge evoked from its boundaries. This is one way in which a high degree of discrimination can be built into the visual system, and its operation allows the cortex to determine exactly the shape and contours of the stimulus.

Columnar arrangement

There is good evidence from microelectrode studies that the various cell types of the visual cortex are arranged in perpendicular columns extending downward from the surface. Impulses arriving in the cortex from the lateral geniculate body spread upward or downward in each column and then to adjacent columns. All the cells within a single column share a common axis orientation of their receptive fields. Simple cells in a column have similar distributions of excitatory and inhibitory regions; complex cells have identical preferences for movement direction. Other properties shared by a column may be concerned with ocular dominance, when the cell is influenced by both eyes, and color coding.

The distribution of striate cells into discrete columns offers certain advantages for decoding information from the retina. To take one example, that of axis orientation, it can be supposed that if one angle of orientation excites a particular column of cells, a slightly different angle will excite an adjacent column of cells. When several columns are activated together, a mechanism is built by which the cortex can detect the angles, form, edges, and contours of the stimulus. As the retina is the main source of input to the striate cortex, the columnar arrangement suggests how the two coordinates of the visual field can be used to deal with three-dimensional aspects of vision.

Summary

The striate cortex contains a number of functionally different cell types arranged in six horizontal layers. The simple type represents the first stage in the processing of impulses from the lateral geniculate bodies and its receptive fields have distinct excitatory and inhibitory subdivisions. The complex type belongs to a higher order of neurons with more elaborate properties; this type exhibits certain preferences for orientation and the direction of movement. The hypercomplex type represents a much later stage of processing and appears to be influenced more by intracortical connections than by afferents from the geniculate projections; although these cells do respond to stimuli of specific lengths. All the different cell types are arranged in a series of vertical columns, each of which serves to extract some specific aspect of the qualities that make up a visual pattern. One quality is the angle of orientation, another is the direction of movement. The effect is similar to that which occurs in other areas of the cerebral cortex whereby information on all the dimensions of a stimulus is selectively decoded.

Abnormalities in the fields of vision

Interruption of fibers in any part of the visual pathway causes loss of sight in the corresponding portion of the visual field. Abnormalities range from a slight loss of visual acuity to total blindness in one eye. The type of defect produced is usually determined by the location of the lesion (Fig. 129).

1. Retina. A retinal lesion may produce a blind spot in the affected eye. Such blind spots are called "scotomata"; they frequently result from toxic or degenerative conditions or from detachment of the retina.

2. Optic nerve. A complete section obviously results in total blindness of the affected eye (unilateral anopia).

3. Optic chiasm. A lesion affecting the lateral portion of the chiasm—e.g.,

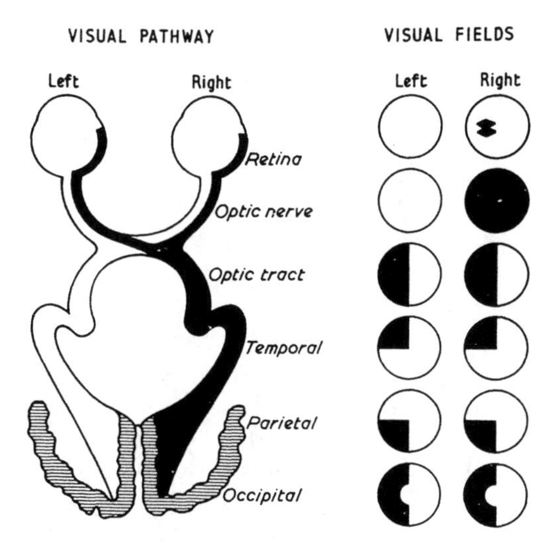

Fig. 129. Diagram of the visual pathways passing to the right cerebral cortex. The dark areas in the visual fields indicate the defects caused by lesions at different sites, as described in the text. A complete lesion of the right optic tract results in a left homonymous hemianopia. Lesions of the right optic radiations produce left homonymous quadrantic defects. A lesion in the right occipital lobe results in a left homonymous hemianopia with macular sparing.

aneurysm of the carotid artery—may interrupt the uncrossed fibers from the temporal half of the retina without damaging the central fibers. This causes loss of vision in the nasal half of the visual field on the same side (unilateral nasal hemianopia). A central lesion of the chiasm—e.g., cyst or tumors of the pituitary gland—may interrupt the decussating fibers from the nasal half of each retina without damaging the uncrossed fibers. This causes loss of vision in both temporal fields (bilateral temporal hemianopia).

4. Optic tract. A complete lesion of one optic tract interrupts the uncrossed fibers from the temporal half of the ipsilateral retina and the crossed fibers from the nasal half of the opposite retina. This causes loss of vision in corresponding halves of the visual fields (homonymous hemianopia).

5. Optic radiations. Lesions in this part of the visual pathway usually produce quadrantic defects. An abnormality in the temporal lobe may interrupt fibers from the lower portions of each retina. This causes loss of vision in the upper quadrants of both visual fields on the side opposite the lesion. An abnormality in the parietal lobe may interrupt fibers from the upper portions of each retina. This causes loss of vision on the contralateral side in the lower quadrants of both visual fields. These partial defects always produce homonymous loss of vision because corresponding fields of vision are affected.

6. Occipital lobe. A lesion of one occiptal lobe–e.g., thrombosis of the posterior cerebral artery–may cause contralateral loss of vision in the corresponding half of each visual field, but central vision remains intact. This is known as "macular sparing." Thus a lesion in the right occipital lobe results in a left homonymous hemianopia except for the macular area.

FUNCTIONS OF THE VISUAL CORTEX

Most of our ideas about the functions of the visual cortex have come from experimental analysis of single unit activity in the vertical columns already described. The properties of these columns will now be examined in terms of the functions they perform–that is, the detection of brightness, detail, movement, and depth. (The processes involved in color vision will be considered in a separate section at the end of this chapter.)

Brightness

The relative brightness and darkness of objects in the visual field is expressed as differences in the rate of firing of the columns whether continuous or intermittent light is used.

Continuous light. A change in the level of illumination affects both the excitatory and inhibitory inputs to the cortical columns. An increase of light intensity causes an increase in the firing rate of on-center excitatory regions and a decrease in the rate of off-center surrounds. At very high intensities, the firing rate may begin to fall owing to light adaptation. Conversely, a decrease of light intensity causes excitation of the surround and inhibition of the center. Thus the contrasts formed by edges and borders and the highlights and shadows of the visual scene are detected by reciprocal interactions between two antagonistic systems–the excitatory and inhibitory regions of the receptive fields. Both systems operate in the retina and lateral geniculate body and continue at least to the level of the simple cells in the striate cortex.

Intermittent light. Neurons in the visual cortex respond to successive flashes of light when the subject looks at a rotating disc, with alternating black and white bands. A stroboscope is more useful for testing because the brightness as well as the duration of the flashes can be finely controlled. At low rates of stimulation the neurons respond to each flash, but at higher repetition rates, the successive flashes become fused to give the impression of being continuous. The transition from intermittent light or flicker to continuous light is known as the critical fusion frequency (cff) and depends upon a number of factors:

1. Level of illumination. Fusion occurs more readily at low intensities of illumination when the rate of flicker is about 10 Hz. At high intensities, the cff rises to about 50 Hz.

2. Light and dark fractions. Fusion occurs more readily when the light and dark fractions are varied and less readily when the alternating phases of each cycle are equal.

3. Part of retina stimulated. The rods are less sensitive to rapid changes in illumination than the cones. Consequently, the cff values for the scotopic and photopic systems can be differentiated. Fusion occurs more readily with flashes directed on the periphery of the retina than on the fovea (Fig. 130).

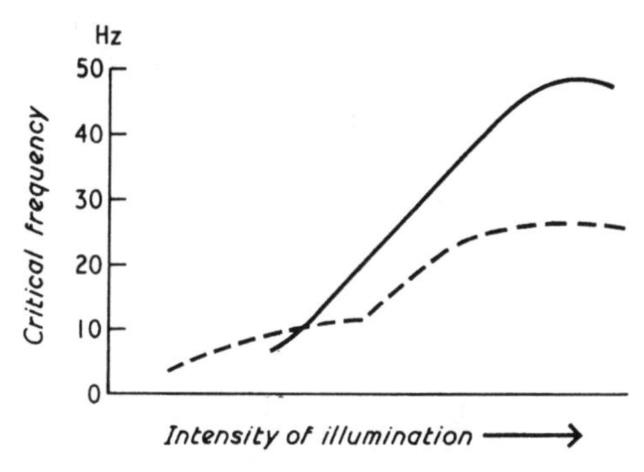

Fig. 130. Critical fusion frequency. Curves show relationship between level of retinal illumination and frequency of light flashes., Continuous line indicates the response to stimulation of fovea; interrupted line, the response to stimulation of periphery. Note that fusion occurs more readily at low intensities of illumination.

Rapid flicker appears as continuous light because excitation persists over a fixed period of time. This persistence effect is used in the technique of motion pictures. Flicker is eliminated when the rate of presentation exceeds a certain value.

Detail

The ability of the visual cortex to detect fine detail is dependent partly on the spatial organization of the macular projections and partly on the selectivity functions of the vertical columns. The highest visual acuity is found in the fovea where the cones are densely packed, each cone possessing a distinct excitatory and inhibitory receptive field. Many of the cones have their own private pathway to the cortex, connecting with a single ganglion cell, in contrast to the considerable convergence of the rods. Cones placed side by side are sensitive to the different levels of brightness created by the sharp borders of the visual pattern.

Macular projections. The topographical relationship between the macula and the optic radiations is faithfully preserved. In particular, the amount of cortex devoted to the foveal region is relatively large. The ratio between the number of ganglion cells and the volume of cortex for each degree of the visual field is called the "magnification factor."

Selectivity functions. Each vertical column of the striate cortex represents a restricted dimension of the retinal image. The column responds to a specific orientation of a line or border when that line lies along the axis of the excitatory region of its receptive field. If the line or border extends beyond this into the inhibitory region of the field, the firing rate of the column is reduced or suppressed. Different orientations are found for different columns. When several columns of cells act together the resultant activity produces various profiles of excitation and inhibition from which the cortex analyzes the fine detail and contours of the retinal image.

Movement

Information about the direction and velocity of movements is transmitted to the cortex under two sets of circumstances—when the eyes are fixed on an object and when the eyes are moved.

Eyes fixed. The detection of movement depends upon the gliding of the image over the retina when the object changes its position in the visual field, but stays at the same distance from the eyes. If the object comes nearer, an increasing area of each retina is stimulated. Most of the cells in the visual cortex are influenced by moving stimuli and appear to be sensitive to both the direction and velocity of the movement; they may be classified as follows:

1. Non-oriented. Cells respond to a moving stimulus irrespective of direction.
2. Oriented. Cells respond to a stimulus moving in a preferred direction.
3. Axis-oriented. Cells respond to a movement along a preferred visual axis.
4. Contrast-oriented. Cells respond to a movement between light and dark borders.
5. Size-oriented. Cells respond to an object approaching the eyes or moving away from them.

Movement is an effective stimulus to all types of cell in the striate cortex. Simple cells give "on" and "off" responses when a slit or edge moves respectively across the excitatory and inhibitory regions of their receptive fields. Some cells respond equally to opposite movements, while others respond only to movement in one direction. Complex cells have a preferred axis orientation, responding to movements along a particular axis but often failing to discharge if the angle is changed slightly. Hypercomplex cells also have a preferred axis for movement as determined by the distribution of their receptive fields. As a general rule, complex and hypercomplex cells can follow movements at high speed without loss of visual acuity, but simple cells respond best to slowly moving objects.

Eyes moving. The extraocular muscles play an important role in maintaining the image on the central portion of the retina. The movements of the eyes are extremely rapid and precise. Voluntary movements are controlled by the cortical eye fields of the frontal lobes. They allow the gaze to be fixed at will on any point in the visual field and then to move to another point. In this way the eyes can follow the course of a moving object and keep it in focus with the minimum of error. This ability of the extraocular muscles is called "pursuit movement" and no doubt requires the cooperation of the cerebellum.

In addition, the eyes are capable of viewing details of the visual scene by a process of involuntary scanning. Involuntary movements are controlled by the occipital lobe. The impulses pass to the superior colliculus and from there are distributed to the oculomotor nuclei. During reading from lines of print, or when a large painting is inspected, the eyes move in brief jerks called *saccades.* As each line is scanned, the eyes drift to a new position and automatically flick to the next fixation point. Saccadic eye movements are executed smoothly and with constant velocity owing to the series elastic component of the muscles.

The visual cortex learns to discriminate between impulses from the retina and impulses from the extraocular muscles, although illusions of movement are common enough in everyday experience. Thus a person sitting in a stationary train may get the impression that he is moving when he observes the departure of another train from the platform. Nevertheless, the primary aim of all these mechanisms is to maintain a stabilized image at the fovea whether the movement is real or apparent and whether the eyes remain fixed or are moved in pursuit.

Depth

The appreciation of three-dimensional space is possible with monocular vision to a limited extent. This is largely due to a knowledge of the true size of the objects observed, from which one can tell the relative distances between them. Depth can also be estimated from the geometrical arrangement of lines that appear to converge to a point on the horizon. The rapid movement of objects nearby can be compared with the slower movements of objects on a distant plane, an effect known as parallax. Another cue that can be used is that distance tends to modify the color of the object. Monocular vision, however, is far from perfect as seen, for example, when attempting to thread a needle with one eye closed or when estimating the distance and speed of an approaching vehicle. The accurate judgment of distance, depth, and movement in space is called stereoscopic vision. It is made possible by overlapping of the fields of vision of the two eyes causing slight disparities between the images formed on each retina.

Principles of binocular vision

1. Fusion of retinal images. When the two eyes are used in normal vision, the extraocular muscles converge the visual axes on a fixed point in space. Within this common visual angle the two retinal images are partly overlapped. A curve passing through the optical centers and the fixation point is called the *horopter* and light passing from any point on the curve will stimulate corresponding points on each retina. Consequently, the images formed in the two eyes appear as a single image. The limits of the area for single vision is known as Panum's fusional area, beyond which diplopia is experienced. Points outside the horopter, either closer to the observer or further away, will project on non-corresponding points on each retina and thus be seen as double.
2. Binocular rivalry. This term is usually applied to the situation in which the images of the two eyes are sufficiently different to make fusion impossible. Thus if vertical lines are seen by one eye and horizontal lines by the other, only one field is seen at a time; or if a red filter is used for one eye and yellow for the other, the colors do not fuse to give orange. The suppression of one image generally results in loss of the ability to discriminate differences in depth.
3. Retinal image disparity. Because the two eyes are horizontally separated by the bridge of the nose, the images formed on each retina are not precisely the same. This can be proved quite easily by looking into the distance while holding an object in front of the nose. It will be noticed that

the object appears double. If the right eye is closed, the object is seen on the right and when the right eye is opened and the left eye is closed, the object appears to move to the left. The experiment shows that there are disparities between the two retinal images due to the fact that non-corresponding points are stimulated. Diplopia is experienced when the disparity extends beyond the dimensions of Panum's area. It is thought that the detection of small differences in the images formed on each retina is the basis of stereoscopic vision.

Neural mechanisms in binocular vision

Our present understanding of depth discrimination began with Hubel and Wiesel's observation that single units in the striate cortex could be driven from both eyes and that is was possible to plot separate receptive fields for each eye. Previously, Hering had postulated that the fusion of dissimilar images might be the basis of distance perception but he did not attempt to apply the theory to receptive field organization.

1. Anatomical findings. The decussation of the optic nerve fibers at the chiasm brings together in the same visual pathway the central projections of corresponding fields of vision. Thus inputs from the two eyes passing through the lateral geniculate body ensure that corresponding retinal areas are kept closely associated with one another in the striate cortex. In other words, the image from each part of one retina is accurately paired with the image from the other retina. Therefore, it can be expected that at least a proportion of the cells in the striate cortex will be influenced equally by the two eyes.

2. Receptive field disparity. When the receptive field of a cortical unit is mapped, first for one eye, then for the other, it is found that the maps are identical for some units and slightly different for others. Accordingly, it must be supposed that some units are unequally stimulated and that the mixing of influences from the two eyes is far from complete. One population of units has exactly the same receptive fields and another population has fields that are out of alignment. The amount of non-correspondence between the two fields is termed "receptive field disparity."

3. Optimal stimulus. An object in the plane of the horopter forms two identical images on each retina and therefore provides the optimal stimulus for those cortical units whose receptive fields are in perfect alignment. When the corresponding parts of the two receptive fields are stimulated, summation occurs and the units discharge at their maximal rate.

An object outside the plane of the horopter forms two slightly different images which provide the optimal stimulus for those units tuned for non-corresponding receptive fields. Such units also discharge at their maximal rate. Thus distance is encoded in the discharge patterns of the cortical units responding to small differences in the images formed on each retina. Each unit can be regarded as a disparity detector, responding best to an object seen at a particular distance from the eyes.

4. Columnar organization. There is some evidence that cortical units showing the same disparity preferences may belong to the same cortical column. That is to say that all the cells in a vertical column are optimally stimulated by objects at the same distance from the eyes. Blakemore suggests that there may be two types of binocular columns–constant depth and constant direction. In a constant depth column, all the cell types have identical preferences for disparity between the two retinal images that occurs as a result of horizontal separation of the eyes. In a constant direction column, horizontal disparities vary from cell to cell, but all the cells have identical axis orientations and identical preferences for the direction of a moving object. The evidence suggests that the two kinds of column are capable of detecting depth and direction of moving targets in their receptive fields. In order for the brain to discriminate between different depths and orientations of the visual scene, it is only necessary to have enough columns to represent a larger binocular field. Interactions between the different types of cell within a column and interactions between neighboring columns may provide the neural basis for building up many of the features of three-dimensional space.

5. Binocular depth neurons. Certain units in the striate cortex fail to respond to monocular stimulation, but give a very brisk response to simultaneous stimulation of the two eyes. Such units are more commonly found in complex and hypercomplex populations; they probably serve the more general functions of depth discrimination–for example, the judgment of distance and speed of on-coming traffic exercised by a driver before overtaking.

Summary

While judgment of depth is possible with monocular vision, true stereoscopic vision requires a degree of overlap in the visual fields to produce slightly different retinal images. Information about the three dimensions of space is transmitted to the visual cortex in terms of these retinal disparities. Binocular units all respond specifically to the distance of an object from the eyes, and moving objects must also be at a particular orientation. If the optimal distance or orientation is changed, unit firing falls off or ceases completely. Binocular units are arranged mostly in vertical columns each of which is selectively tuned for the recognition of disparity. Interaction between the mosaic of columns is the probable neural basis for discriminating the details of visual space.

COLOR VISION

The detection of color is performed partly by the retina and partly by the brain. Persons with normal color vision can discriminate between two lights of different spectral composition that are equal in brightness, whereas color-defective individuals rely on a brightness difference. This fact suggests that color information is processed by two separate systems, one based on spectral sensitivity for a narrow band of wavelengths and the other depending on brightness information from a broad band of wavelengths. In describing the mechanisms of color vision, it is useful to begin at the photoreceptor level in the retina, where light rays are aborbed selectively by different cone pigments. An account will then be given of neural processing in the visual pathway with particular reference to the cells of the lateral geniculate body, as these have been subjected to detailed investigation. Finally, it is important to consider the causes of defective color vision in man and the tests that are used for its detection.

Retinal mechanisms

Wavelengths of light

The visible spectrum is composed of a band of wavelengths ranging from 700 nm to 400 nm. Newton was the first to show that if a beam of sunlight is allowed to pass through a glass prism on to a screen, the rays will be bent differently and appear to the eye in different colors (Fig. 131). The long wavelengths are bent the least and appear red, the short wavelengths are bent the most and appear blue, while the intermediate ones appear green. Outside this range, the wavelengths are invisible to the human eye. Artists do not need an infinite variety of paints in order to obtain a desired color, but achieve their results by delicate mixtures. Theoretically, white can be represented by a mixture in equal proportions of red, green, and blue, and these are called the primary colors. By varying the brightness as well as the relative proportions of the three primary colors, it is possible to reproduce any shade or hue, and by adding white to a particular mixture, the color

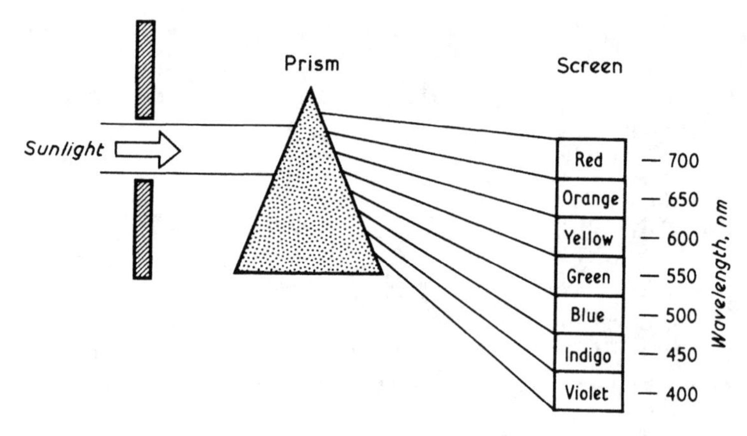

Fig. 131. Colors of the visible spectrum. Light passed through a prism is split into individual wavelengths, each of which appears as a color when projected on a screen.

becomes paler or less saturated. Accordingly, we now accept the ideas that color vision is trichromatic and that the changes set up in the retina are due to specific wavelengths of the visible spectrum.

Cone pigments

It is generally assumed that the rod pigment, rhodopsin, is not involved in color vision despite arguments that it may contribute to processing at the blue end of the spectrum. On the other hand, the cone pigments, which operate mainly at high levels of illumination, possess sensitivities for monochromatic light with peak absorptions limited to a narrow part of the spectrum. The absorption bands can be demonstrated by spectrophotometry. If an intense red light is shone on the retina, the red pigment alone is bleached, leaving a characteristic deficiency at the long end of the spectrum. Similarly, an intense blue light deactivates the short end of the spectrum. It can be seen from Figure 132 that there is a peak sensitivity for each kind of cone pigment–red, green, and blue at 570, 535, and 445 millimicrons, respectively.

Theories of color vision

1. Trichromatic theory. The Young-Helmholtz theory is based on the assumption that the three primary colors–red, green, and blue–are matched by three types of cones and that color sensation is determined by

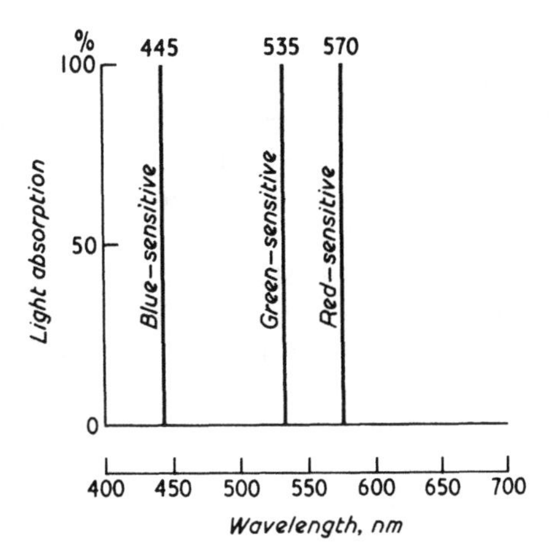

Fig. 132. Light absorption of the respective pigments of the three color-receptive cones in the human retina. Peaks of light absorption are shown as a maximum percentage occurring at specific points in the spectrum. (Modified from Rushton.)

the ratio of these three cone outputs. The sensation of white arises when all three are equally excited; intermediate shades of color are produced by different sets of ratios. Thus red light stimulates the cones in the ratios 75:13:0, green light stimulates in the ratios 50:85:15, and blue light in the ratios 0:14:86. Equal stimulation of red and green cones produces the sensation of yellow. The trichromatic theory satisfies most aspects of the spectral sensitivities of color defective vision.

2. Opponent theory. In its original form the theory assumes that photochemical pigments exist as three complementary pairs—red-green, yellow-blue, and white-black. Hering proposed that each pair responds to different wavelengths of light by producing opposite effects on the photoreceptors. Impulses are set up when the pigment is bleached and again when resynthesis takes place. For example, in the red-green pair, when breakdown of pigment is in excess of resynthesis, the effect produced is red; when resynthesis is in excess of breakdown, the effect is green. Comparable interactions take place for the yellow-blue and white-black complementary pairs. Although there are a number of objections to this theory, particularly the need for a white-black substance, Hering's proposal clearly anticipated modern ideas of color processing by spectrally opponent elements in the visual pathway.

Slow potentials

Further evidence that color processing begins in the retina has come from intracellular recordings of horizontal cells during stimulation by light. Horizontal cells give slow hyperpolarization potentials that spread laterally to the bipolar cells, causing an inhibitory surround at each excited point. Some horizontal cells are sensitive to specific wavelengths of light, giving a large-amplitude response to red light and a smaller response to green. Other horizontal cells show similar amplitude changes for the yellow and blue parts of the spectrum. These responses are called S-potentials, named after their discoverer, Svaetichin.

Dominator-modulator curves

In extensive work on a large number of animals, Granit confirmed the view that the retina contains three main types of color-sensitive elements. Recording with microelectrodes from single ganglion cells and optic nerve fibers, he was able to construct sensitivity curves for white and monochromatic flashes of light. The curves were of two kinds: (1) A dominator curve, resembling the photopic visibility curve, showed the response to stimulation by a broad band of wavelengths. (2) Modulator curves were each sensitive to a narrow band of wavelengths in three preferential regions of the spectrum—red, green, and blue (Fig. 133).

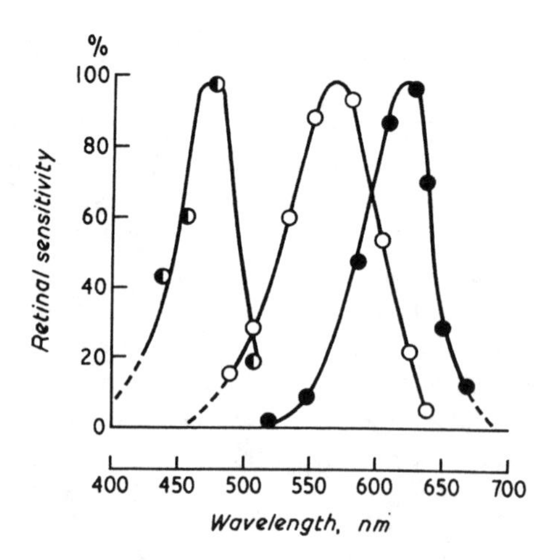

Fig. 133. Sensitivity curves of the dark-adapted cat's retina plotted as the percent of the maximum against the wavelengths of light. Filled circles represent red modulators; open circles, green modulators; half-filled circles, blue modulators. (From Granit.)

Although modulators have been reported in many animal species by other investigators, Granit's work has been criticized on the grounds that it did not take into account intensity differences in the flashes of light. The response curves could have been carrying information on brightness, and not just on the color of the light. Another source of error was that the inhibitory components of the responses were not considered. Thus color-sensitive elements may be excited by one wavelength and inhibited by other wavelengths, and many complex interactions must take place before the signal is transformed into the all-or-nothing spike potential.

Color matching

The color of one light can be matched to resemble the color of another light by mixing the output of the three cone pigments. If the brightness of one light is changed, the mixture of the pigments must also be changed. Mixtures that match each other are all represented by the relative proportions or "catch" of the individual cones. When two or more lights are mixed, the resultant color is formed from the total quantum catch. White light will produce equal catches in all three pigments.

In summary, the peripheral apparatus for color vision is to be found in the retina itself. The suggestion originally made by Young that there are three different kinds of cones is generally accepted as the mechanism on which our sense of color depends. The spectral sensitivities of the different cones are essentially the same as the light absorption curves of their respective pigments. Color therefore depends upon the relative quantum catch in the three cone pigments, which in turn is determined by their brightness thresholds.

Central mechanisms

The extent to which a person is capable of recognizing color depends upon the impulse patterns that are delivered by the optic nerves to the brain. Activity originates in the cones and spreads as amplitude-variant potentials when light goes on or off. Horizontal cells hyperpolarize to one color and depolarize to another. Amacrine cells give a transient depolarization in response mainly to changes in brightness. The bipolar cells act as a link between the photoreceptors and the ganglion cells, balancing one potential against another; but only the ganglion cells encode the net result into a frequency-variant spike signal (Fig. 134). The retina may therefore be regarded as the first stage in the coding of color events, and the optic

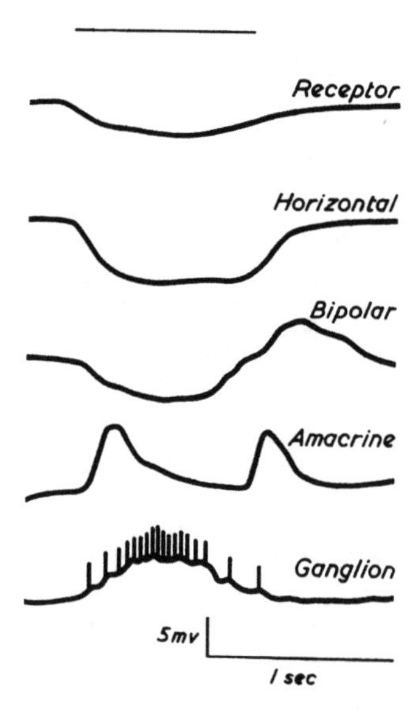

Fig. 134. Intracellular potentials recorded from different layers of the retina. Rods and cones and horizontal cells hyperpolarize (positivity downward). The bipolar cell hyperpolarizes when light is on and depolarizes when light goes off. The amacrine cell gives a brief depolarization to "on" and "off." The ganglion cell responds by a steady stream of nerve impulses. Line above tracings indicates when light is on. (Modified from Rushton.)

nerve fibers as the second stage. Further transformations take place in the lateral geniculate bodies where color information is appropriately channeled before being transmitted to the visual cortex.

Lateral geniculate body. Microelectrode studies of the lateral geniculate body indicate that there are two classes of cells—broad band and spectrally opponent cells. The former transmit white light signals and the latter color signals.

1. Broad band cells. These may be divided into two general types: *Excitatory* cells respond to light of all wavelengths; their firing rate is increased when the intensity of light is increased. *Inhibitory* cells respond to light of all wavelengths; their firing rate is decreased when the intensity of light is increased. Broad band cells are comparable to Granit's dominators and appear to carry brightness information. They are probably not involved in color discrimination.

2. Spectrally opponent cells. These cells have both excitatory and inhibitory components, responding to some wavelengths by an increased discharge and to other wavelengths by a decreased discharge. They may be subdivided into red-green and yellow-blue pairs:

Red-green opponent cells

• Red excitatory and green inhibitory (+R–G). This type of cell gives a maximal discharge, with wavelengths around 630 nm, and a minimal discharge with wavelengths around 500 nm.
• Green excitatory and red inhibitory (+G–R). This type of cell gives a maximal discharge, with wavelengths around 500 nm, and a minimal discharge with wavelengths around 630 nm.

Yellow-blue opponent cells

• Yellow excitatory and blue inhibitory (+Y–B). This type of cell gives a maximal discharge, with wavelengths around 600 nm, and a minimal discharge with wavelengths around 440 nm.
• Blue excitatory and yellow inhibitory (+B–Y). This type of cell gives a maximal discharge, with wavelengths around 440 nm, and a minimal discharge with wavelengths around 600 nm.

The responses of opponent geniculate cells are the result of interaction between the excitatory and inhibitory components of two spectrally antagonistic systems originating at the ganglion cell level. They are similar to the modulators reported by Granit, although he failed to detect the inhibitory input from the opposite spectral region. It may be concluded that while broad band cells signal increases and decreases of light intensity by means of proportional changes in firing rate, spectrally opponent cells signal changes to longer or shorter wavelengths of light and therefore represent the mechanism that carries information about color.

Striate cortex (area 17). Surprisingly, cells in the striate cortex with color-selective properties are relatively difficult to find. Hubel and Wiesel found columns of simple cells responsive to color, but only a few complex and hypercomplex cells having opponent color qualities. The scarcity of color-sensitive neurons in the striate cortex suggests that much of the information on narrow wavelength bands has already been processed at the geniculate level. Alternatively, the information received by the cortex may be used to detect form and movement independent of the color patterns being carried. On the evidence at hand, it seems likely that all channels of visual information combine at the cortical level, where the analysis of color has already been largely completed.

Prestriate cortex (areas 18 and 19). In contrast to the primary visual area, color-coded cells are relatively abundant in the prestriate cortex, the highest order of detection. The cells belong to the complex and hypercomplex types and many of them have preferred stimulus orientations. Their receptive fields have both excitatory and inhibitory components. One group of cells responds vigorously to light of a preferred wavelength; another group responds only to a change of wavelength, for example, from red to blue or from green to red. The important point to emphasize here is that the cells are grouped together according to their preferences for specific aspects of the visual stimulus, particularly the subtle changes of color between the boundaries and edges of three-dimensional patterns.

DEFECTIVE COLOR VISION

In the great majority of cases defective color vision is congenital, resulting from absence of appropriate color genes, and cannot be corrected. The recessive trait is passed by the father to his daughters who will be carriers, although few females are themselves color-defective. The total number of hues that can be recognized by an individual with defective color vision is much smaller than the number recognized by a person with normal color vision. The affected individual can match colors accurately only when there is a brightness difference. In poor conditions of light, the number of errors made is greatly increased. The disability arises from a lack of one or more cone pigments, most defects occurring in the red and green parts of the spectrum. If the red pigment is missing, the ratio of stimulation of the different cones does not change as the color changes from green to the red end of the spectrum. Defects in the yellow and blue regions are less common. Total color blindness is a rare condition in which the three cone pigments are absent and the subject sees the world virtually in black and white.

Classification

1. Trichromats. All three cone pigments are present, but the individual uses them to match a color in different ratios from those used by the person with normal color vision.

- Protanomaly. The individual has a deficiency of the red pigment and consequently requires more red than normally required when mixing red and green to produce yellow. His match will appear orange to the person with normal color vision.
- Deuteranomaly. The individual has a deficiency of the green pigment and consequently requires more green than normally required for color

matching. Such a person cannot recognize all the different hues in the intermediate part of the spectrum.

• Tritanomaly. This is a rare condition in which the individual confuses colors at the short end of the spectrum.

2. Dichromats. One of the cone pigments is lacking in these individual so that they must depend upon the other two for color discrimation.

• Protanopia. The individual has only one pigment for the red-green part of the spectrum and requires red light to be very intense to match a green color. He is unable to recognize hues between red and yellow and thus is aptly termed "red-blind." The protanope confuses all colors in the red-green range.

• Deuteranopia. Absence of the green pigment means that the individual must use red and blue pigments for all his color matching. Since green occupies the intermediate part of the spectrum, it follows that the overall range of wavelengths is not very different from the normal range.

• Tritanopia. Absence of the blue pigment means that the visual spectrum is noticeably shortened, and all color information is carried by the red-green range.

3. Monochromats. When only one pigment is present, color mixing is impossible. The spectrum is recognized only by variations in brightness intensity.

Tests for color vision

The tests should be given at school age before children may be thinking about careers requiring a high standard of color vision. Early detection of a defect may save much disappointment later. Examples of tasks in industry requiring color vision are the matching of cables and paints, the sorting and blending of textiles, and the recognition of color signals in transportation.

Most tests of color vision depend upon a person's ability to judge correctly the degree of contrast between colors and, of course, his ability to match one color with another. The cone pigment triangle illustrated in Figure 135 shows how a particular color may be confused with other colors when one pigment is defective. The three cone pigments R, G, and B represent the three corners of the triangle, and the position of the spectral wavelengths is marked in it with white at the center. A person with normal color vision can distinguish any color P from all the colors plotted on a straight line from P to each corner of the triangle. The protanope confuses P with the colors found on the line PR; the deuteranope confuses P with the colors on the line GP, and the tritanope confuses P with the colors on the line BP.

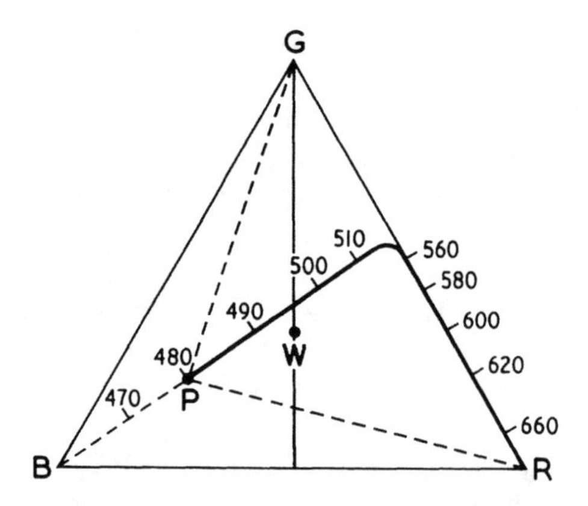

Fig. 135. Cone pigment triangle representing the three cone pigments R, G, and B at each corner. The position of white (W) is determined from the quantum catches of all three pigments. The normal eye sees the color P at the point on the spectral wavelengths marked in the triangle and also sees all the colors represented on the dotted lines PR, PG, and PB. The protanope lacks the red cone pigment and confuses P with the colors represented on PR. The deuteranope lacks the green cone pigment and confuses P with the colors represented on PG. The tritanope lacks the blue cone pigment and confuses P with the colors represented on PB. (After Rushton.)

Thus for each kind of dichromat, the lines of confusion radiate from the corner of the missing pigment.

1. Ishihara charts. This is the simplest and most rapid of many tests used for initial screening. The charts are arranged with numbers or designs composed of colored spots printed on a background of spots made up of confusion colors. Several shades are used to reduce the effect of brightness differences. A person with normal color vision reads all the charts correctly, while a color-defective individual sees a different number or design.

2. Holmgren wool test. The individual is given an assortment of skeins of different colors and asked to place those of the same color in separate piles. The person with normal vision recognizes all the various hues. However, if the individual is red-blind, he places red and orange in the yellow and green piles.

3. Anomaloscopes. These are tests in which the individual is required to match two fields of light for both hue and intensity. The observer rotates a drum, which presents a color mixture, and the subject attempts to match the field by rotating a second drum. For example, if red and green lights are adjusted in intensity to look like yellow, the deuteranope will match the field with a mixture that appears green to the observer.

4. Edridge Green lantern. The instrument can be used in two ways:

- As a spectroscope. The individual moves a shutter to the side of the spectrum where he sees red begin, and the position on the wavelength scale is noted. The shutter is then moved along the spectrum, and the wavelength noted for each color recognized. A color-defective person will map out fewer spectral strips than the person with normal vision, and his spectrum may be shortened at the red or blue end.

• As a filter. Various colors are presented to the subject through apertures of different size. The colors can be modified by filters to represent color identification under certain difficult conditions—e.g., haze. This method is useful for testing individuals in circumstances in which safety and accuracy are important factors—recognizing signals from a distance, for example.

SELECTED BIBLIOGRAPHY

Blakemore, C. The representation of three-dimensional visual space in the cat's striate cortex. *J. Physiol.* 209:155-178, 1970.

Brindley, G.S. *Physiology of the Retina and the Visual Pathway.* London: Edward Arnold, 1970.

Cowey, A., and Gross, C.G. Effects of foveal prestriate and inferotemporal lesions on visual discrimination by Rhesus monkeys. *Exp. Brain Res.* 11:128-144, 1970.

Creutzfeldt, O., and Ito, M. Functional synaptic organization of primary visual cortex neurones in the cat. *Exp. Brain Res.* 6:324-352, 1968.

Gouras, P. Trichromatic mechanisms in single cortical neurons. *Science* 168:489-492, 1970.

Hubel, D.H., and Wiesel, T.N. Integrative action in the cat's lateral geniculate body. *J. Physiol.* 155:385-398, 1961.

Hubel, D.H., and Wiesel, T.N. Receptive fields and functional architecture of monkey striate cortex. *J. Physiol.* 195:215-243, 1968.

Nikara, T., Bishop, P.O., and Pettigrew, J.D. Analysis of retinal correspondence by studying receptive fields of binocular single units in cat striate cortex. *Expl. Brain Res.* 6:353-372, 1968.

Rushton, W.A.H. Pigments and signals in colour vision. *J. Physiol.* 220:1P-31P, 1972.

Schneider, G.E. Two visual systems. *Science* 163:895-902, 1969.

CHAPTER XVIII

Central Processes of Hearing

When recording from any neural system, it is common laboratory practice to switch off the loud-speaker in order to ensure that an evoked response is not being influenced by an auditory stimulus. This precaution arises from the fact that many parts of the central nervous system receive inputs of auditory origin that are not directly concerned with the analysis of hearing. As a corollary, the auditory pathway itself is very complex, for in addition to its main sensory projections, there are many ascending and descending tracts whose functions are primarily reflexive in nature.

Sound plays a very important role in the environment. As a means of communication, it reaches its highest expression in man with the development of speech, language, and musical appreciation. Further, sound has a considerable influence on behavior. It can produce a state of contentment or annoyance or provoke widespread bodily reactions that sometimes may be essential for survival. It is not surprising, therefore, that the auditory input is carried over a multiplicity of nervous channels. The central mechanism attempts to narrow down this flow of information to relevant and useful signals which, in the end, provide a hearing ability of immense range and sensitivity.

THE AUDITORY PATHWAY

The internal ear is supplied by the auditory division of the eighth cranial nerve. On entering the pons, the nerve divides into two branches, one terminating in the ventral cochlear nucleus and the other in the dorsal cochlear nucleus. Second-order neurons from these two nuclei ascend in the lateral lemniscus to the inferior colliculus of the midbrain. As they ascend, many of the fibers give off collaterals to the reticular formation and cerebellum. From the inferior colliculus the pathway passes to the medial geniculate body and then via the auditory radiations to the auditory cortex in the temporal lobe.

Lower auditory centers

Impulses from the cochlear nuclei are transmitted to a number of bilateral structures in the brain stem from which collaterals pass to the cerebellum and spinal cord. They constitute a lower level of neural integration and serve such features of the acoustic organization as sound direction, filtering effects, and reflex muscular reactions.

1. Coclear nuclei. The fibers of the auditory nerve maintain their tonotopic arrangements within the subdivisions of these nuclei. Many of the cells are spontaneously active, but respond to tonal stimulation by discharging spikes at a frequency proportional to the sound intensity. Other cells exhibit strong inhibitory tendencies.

- The cells of the ventral cochlear nucleus respond more or less faithfully to inputs from the cochlea with respect to changes in sound frequency and amplitude. Their axons run medially in the pons as the trapezoid body, decussating with those of the opposite side before entering the lateral lemniscus.
- The cells of the dorsal cochlear nucleus have more complicated response characteristics, many exhibiting a reduction in firing rate with increase of intensity and others giving a mixture of excitatory-inhibitory patterns. These observations strongly suggest that dorsal cells have filtering properties in which wide band noise signals are actively suppressed. Their axons cross the midline along the floor of the fourth ventircle to join the lateral lemniscus.

2. Nuclei of the trapezoid body. Cell groups within the trapezoid body give off collaterals to the medial longitudinal bundle through which impulses are transmitted to the motor nuclei controlling movements of the head and neck.

3. Superior olivary complex. These cell groups comprise a number of well defined masses, including a prominent S-shaped segment or lateral nucleus, the accessory olive, and the nuclei of the lateral lemniscus. They form relay stations on the main afferent pathway from the cochlear nuclei to the inferior colliculus. As many olivary neurons are affected by stimulation of either ear, they may play an important role in the localization of sound sources. Other cells in this group give origin to the olivocochlear tract—a descending system of fibers running back in the auditory nerve to the sensory hair cells of the organ of Corti, on which they exert an inhibitory influence. In this way, a mechanism is provided by which the cochlea may achieve some degree of auto-regulation.

4. Cerebellum. Auditory representation in the cerebellum was first demonstrated by Snider and Stowell, who recorded surface responses to click stimuli in the decerebrate cat. Single unit analysis later revealed that the

responses were localized in the lobulus simplex and tuber vermis of the posterior lobe, a region overlapped by visual and cutaneous representations. The cerebellum derives its information from a wide variety of receptors that are directly or indirectly concerned with the control of the locomotor system.

Functions of lower auditory centers

This section will cover the more general features of the auditory pathway, by which sound impresses itself on other systems of the body. Sound is capable of eliciting reflex actions in the motor systems. The source of a sound can be localized with a remarkable degree of accuracy. The signals generated by sound can be attenuated by centrifugal control.

1. Auditory reflexes. The muscles operating the external ear are well developed in animals, although they possess very little action in man. An animal will respond to sound by raising its ears, probably to help it to localize the source and direction of the sound. Loud sounds tend to discharge a large number of auditory nerve fibers. These constitute the contribution to the neural network of the brain stem, through which impulse activity is transmitted to the spinal motoneurons. Movement patterns elicited by sound are coordinated by the cerebellum, while the reticular formation plays its part in regulating changes in muscle tone and posture. The lower auditory centers are thus responsible for mediating instantaneous reactions to sudden and unexpected sounds.

2. Localization. Head and ear movements undoubtedly play some role in localization, but it is possible to detect the source of a sound without any movement at all. A person with normal hearing uses differences in the time and intensity of the sound to derive information about its source. The superior olive can detect differences on the order of 10 msec to 15 msec between the arrival of impulses from the two ears. The impulses arrive earlier on the contralateral side owing to crossing of the fibers from the cochlear nuclei. The effect of intensity differences will also influence the probability of response of the olivary neurons. If the sound comes from the contralateral side, the probability of firing is high; if the sound comes from the ipsilateral side, the probability is lower. Other evidence suggests that olivary neurons may be able to discriminate between different tones, some being excited by a tone to one ear and inhibited when the same tone is presented to the other ear.

3. Centrifugal control. An important feature of many sensory systems is that information from the receptors may somehow be controlled by the central nervous system itself. The cochlea is no exception. Recurrent

efferent fibers reach the hair cells of the organ of Corti on which they have both excitatory and inhibitory influences:

• Excitatory pathway. Single cell recordings from the ventral cochlear nucleus indicate that the discharges are increased following stimulation of the superior olive. At the same time their threshold to sound stimulation is lowered. The olivary pathway to the ventral cochlea nucleus mediates an excitatory centrifugal influence by which the response of the hair cells to weak sounds may be enhanced.

• Inhibitory pathway. Rasmussen was the first to describe an efferent component in the auditory nerve originating from the superior olive and known as the olivo-cochlear bundle. The fibers emerge from the dorsal cochlear nuclei and terminate on the organ of Corti. Fex recorded impulse activity in single axons and found that an inhibitory effect was exerted on the auditory input. A crude form of tonal localization matching the afferent input was also observed. Galambos showed that electrical stimulation of the bundle suppressed the auditory responses elicited by click stimulation. These findings suggest that the function of the inhibitory pathway is to raise the threshold of the hair cells to sound. This would have the effect of diminishing background noise and thus improve the discriminating properties of the hearing mechanism.

Higher auditory centers

The majority of fibers in the lateral lemniscus terminate in the inferior colliculus, a major relay station of the auditory pathway. At this level there is a fair amount of intermingling of connections from the two sides, but tonotopic organization of the auditory projections is maintained. Axons from the collicular cells pass to the medial geniculate body from which the terminal radiations extend over a wide area of the temporal lobe.

1. Inferior colliculus. Most of the cells in the inferior colliculus can be activated by stimuli presented to either ear. This suggests that the lateral lemniscus carries a proportion of both crossed and uncrossed fibers from the cochlear nuclei. Bilateral contributions also reach the inferior colliculus from the reticular formation of the brain stem.

2. Medial geniculate body. This nucleus has a small-celled lateral or principal division and a large-celled medial division. The former lies on the main auditory pathway to the cerebral cortex; its projections terminate largely, though not exclusively, on the primary auditory area. The cells of the medial division have many synaptic connections with dendrites of adjacent thalamo-cortical neurons, suggesting that a high order of auditory processing may occur at this level.

3. Auditory cortex. The boundaries of the auditory cortex are not easily

defined owing to an abundance of intracortical connections. Mapping experiments have been performed in animals either by the method of retrograde degeneration following localized cortical ablations or by means of evoked potential recordings.

• Primary auditory area. This area (A1) receives its projections from the anterior part of the principal division of the medial geniculate body. In the cat, it lies in the middle ectosylvian gyrus. In the human being, it is located in the anterior transverse temporal gyrus on the inferior lip of the Sylvian fissure.

• Secondary auditory areas. On the basis of evoked potential studies in the cat, Rose and Woolsey defined a belt of cortex occupying most of the posterior ectosylvian gyrus below A1. The belt was subdivided into several zones corresponding to the projection patterns obtained by retrograde degeneration (Fig. 136). The findings gave rise to the concept of "sustaining projections," in which collaterals from medial geniculate neurons diverge on several zones of cortex and so "sustain" the auditory pathway when one of the zones is removed. In man, the cortical areas that can be influenced by auditory inputs are very widespread, extending beyond the temporal lobe into the insula and somatosensory area II.

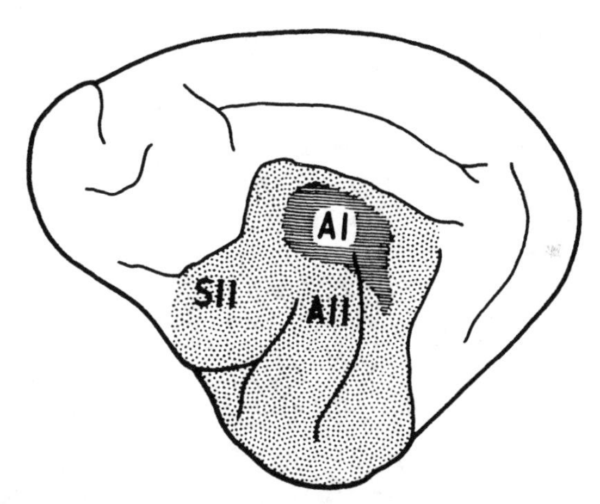

Fig. 136. Diagram of left cerebral hemisphere of cat showing position of the auditory areas. The dark zone A1 lies in the middle ectosylvian gyrus. This is the primary auditory area receiving fibers from the principal division of the medial geniculate body. The surrounding zones include the secondary auditory area A11 and the secondary sensory area S11.

Functions of higher auditory centers

By the time the tonal patterns from the cochlea have reached the higher levels of the auditory pathway, a greater proportion of neurons respond to only a narrow range of frequencies, and a smaller proportion to a broad range of frequencies. It seems that sound information is sharpened by lateral inhibition exerted through collaterals on the primary auditory pathway. The principal function of the higher centers is to discriminate the different pitches of sound. The correct interpretation of the signals and their subsequent storage are functions of later stages in processing in which the auditory association areas obviously play an important role.

Inferior colliculus

About half of the cells in the inferior colliculus respond to standard click stimulation. The remaining half of the population show some form of tonal discrimination. Tuning curves have certain best frequencies at threshold levels, although the same cell may show a different range of responsiveness at higher stimulus intensities. In general, the discharge rate of all collicular cells increases with increasing sound intensities. Some cells respond to inputs from either ear, while others are more selective to sound direction.

Medial geniculate body

All acoustic impulses destined for the cerebral cortex have their synaptic relays in the medial geniculate body. There appears to be an orderly arrangement of frequency representation in the principal nucleus from which a point-to-point projection exists to the primary auditory cortex. Short- and long-latency responses have been reported to both clock and tonal stimulation. In some cells, activity is followed by a period of depression, suggesting that strong excitatory-inhibitory interactions occur at this level.

Auditory cortex

The functions of the auditory cortex have been investigated by electrical recordings in anesthetized and conscious animals and by estimations of hearing loss following localized cortical ablations.

1. Electical activity. The potential changes evoked by sound stimuli can be demonstrated by simultaneous recordings from a number of different points on the cortical surface. Mapping of the auditory areas is then carried out by plotting the sites of maximal amplitudes. The method demonstrates in a crude way the tonotopic representation of the cochlea for areas responding to low, intermediate, and high pitched sounds. Woolsey stimu-

lated discrete bundles of auditory nerve fibers along the cochlea and showed that different parts of the spiral activated vertical bands of cortex, with the base of the cochlea represented anteriorly and the apex posteriorly (Fig. 137). From this evidence it would seem that there exists an orderly spatial projection of the cochlea on the primary auditory area.

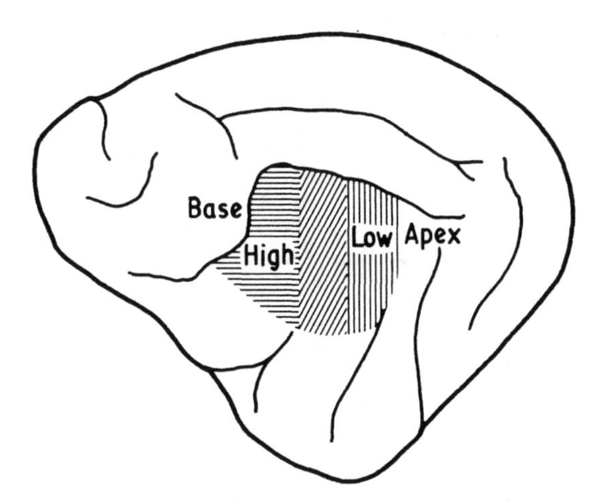

Fig. 137. Sites of maximal amplitude of evoked potentials demonstrating tonal localization in the auditory cortex of a cat. The base of the cochlea is represented anteriorly, the apex posteriorly. Horizontal lines show the area responding to high tones; diagonal lines, the area responding to intermediate tones; and vertical lines, the area responding to low tones.

2. Single unit activity. Microelectrode studies of single unit discharges provide a more detailed picture of the functional aspects of the auditory cortex. Surprisingly, it has become clear that a strict tonotopic organization does not exist for individual units, which, on the contrary, show diverse response patterns in the same vertical penetration. Since many cortical units respond to a wide range of sound frequencies, it is impossible to assign a particular auditory input to a specific locality.

The responses of auditory units to sound stimulation are extremely variable. Examples are given in Figure 138. The majority display spontaneous firing and have low thresholds to noise or clicks. About two-thirds respond to tones either by giving a sustained discharge or by inhibition of their spontaneous activity. A small proportion of units respond only when the sound is changed from one frequency to another. All tend to show habituation to repeated stimuli, and some units have prolonged recovery times before the response can be reproduced. The conclusion to be drawn from the evidence at hand is that the primary cortical area is not rigidly bound to a frequency modulated projection system, as was previously thought, but

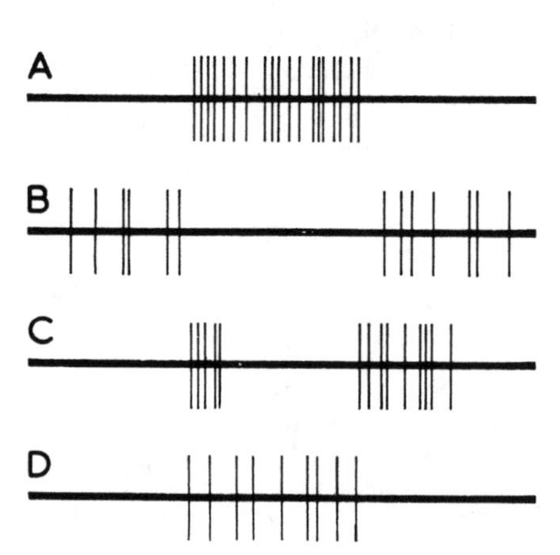

Fig. 138. Examples of unit responses in the auditory cortex of the unanesthetized cat. A represents sustained excitation; B, sustained inhibition; C, "on"-"off" response; and D, response to changing frequency. Line above traces indicates when sound is on. (After Evans and Whitfield.)

that it serves as a pattern analyzer, responding critically to the shifting character of the auditory input in a world of constantly changing sounds.

3. Bilaterality. Nearly all units in the primary auditory area can be influenced by stimuli from both ears. The contralateral side appears to be more strongly represented and has lower stimulus thresholds. If a click stimulus is presented to each ear in a conditioning-testing sequence, the contralateral ear usually excites the unit and the ipsilateral ear inhibits it. Bilateral representation for hearing follows from the knowledge that collaterals in the auditory pathway intermingle as they ascend through the brain stem, most of the crossing being completed below the level of the medial geniculate body. Evidence of bilaterality is also provided from ablation experiments in animals and from clinical assessment of hearing loss following temporal lobe lesions in man:

- Cats are trained to make an avoidance response to clicks presented to one ear through headphones. After learning this discrimination, the signal is transferred to the other ear, when the animal immediately makes the correct response. Continuing the experiment, the contralateral auditory cortex is ablated. An initial deficit in performance is found, although relearning is rapid. After bilateral ablation of the auditory cortex, there is a severe loss of the discrimination response with slow recovery by retraining.
- Unilateral temporal lobe lesions in man cause only slight hearing loss but poor discrimination abilities if the lesion extends into the secondary audi-

tory areas. As speech is developed in the left cerebral hemisphere, the deficits seen in tests based on speech discrimination are significantly greater for lesions on the left side.

The results of these tests indicate that hearing loss may not be marked when one auditory cortex remains intact and that the crossed and uncrossed components of the auditory pathway are about equal. On the other hand, if the damage extends beyond the primary auditory cortex, discrimination performances may be severely impaired. For human subjects, the defects of language found in left-sided lesions emphasize the importance of the temporal cortex in the recognition of auditory patterns.

DEAFNESS

Types of deafness

A child born totally deaf or with a marked hearing defect may have arrest of speech development. If the impairment is acquired early in life it may interfere with the quality of the voice and the clarity of articulation. In older children and adults, loss of hearing may result from interruption of the auditory pathway at any site from receptor to cortex. Deafness is usually divided into three types—first, that caused by impairment of the mechanisms for transmitting sound to the cochlea; second, that caused by damage to the receptor sites and auditory nerve; and third, that caused by lesions in the central nervous system.

Transmission deafness

This may be due to a number of causes. One possible cause is interference with the conduction of sound waves through the external auditory canal. Wax is the commonest reason for this, but occlusion may occur from foreign bodies or collections of pus.

Another possible cause of transmission deafness is blockage of the auditory tube. In nasopharyngeal catarrh, the air in the middle ear cannot be renewed, and this leads to inequalities of pressure on the two sides of the tympanic membrane. The condition is aggravated when the individual is airborne, and particularly on rapid descent of the aircraft. Interference with hearing results from bulging of the tympanic membrane, but the discomfort can be alleviated by chewing or yawning to help open the orifice of the tube.

Otitis media can also produce transmission deafness. Infections of the middle ear may cause inflammation of the drum, which appears indrawn and red on examination. A more serious condition is suppurative otitis media, often leading to perforation of the drum and permanent loss of hearing. The commonest cause is bacterial infection, although there is a high incidence of the disease in children due to the complications of measles, scarlet fever or influenza.

Finally, otosclerosis is another possible cause of transmission deafness. Adhesions of the ossicular chain or degenerative thickening of the stapes diminish the efficiency by which sound vibrations are conducted through the middle ear to the membrane of the oval window. If the stapes is rigidly fixed, the vibrations are conducted to the round window with consequent loss of mechanical sensitivity.

Nerve deafness

- Damage to the hair cells of the organ of Corti. Slow, progressive high-tone deafness comes with age. More abrupt causes are from sudden explosions or bomb blasts, which may result in irreversible damage and permanent deafness. The hair cells may suffer damage from exposure to excessive noise; this often occurs among industrial workers although, fortunately, the condition is usually a temporary one. Prolonged exposure to intense high-frequency sounds is sometimes an occupational hazard that may lead to permanent tone deafness.
- Damage to the auditory nerve. Impairment of hearing is confined to the affected ear when there has been a physical head injury or there is an acoustic nerve tumor. The patient may also show signs of vestibular involvement.

Central deafness

Any lesion of the central auditory pathway may interrupt the conduction of impulses from the cochlear nuclei to the auditory cortex. However, as already mentioned, the loss of hearing may not be more than a few decibels owing to the bilaterality of the auditory pathway; this is especially true of lesions confined to one cerebral hemisphere.

Methods of testing hearing ability

The object of applying hearing tests is to determine whether each ear functions normally with respect to a standard measurement of audibility. It is essential to examine the extent of a hearing defect at different sound frequencies in order to ascertain the most suitable form of treatment or, in the case of children, to determine whether there is need for special educational provision.

1. Simple tests. The "forced whisper" or spoken voice can be used for testing hearing loss in the speech frequency range. Tape recordings of numbers and words transmitted to earphones are useful for routine group testing.

2. Audiometry. The instrument consists of an oscillator capable of emitting pure tones ranging from low frequencies to high frequencies, an attenuator to regulate the loudness of each tone, and headphones that can test one ear at a time. The operator selects a frequency and changes the volume control step by step until the sound is just audible. The subject indicates that the sound has been heard by pressing a button to light a signal lamp. His hearing at that particular frequency is noted and another frequency is selected.

The results of the test are plotted in the form of a graph, with the frequencies on the abscissa and the intensity level on the ordinate. In practice, the instrument is calibrated so that the zero intensity level represents the threshold of audibility for persons with normal hearing; this forms the base line on the chart. If, for example, the loudness of a tone must be increased to 20 decibels above the normal threshold before the subject can hear it, it indicates that he has a hearing loss of 20 decibels for that tone. The chart in Figure 139 depicts a high-frequency tone deafness with reference to the base line.

3. Clinical tests. Bone conduction plays little part in normal hearing. If

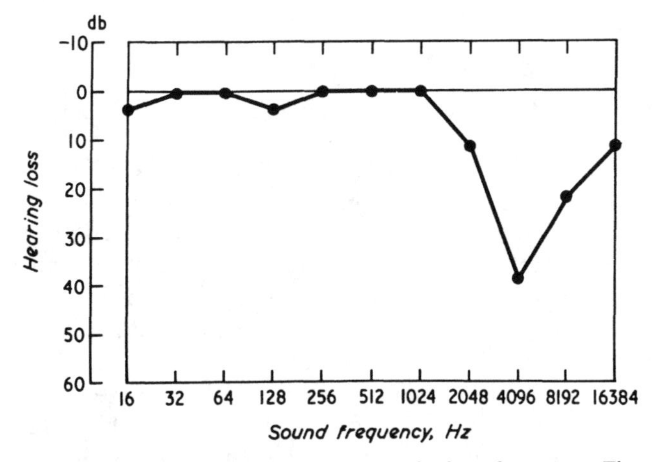

Fig. 139. Audiogram showing a high frequency tone deafness for one ear. The normal threshold of hearing is taken as 0 db, and a negative loss above the base line indicates increased auditory sensitivity.

the base of a tuning fork is applied to the mastoid process, its tone can be heard but soon ceases; if the fork is now placed close to the ear, the tone is heard again. Air conduction is therefore better than bone conduction because of the great loss of energy that occurs as the vibrations pass to the cochlea through the bones of the skull.

• Rinne's test. In middle ear deafness, transmission by the air route is impaired. A vibrating fork is placed close to the subject's ear until he can no longer hear the sound. The base of the fork is then applied to the mastoid process when he will again hear the sound. Obviously bone conduction is better than air conduction.

• Weber's test. When the base of a vibrating tuning fork is applied to the center of the forehead, sound is conducted equally to both ears. If one ear is plugged, the tone appears louder in that ear. In like manner, a subject with middle ear deafness will hear the tone better on the affected side.

NOISE

Unwanted sounds are classified as noise. The sounds need not necessarily be loud, but the disturbance can be sufficiently annoying to upset bodily health and performance at work. Persons exposed to continuous noise may sustain a hearing loss for certain frequencies without actually being aware of it. This may occur when normal conversation can be heard, and the hearing loss comes to light only when such a person is given an audiometric test. Excessive noise is encountered not only in industry but also in residential areas, especially near airports where effective control has now become a major problem.

Noise-induced hearing loss

Brief exposure to noise is unlikely to cause permanent ill effects unless the sound pressure level is high enough to produce physical damage to the ear drum. Continuous exposure to noise at levels over 80 decibels—for example, the clatter of machinery, gunfire, or jet aircraft—may cause elevation of the hearing threshold of 30 decibels or more, but recovery can be anticipated with appropriate rest periods. Hearing of high frequency sounds is lost first because their reception is at the origin of the basilar membrane. The effects of excessive noise are always more serious in persons who have a history of middle ear disease with scarred drums from old perforations. Then, if little is done to eliminate the hazard, hearing loss may become permanent and adversely affect the general health and temperament of the sufferer.

Measures of control

Sound level meters handle all input signals at the noise source and give accurate measurements, automatically recorded over selected frequency bands. A system designed to measure the sound levels at a number of locations is used for monitoring airport noise. Generally speaking, sound pressure levels of less than 85 db are unlikely to damage hearing, although intensities greater than 95 db will almost certainly do so.

Measures adopted to prevent the hazards of noise include attempts to reduce sound levels at the source and measures to protect exposed personnel. Obviously, the quieter the working conditions, the less risk there is of developing a hearing disability. Machinery can be fitted with silencers or enclosed in soundproof coverings, and the surounding buildings can be insulated. Earmuffs or helmets can be worn by workers, and periodic audiometric examinations carried out to detect early changes in hearing acuity. Yet despite such precautions and the strict application of standards, the problem of noise control remains a serious challenge to modern society.

SELECTED BIBLIOGRAPHY

Comis, S.D., and Whitfield, I.C. Influence of centrifugal pathways on unit activity in the cochlear nucleus. *J. Neurophysiol.* 31:62-68, 1968.

Desmedt, J.E. *Auditory Evoked Potentials in Man.* Basel: John Wiley, 1979.

Evans, E.F., and Whitfield, I.C. Classification of unit responses in the auditory cortex of the unanaesthetized and unrestrained cat. *J. Physiol.* 171:476-493, 1964.

Evarts, E.V. Effect of auditory cortex ablation on frequency discrimination in monkey. *J. Neurophysiol.* 15:443-448, 1952.

Heffner, H., and Masterton, B. Contribution of auditory cortex to sound localization in the monkey. *J. Neurophysiol.* 38: 1340-1358, 1975.

Hind, J.E. An electrophysiological determination of tonotopic organization in auditory cortex of cat. *J. Neurophysiol.* 16:475-489, 1953.

Katsuki, Y. Neural mechanisms of hearing in cats and monkeys. *Progr. Brain Res.* 21:71-97, Amsterdam: Elsevier, 1966.

Moller, A.R. Coding of amplitude and frequency modulated sounds in the cochlear nucleus of the rat. *Acta Physiol. Scand.* 186:223-238, 1972.

Morest, D.K. The neuronal architecture of the medial geniculate body of the cat. *J. Anat.* 98:611-630, 1964.

Neff, W.D. Neural mechanisms of auditory discrimination. In Rosenblith, W.A. (Ed.): *Sensory Communication.* New York: John Wiley. 1961.

Powell, E.W., and Hatton, J.B. Projections of the inferior colliculus in cat. *J.Comp. Neurol.* 136:183-192, 1969.

Whitfield, I.C. *The Auditory Pathway.* London: Edward Arnold, 1967.

CHAPTER XIX

Disturbances of Sensory Function

Lesions may be present in the sense organ itself, in the peripheral nerve, or in the central nervous system. While the patient's symptoms and physical signs may indicate a disorder of the sensory system, pathological changes involving other systems of the body may also be present and require careful examination. Destructive lesions generally cause bluntness or loss of sensation; irritative lesions may produce hypersensitivity or abnormal sensations. The principal symptoms of a sensory disturbance are tingling, numbness, tenderness, and pain. The principal physical signs are anesthesia, analgesia, diminished or absent reflexes, loss of position sense, and joint deformities. Lesions in the cerebral cortex usually affect the discriminating functions of sensation or produce complex disturbances involving evaluation and interpretation of the sensory signals.

Tests for the sensory system

A good deal of information can be obtained from a few simple clinical tests for each form of sensation, although much depends upon the cooperation of the patient. Electrodiagnostic techniques provide additional, often unequivocal evidence of functional integrity or of the existence of abnormalities in the sensory pathway.

1. Clinical tests. The patient should keep his eyes closed. The sensitivity of the proximal and distal parts of each extremity is determined and also whether the sensory changes are confined to one side of the body. If a sensory change is detected, its dermatomal distribution is evaluated and compared with the findings on the opposite side.

Superficial sensations

Tactile sensation. The sense of touch is tested by a wisp of cotton wool applied lightly to the skin. Quantitative estimations are made by using Von

Frey hairs, graduated in thickness so that different pressures are needed to bend them. The threshold varies considerably on different areas of the body, according to the thickness of the epidermis and the presence or absence of hairs.

Thermal sensation. This is tested by applying test-tubes to the skin, one filled with ice, the other with hot water. The patient must be told to respond to the feeling of cold or heat and not to the sensation of touch or pressure.

Pain sensation. This is usually assessed by giving a pinprick or by using an algesiometer, which can grade the severity of the stimulus. In tender parts of the skin a mild stimulus may provoke a marked reaction.

Deep sensations

Pressure sense. This is tested by squeezing the calf and forearm muscles or the deep tendons and noting sensitivity.

Vibration sense. The base of a vibrating tuning fork is applied to the bony prominences–e.g., elbow, shoulder, knee, shin, and ankle. The patient's ability to detect when the vibration stops is noted.

Position and joint sense. The fingers and toes are moved passively by the examiner, and the patient is asked to indicate the direction of the movements. In another test, the examiner alters the position of a limb in space and the patient points to each new position.

Discriminating sensations

Two-point discrimination. To judge whether a tactile stimulus is single or double, various parts of the body are touched simultaneously with the blunt points of a compass or other applicator. The patient is asked each time if he is being touched by one or two points. The normal threshold of discrimination is 1 mm on the tip of the tongue, 2 mm to 3 mm on the fingers, 10 mm on the palm of the hand, and as much as 60 mm on the back.

Stereognosis. The appreciation of texture, weight, size, and shape of objects is tested by feeling the materials in the hands. Familiar objects like coins and keys are identified by rolling them between the fingers.

Graphesthesia. The patient is asked to identify a letter or number traced on his skin with a blunt edge; the two sides of the body are then compared.

2. Electrodiagnostic tests. Latency measurements of sensory nerve action potentials and studies of reflex conduction times are useful in differentiating disorders of the peripheral nerve from those of central origin. These tests are also used to demonstrate the progress of recovery in regenerating fibers after nerve damage.

Sensory action potentials. In this technique, a stimulus is delivered through a pair of ring electrodes placed round the thumb or little finger, and action potentials are recorded through electrodes strapped to the wrist. In the lower limb, the anterior tibial nerve is stimulated at the ankle while recordings are taken from the lateral popliteal nerve at the knee. Figure 140 shows the stimulating and recording arrangements for the median nerve. The stimulating electrodes embrace the thumb; a metal plate over the thenar eminance is connected to ground. The recording electrodes consist of a pair of silver discs, lightly covered with electrode jelly and applied firmly over the median nerve at the wrist. The tracing in the figure shows the stimulus artifact followed by the sensory nerve action potential. The latency and amplitude of the action potential can be measured directly from the tracing or else computed from the average of successive superimposed sweeps.

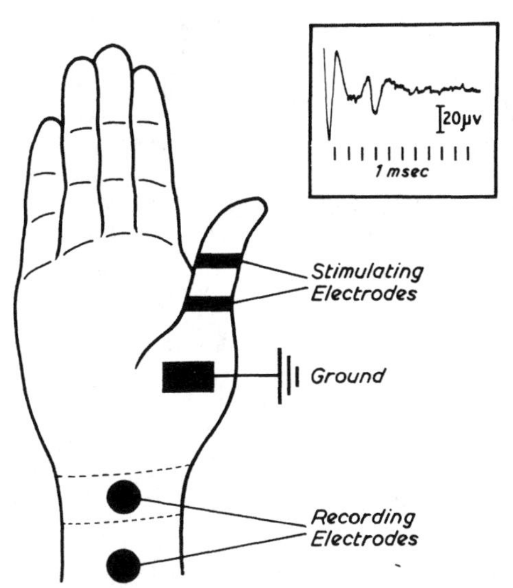

Fig. 140. Method of recording sensory action potentials from the median nerve. Ring electrodes are placed around the thumb and connected to a stimulator; a metal plate over the thenar eminence serves as grounding. Recording electrodes are placed over the median nerve at the wrist. The tracing shows atimulus artifact followed by a nerve action potential. Calibrations are 1 msec and 20 μ v.

Reflex conduction times. In this technique, a tendon jerk is elicited, and the reflex action potential recorded from the appropriate muscle. For the knee reflex, the recording electrodes are placed on the skin overlying the quadriceps muscle and the patella tendon briskly tapped. For the ankle reflex, the subject kneels on a chair, the recording electrodes are placed over the calf muscles, and the tendon Achilles tapped. The hammer contains a battery and contact switch to trigger the sweep on the oscilloscope; a suitable time base is displayed simultaneously on a second channel (Fig. 141).

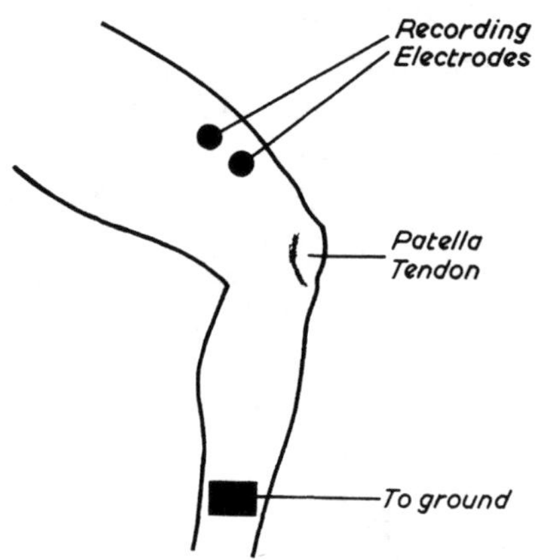

Fig. 141. Method of recording knee jerk. The patella tendon is tapped by a hammer which triggers the sweep of an oscilloscope. The reflex muscle action potential is displayed through recording electrodes placed over the quadriceps above the knee. A metal plate on one ankle is connected to the ground. For the ankle jerk, the recording electrodes are placed over the calf and the tendon Achilles tapped. (See Fig. 89 for relevant tracings.)

Lesions of the peripheral nerve

Certain cutaneous nerves are particularly vulnerable to damage by local pressure or trauma. In addition, afferents from muscles, tendons, and joints make a significant contribution to the fiber spectrum of nerve trunks and such lesions can be expected to give a combination of sensory and motor deficits.

1. Cutaneous nerves. Partial denervation causes tingling, numbness, and impaired sensation, while complete section results in the loss of all forms of sensation in the area supplied. If, however, only a small branch is involved, the loss may be unnoticed because of overlap. Nerve conduction studies may provide evidence of early denervation, and rapid recovery is possible if the pressure is relieved. Thus compression of the median nerve in the carpal tunnel leads to a decrease in amplitude of the nerve action potential and prolongation of its latency.

2. Muscle afferents. Section of a nerve trunk causes muscle paralysis as well as loss of joint sense, deep pressure, and vibration. Recovery of function occurs if the nerve is resutured, but the fibers growing distally may take some time to find their original receptors. Accurate localization and tactile

discrimination remain unsatisfactory for several years; muscle tone is diminished and the deep reflexes are difficult to elicit.

3. Dorsal nerve roots. Lesions of sensory roots are of two kinds. If the lesion is distal to the ganglia, it results in degeneration of the peripheral fibers and loss of nerve action potentials. If the lesion is proximal to the ganglia, the peripheral nerves remain in contact with their cells of origin and do not degenerate. Accordingly, the absence of nerve action potentials implies a more favorable prognosis since regeneration of the central processes is most unlikely. Lesions of the brachial plexus are relatively common. All sensibility is lost in the distribution of the damaged roots, but the extent of the damage may be masked because of overlap between adjacent roots. Pain is a common symptom of dorsal root compression. It can be very severe in sciatica due to irritation of the root fibers as they pass into the cord or to compression resulting from a diseased intervertebral disc.

Lesions of the spinal cord

A knowledge of the conduction pathways in the spinal cord is essential for the diagnosis of sensory abnormalities. It is also important to know the segmental distribution of the dermatomes in order to determine the level of the lesion when this happens to be discrete. Unfortunately, many diseases give rise to multiple lesions that may not be confined to the spinal cord, and the clinical picture is often complicated by motor as well as sensory defects. Another difficulty in the diagnosis and treatment is that there tend to be periods of remission of the symptoms followed by periods of relapse.

1. Tabes dorsalis. The lesion in this syphilitic disease affects the dorsal nerve roots and dorsal columns of the cord. All forms of sensation are impaired or lost except that attacks of lightning pains come on at intervals due to irritation in the root entry zone. The clinical manifestations are as follows:

- There is loss of deep sensibility including deep pressure, vibration, and position sense. The patient cannot indicate the exact position of the limbs or joints when altered by passive movements.
- Muscle tone and the tendon jerks are diminshed or absent since tone depends upon the integrity of the stretch reflex.
- Visceral sensibility is disturbed, especially in the bladder and rectum. The patient is unaware of when the bladder is full and allows it to become over-distended.
- The appreciation of pain becomes impaired. The Charcot joint is a deformity produced by repeated injuries since no pain is aroused and the protective reflexes abolished.

• There may be considerable disturbance of voluntary movement, and abnormal gait may result from loss of proprioceptive impulses from the muscles and joints.

2. Subacute combined degeneration. In this complication of pernicious anemia, there is a degeneration of fibers especially affecting those involved in deep sensibility in the dorsal columns. The patient suffers from ataxia and walks unsteadily with his legs apart.

3. Syringomyelia. The pathological changes in this condition result in cavity formation in the central gray matter of the cervical cord. Here the decussating pain and temperature fibers are interrupted with loss of these sensations, but dorsal column sensibility is preserved. The clinical findings are described as "dissociated anesthesia"–i.e., loss of pain and temperature sensations, with the sense of touch and deep sensibility remaining. The patient may show signs of injury or burns on his fingers due to loss of the nociceptive reflexes.

4. Multiple sclerosis. The disturbances attributable to this disease are due to damage to the fiber tracts by plaques of demyelination occurring almost anywhere in the cord or brain. Beginning usually in early adult life, the disease progresses slowly over many years with episodes of acute disability followed by long periods of remission. Involvement of the motor tracts gives rise to weakness in the limbs with difficulty in walking and lack of muscular coordination. Visual failure and diplopia may also occur. Sensory deficits may be widespread. Demyelination in the dorsal columns causes impairment of position and joint sense, producing the so-called "useless hand" syndrome, in which the patient is virtually unable to use one hand, though motor power remains intact. Pain and temperature sensations may also be impaired in one or more limbs.

Lesions of the thalamus

Certain nuclei of the thalamus may be damaged by thrombosis of the arterial supply, and the patient suffers a series of abnormalities as follows:

1. Ventrobasal complex. Loss of all forms of sensation from the opposite side of the body due to destruction of the lemniscal relay fibers. Discrimination is particularly affected, as is the case in patients with lesions of the sensory cortex itself.

2. Ventrolateral nucleus. Muscular incoordination and ataxia may result from damage to afferent relays from the cerebellum.

3. Posterior group. Severe burning pain is felt particularly around the angle of the mouth and cheek or in the hand and foot. It is often accompanied by other unpleasant sensations associated with irritation of the anterolateral system.

4. Reticular nuclei. Altered emotional states and other affective disturbances may be caused by lesions of the reticular and intralaminar nuclei of the thalamus with consequent interruption of the diffuse projections from the nuclei to almost all parts of the neocortex.

Lesions of the sensory cortex

Vascular lesions are rarely confined to the primary sensory area and generally extend into the parietal lobe or motor cortex. Indeed, ataxia may be more pronounced than the loss of sensory modalities.

1. Injuries to the postcentral gyrus cause impairment of sensation on the opposite side of the body. Accurate localization for tactile and position sense is lost as are all forms of discriminative abilities including two-point discrimination and the recognition of small joint displacements. Pain is not specifically affected owing to its bilateral distribution.

2. Sensory epilepsy may arise from irritation of cortical neurons in the region surrounding a tumor or abscess. The patient may complain of tingling and constricting sensations in representative areas of the body.

Lesions of the parietal lobe

A tumor of the parietal lobe may be present without any obvious sensory defect, although disorders of recognition are commonly revealed on clinical examination:

1. Astereognosis. The patient is unable to identify familiar objects by palpation. He cannot distinguish the value of different coins by their size or form.

2. Agraphesthesia. The patient fails to recognize a letter or number traced on the skin.

3. Apraxia. The patient is unable to carry out simple motor skills on request. He cannot construct letters of the alphabet with matches.

4. Aphasia. The patient may not be able to reply coherently to simple questions, and the appreciation of musical sounds may be lost.

5. Disorientation. The patient may fail to recognize the relationships of his body image. He may not be able to distinguish between the left and right sides.

Lesions of the prefrontal lobes

The integrity of the prefrontal lobes is important for the critical assessment of information supplied by the sense organs in relation to previous experience. The degree of attention given to a sensory stimulus no doubt determines the type of reaction it evokes and influences the normal restraints on behavior. Emotional changes are common in patients suffering from a prefrontal lobe tumor, while bilateral injuries are known to produce marked deficits of personality and intellect. On the other hand, selective bilateral lesions tend to increase the tolerance of pain. Or, by relieving the associated anxieties and tensions, these lesions may reduce painful sensations to an acceptable level. Further details on this subject will be found in Chapter XXIII.

CHAPTER XX

Nervous Control of the Viscera—Afferent System

The visceral functions of the body are under the control of the brain although little influenced by volition. The functions include cardiac output, pulmonary ventilation, gastrointestinal secretions, body temperature, and many other activities. The brain also regulates the secretions of most endocrine glands either directly by nervous action or indirectly by means of hormones released into the blood stream. Nervous and chemical processes thus act in combination to maintain a constant internal environment, a concept first developed by Claude Bernard and described as *homeostasis* by Cannon. Nervous and chemical processes are involved in feedback mechanisms by which the body samples the environment and compensates for its variations. Finally, they play a major role when the body is called upon to meet excessive demands or to deal with a sudden emergency.

In its general plan, the visceral system operates very much like the somatic nervous system, with input and output mechanisms and a central control. Visceral afferents transmit signals from a variety of sense organs. The information is used at the spinal level to cause an immediate reaction or to modify local reflex networks. Above the spinal cord, visceral afferent pathways proceed to the thalamus and cerebral cortex along well defined projections which have their counterpart in the lemniscal and anterolateral systems. These afferent projections are subjected to influences imposed upon them by the cerebellum and reticular formation. Relatively discrete pathways terminate in the primary and secondary sensory areas of the cortex, while non-specific elements are distributed more widely to a group of cerebral structures frequently called the *limbic complex*. As in the somatic nervous system, there are many corticofugal pathways transmitting signals to lower relay stations; they modify synaptic transmission, usually by exerting an inhibitory influence at each important level of sensory inflow.

The limbic complex can be regarded as a neuronal link connecting visceral input to visceral output. It is, in fact, a network of related cortical and subcortical structures where various channels of information are integrated and impulse patterns elaborated for controlling many of the internal functions of the body. For these reasons it is sometimes referred to as the "visceral brain." A plan representing the principal input-output relationships of the visceral nervous system is shown schematically in Figure 142. It will be seen that the output begins at the hypothalamus and that from this structure emerge two major output mechanisms, nervous and humoral. The nervous outflow constitutes the autonomic nervous system, and the humoral outflow arises from neurosecretory substances acting on the pituitary gland.

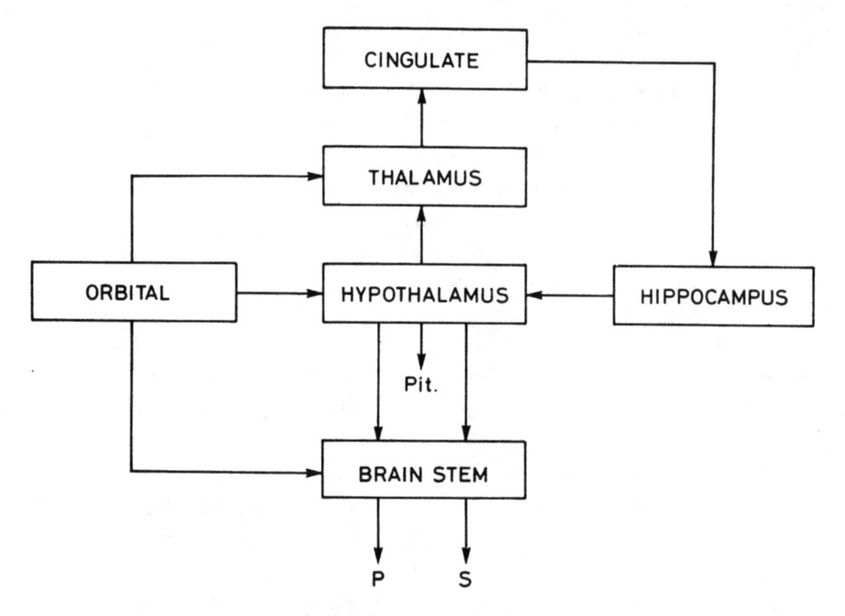

Fig. 142. Block diagram showing input-output relations of hypothalamus. The orbital cotex and hippocampus have direct links with the hypothalamus; the cingulate is connected via the thalamus and hippocampus. Output of hypothalamus is distributed to the pituitary gland (Pit) and autonomic centers of brain stem. A link between orbital cortex and brain stem is also shown. P represents parasympathetic outflow; S, sympathetic outflow.

Visceral sensibility

The internal organs and their attachments are supplied with receptors that respond to a variety of stimuli including pain. The afferent discharges are conveyed centrally along pathways in the distribution of the autonomic nerves, and their effects can be seen at all functional levels. The activity set up by natural stimulation accounts for a vast number of visceral reflexes that serve the individual organs. Strong stimulation may give rise to wide-spread reactions involving the somatic reflexes as well. These effects permit the performance of more strenuous physical activity than would otherwise be possible. In certain conditions due to local irritability or to over-distension, a feeling of discomfort is aroused, and any organ stimulated excessively may give rise to the sensation of pain. Inflammatory conditions often spread to the serous coverings of the viscera which are supplied by somatic nerves; pain is then accompanied by deep tenderness and reflex muscular rigidity.

Types of receptor

Apart from Pacinian corpuscles, which are found in large numbers in the mesentery, no definite structures have been described to distinguish the various types of nerve endings. These are the following:

1. Chemical. Cardiovascular reflexes are served by chemoreceptors in the carotid body and aortic arch. Gastric receptors detect changes in pH in all parts of the stomach. Intestinal chemoreceptors are involved in absorption processes.
2. Mechanical. Pacinian corpuscles are readily excited by mechanical stimuli and are rapidly adapting. Tension recorders are found in the muscular walls of the stomach, uterus, and urinary bladder. Stretch receptors are present in the carotid sinus, atria, and walls of the major arteries.
3. Thermal. Cold receptors have been described in the area of the heart and great veins. Warm receptors are located in the posterior abdominal wall and my cause panting in sheep without a corresponding rise of brain temperature.
4. Nociceptive. There are no clear-cut boundaries to distinguish nociceptors from other types of receptor. The viscera are supplied with free nerve endings arranged in a plexus of fine unmyelinated axons in the muscular walls and mucous membrane.

Adequate stimulus

Internal operations performed under local anesthesia may cause little distress or pain. No painful sensations are aroused by crushing, cutting, or burning as the viscera are insensitive to these forms of stimulation. The adequate stimuli are those which result in excessive activity such as chemical irritants, over-distension, or vigorous muscle contractions. Thus, in spite of the apparent insensitivity of the viscera to ordinary forms of stimulation, severe discomfort and pain are only too frequently a common experience in health and disease.

Types of fiber

Visceral afferents cover a wide range of the fiber spectrum. Their identification by histological methods can prove very difficult, as many are of small diameter and are unmyelinated. One way of identifying them is by calculating their conduction velocities. Under the dissecting microscope, the nerve trunk is divided into small strands and subdivided with needles until a single fiber is isolated. Action potentials are then recorded from the fiber by stimulating the nerve endings. Another method is known as the collision technique. This technique depends on the observation that antidromic impulses evoked by electrical stimulation of the nerve are extinguished by impulses initiated at the nerve endings. For example, a stimulus is delivered to the proximal end of the vagus nerve in the neck, and the compound action potential recorded distally. If the stomach is now distended by a balloon to excite the sensory nerve endings, the size of the C potential is markedly reduced.

1. Vagal afferents. The majority of afferent fibers carried by the vagus nerve belong to the smallest diameter group. Their conduction velocities are less than 2 m/sec.
2. Pelvic afferents. Afferent fibers from the pelvic viscera reach the spinal cord in the sacral division of the parasympathetic. A proportion of the fibers are myelinated A delta afferents, but the largest group is unmyelinated with slow conduction velocities.
3. Splanchnic afferents. Afferent fibers in the splanchnic nerves fall into three main groups—A beta, A delta, and C. A beta fibers contribute to the earliest wave of the compound action potential. They have very low thresholds and fast conduction velocities (70 m/sec to 75 m/sec). A delta fibers produce a second wave after the A beta is completed. The amplitude of the wave is smaller, indicating fewer fibers in this group, and conduction velocities range between 30 m/sec and 35 m/sec. Both groups are myelinated. C fibers have the highest threshold, with conduction velocities less than 2 m/sec. They are all unmyelinated.

Peripheral afferent pathways

Impulses of visceral origin are conveyed to the central nervous system along pathways which follow, in the main, the distribution of the vagal, splanchnic, and pelvic nerves.

1. Vagal afferent pathways. Impulses from the thoracic viscera, stomach, and small intestine ascend in the trunk of the vagus to the dorsal vagal nucleus in the medulla. They form the sensory component of several important reflex arcs which operate in the medulla and pons—for instance, the reflexes controlling respiration, blood pressure, and heart rate.

2. Splanchnic afferent pathways. Impulses from the abdominal viscera are conducted into the spinal cord from the third thoracic to the first lumbar segments inclusive. The extent of the splanchnic innervation was first demonstrated by using pupillary dilatation as an index of response to moderate distension of the gut by inflating a balloon with air. The afferent fibers pass through the sympathetic ganglia without any synapse, ascend in the splanchnic nerves or sympathetic chain to the spinal ganglia, and enter the spinal cord through the dorsal roots. Many afferent fibers take a devious course through numerous abdominal plexuses, but the pupillary response is abolished after section of the splanchnic nerves.

3. Pelvic afferent pathways. Impulses from the pelvic viscera travel in both sympathetic and parasympathetic supply routes to gain access to the spinal cord.

The sympathetic route serves sensation from the large intestine and urinary bladder. Afferents from the colon pass through the celiac, superior, and inferior mesenteric plexuses to reach their cell stations in the dorsal root ganglia. Afferents from the bladder pass through the hypogastric plexus in front of the sacrum to reach the dorsal roots of the twelfth thoracic and first lumbar segments.

The parasympathetic route serves rectal sensation and the micturition reflex. The afferent fibers pass through the hypogastric ganglia, and thence via the pelvic nerves to the dorsal roots of the second and third sacral segments.

Visceral sensation

Hunger is a normal sensation aroused by an empty stomach. Other sensations are fullness and discomfort due to mild physiological stimuli, while abnormal conditions and pathological states may give rise to acute pain.

1. Hunger. If food is presented to a hungry animal it will usually feed. A measure of food intake is not, however, a reliable guide to the degree of hunger, as feeding is regulated centrally. The sensation of hunger is the result of a complex interplay of many factors in which the stomach itself may have only a contributory role. Gastric contractions are believed to stimulate sensory nerve endings in the stomach, yet they are not essential in order to experience hunger. Again, impulses from the stomach are conducted centrally in the vagal nerves, but hunger may occur in patients after

bilateral vagotomy. There is some evidence that the hypothalamus is sensitive to changes in blood sugar concentration. Consequently, when the blood sugar is lowered by starvation, hypothalamic activity is increased and suitable adjustments are made to the mechanisms under its control. These may include feeding drives to supplement local reflexes from the empty stomach. Thus loss of one source of stimulation, as occurs after vagotomy, does not necessarily impair the experience of hunger.

2. Discomfort. A sensation of fullness or discomfort is commonly experienced after a large meal, when the bladder is distended or the rectum loaded. It seems that the hollow organs are sensitive to natural distension, although the symptoms are quickly relieved by emptying these organs. Similar feelings of discomfort, often accompanied by deep-seated tenderness, may result from vascular congestion in localized areas, as occurs, for example, in the premenstrual period.

3. Pain. If the stimulus becomes too great, the discomfort becomes acute and the subject experiences pain. Any inflamed organ may give rise to pain. The mesentery is particularly sensitive, and this probably explains the pain produced by adhesions. Cardiac pain is more usually caused by ischemia of the heart muscle due to inadequate blood flow when a branch of the coronary artery is occluded.

VISCERAL REFLEXES

On entering the spinal cord, many visceral afferents take part in some form of reflex activity either in the same segment of the cord or in a number of segments. Other afferents ascend to higher levels of the nervous system—to the brain stem, where the principal autonomic centers are located, or to the midbrain, which is the center for the pupillary reactions.

Spinal cord

Normally, the reflex functions of the spinal cord are under the control of descending influences from the brain. Therefore, to study the isolated cord reflexes, it is necessary to separate the cord from the higher centers. This is usually done in experimental animals by administering a general anesthetic or by using spinal and decerebrate preparations. In the human being, reflex activity can be observed after recovery from shock following a spinal injury, but the clinical picture tends to vary according to the level of the lesion and the extent of damage.

Viscero-vascular reflexes

Local reflex arcs maintain a tonic sympathetic discharge to the blood vessels. This is usually sufficient to elicit a reaction to changes in body temperature or to cause a rise in arterial blood pressure.

1. Effects on body temperature. When radiant heat is applied to the skin of a spinal patient, reflex vasodilation and perspiration may occur in the limbs. Cooling the skin results in vasoconstriction, although shivering is absent because the muscles are paralyzed. The body temperature can be maintained to a limited extent by means of these reflex actions. However, the patient cannot respond to rapid changes in the environment, for if he is exposed to sudden heat or cold the body follows passively the temperature of the surroundings.

2. Effects on arterial blood pressure. It is possible to show experimentally that vascular reflexes can be elicited in the spinal animal by stimulating an abdominal organ. The effect is usually a marked vasoconstriction associated with the discharge of adrenaline and a rise in blood pressure. The tracing shown in Figure 143 is from the femoral artery of the cat. The vagus nerves are cut in the neck to eliminate possible influences from that source. A small balloon is inserted into the gallbladder. On inflation of the balloon there is an immediate rise in blood pressure, and a slower fall persisting for some time after deflation. The response is reduced if the left splanchnic nerve is cut and abolished after cutting the right splanchnic nerve.

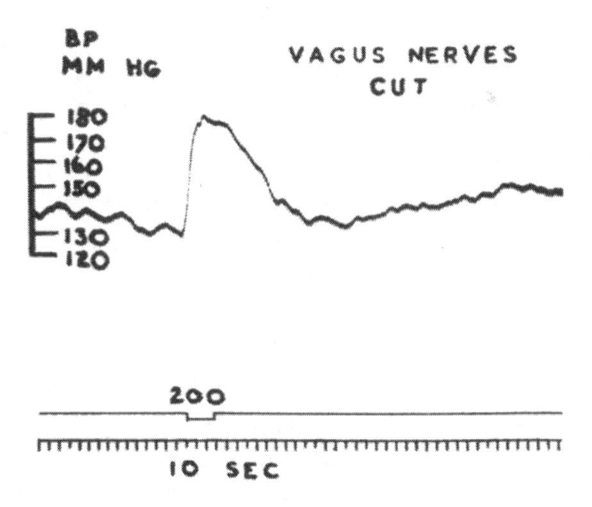

Fig. 143. Blood pressure response to distension of the gall bladder in an anesthetized cat with the vagi cut. On inflation of the balloon there is a marked rise in arterial blood pressure which is partly reflexive and partly due to the discharge of adrenaline.

The experiment suggests that impulses from the gallbladder proceed to the spinal cord where the vasomotor reflex arcs are excited. The effects are exerted bilaterally on the splanchnic blood vessels and adrenal glands. Viscero-vascular reflexes can also be demonstrated by distension of the gut or urinary bladder or by stimulation of the afferent nerves supplying these organs.

Viscero-motor reflexes

Afferent impulses from abdominal viscera can reflexively elicit contractions of skeletal muscle. It is presumed that the incoming splanchnic fibers converge on the cells of the internuncial system whose axons supply the motoneurons of the anterior horn. The evidence is as follows:

1. The muscular rigidity of the abdominal wall is an important clinical sign of underlying disease. The patient frequently draws up his legs to diminish intra-abdominal pressure and relieve pain. These actions may be regarded as protective reflexes.
2. Distension of the gallbladder or a loop of intestine results in disappearance of the knee jerk in the chronic spinal monkey. This effect is analogous to inhibition of extensor reflexes by antagonistic flexors.
3. Viscero-motor reflexes do not involve the same spinal arcs as the viscero-vascular reflexes. Nevertheless, they are dependent upon a comparable spinal organization. Thus, if a threshold stimulus is delivered to the central cut end of a splanchnic nerve, the reflex discharges are found only in the lower intercostal nerves, whereas a stronger stimulus causes irradiation of impulses into upper thoracic levels and even to the opposite side of the cord.
4. Descending influences from the brain stem generally inhibit visceromotor reflexes. This can be demonstrated in the decerebrate animal by subsequent transection of the spinal cord when the reflex is greatly enhanced.

Pelvic reflexes

Surgicál anatomy is of practical importance in operations on the pelvic viscera as the nerves are easily damaged. The reflex control of the bladder and rectum is particularly relevant to the management of patients with lesions of the spinal cord.

1. When the bladder is filled, afferent impulses reach the sacral portion of the spinal cord and cause reflexive emptying by contraction of the detrusor

muscle and relaxation of the sphincter. Likewise, distension of the rectum evokes reflex contraction of the colon and associated muscles of the abdominal wall.

2. Weak stimulation of the afferent fibers from the bladder elicits a reflex discharge in the lumbar sympathetic outflow. With strong stimulation or during active bladder contraction, the reflex is reduced or completely suppressed. If the spinal cord is transected, the inhibitory effect is abolished. These findings indicate that under normal conditions facilitation as well as inhibition of sympathetic discharges accompanies reflex micturition.

3. After complete transection of the spinal cord, voluntary micturition is lost but reflex activity returns. The patient is unaware of the state of the bladder, but reflex evacuation occurs as the pressure rises. In lesions affecting the sacral portion of the cord, the pelvic reflexes are abolished.

Brain stem

Many important structures are located in the medulla and pons which serve in the reflex regulation of visceral functions. These include the central mechanisms controlling the cardiovascular, respiratory, and alimentary systems. In experimental work, the brain stem can be investigated from either a dorsal or ventral approach. The former allows the insertion of electrodes through the substance of the cerebellum and is probably less disturbing to the animal. The ventral approach is more complicated since the soft tissue of the neck must first be excised to gain access to bone, and care must be taken to avoid the basilar artery which extends in the mid-line to the upper border of the pons.

Cardiovascular reflexes

1. The *cardiac inhibitory center* is a collection of neurons within the medullary reticular formation. It transmits a stream of impulses to the vagal nerves that slow the rate of the heart, shorten the systolic period of the heart cycle, and reduce the force of contraction. Hence, if the vagi are cut, the heart accelerates and beats more strongly. This tonic activity is responsible for maintaining the resting heart rate at about 70 beats per minute (vagal tone).

Afferent influences on the cardiac inhibitory center arise mainly from cardiovascular receptors:

• Mechanoreceptors signal pressure or volume changes in the heart and great vessels, and their discharge are characterized by a pulsatile pattern related to the cardiac cycle. They can be stimulated by probes or small balloons placed in the atria or ventricles while recordings are taken from single fibers of the vagus nerve.

• Chemoreceptors give intermittent discharges under resting conditions; their effects are negligible and have no relation to the pulse wave. If the animal breathes low oxygen mixtures, the chemoreceptor discharges increase causing reflex hyperpnea and tachycardia. The increased heart rate is a secondary effect due to stimulation of the respiratory mechanism. If respiration is controlled by inflating the lungs or by holding the breath as is done in diving, extreme bradycardia ensues due to stimulation of the cardiac inhibitory center.

2. The *cardiac acceleratory center* lies in the medullary reticular formation. Its descending pathways connect with cells in the lateral horn of the thoracic cord from which the sympathetic outflow arises. Sympathetic fibers influencing the rate of the heart are described as *chronotropic,* and those influencing the force of contraction are called *inotropic.*

In 1915 Bainbridge demonstrated that rapid intravenous infusions of blood or saline in the anesthetized dog brought about an increase in the heart rate, and the mechanism was thought to play a part in the acceleration of the heart during muscular exercise. At one time the Bainbridge reflex was attributed to inhibition of vagal tone due to afferents arising from the venous side of the heart, but it is clear that cardiac acceleration is obtained only when certain conditions are fulfilled. If the heart rate before infusion is more than 130/min, the result is bradycardia, while an increase in heart rate commonly occurs when the initial heart rate is low. The adequate simulus for eliciting the Bainbridge reflex is therefore uncertain. The tachycardia produced by stimulation of atrial receptors is thought to be mediated by the cardiac acceleratory center, since the reflex is abolished after cutting the sympathetic supply to the heart.

3. The *vasomotor center* is situated in the floor of the fourth ventricle. It gives rise to vasoconstrictor fibers that pass out of the spinal cord in the sympathetic nerves to the various parts of the body. Tonic vasoconstrictor impulses maintain the peripheral resistance and resting level of the arterial blood pressure. If the frequency of discharge increases, the blood pressure rises mainly because of constriction of the arterioles in the splanchnic and skin areas. If the frequency is reduced, the arterioles dilate and the blood pressure falls. Thus any change in the level of the systemic blood pressure, as occurs, for example, after a moderate hemorrhage, can be reflexively adjusted by altering the frequency of vasomotor discharge.

Among the many factors that can influence the vasomotor center, the most important are the baroreceptors in the walls of the carotid sinus and aortic arch. These receptors are spray-type nerve endings sensitive to stretching, as occurs when the pressure within the vessels is suddenly increased. Afferent fibers from the carotid sinus form the sinus nerve, which accompanies the glossopharyngeal nerve to the medulla; afferents from the aortic arch and neighboring arteries are transmitted through the vagus nerves to their terminals in the medulla. The sinus and aortic nerves constitute a "buffer" mechanism for maintaining a stable blood pressure level under resting conditions, and these nerves respond to sudden rises or falls by restoring the blood pressure back to normal. Baroreceptor discharges exert an inhibitory influence on the vasomotor center. If the arterial blood pressure rises, baroreceptor discharge is increased and the vasomotor center inhibited; sympathetic activity is correspondingly reduced, the arterioles dilate, and the blood pressure falls again. Conversely, if the arterial blood pressure is suddenly lowered, baroreceptor discharge is diminished, thus causing increased vasoconstrictor activity and a tendency to return the blood pressure to normal. These effects can be readily demonstrated by experiment:

- Perfusion of the isolated carotid sinus of the dog under chloralose evokes a reflex fall of blood pressure when the sinus pressure is raised above the threshold of about 70 mm Hg. The reflex is not abolished after bilateral vagotomy.
- Electrical stimulation of the sinus or aortic nerve also lowers the blood pressure owing to inhibition of the vasomotor center.
- Section of the sinus nerves or occlusion of the carotid arteries raises the systemic blood pressure because the restraining activity on the vasomotor center is removed with resultant vasoconstriction.

4. The *vasodilator center* is an ill defined, scattered region or series of depressor points in the medullary reticular formation. Electrical stimulation of depressor points causes a fall in arterial blood pressure associated with an increased blood flow in the skeletal muscles, splanchnic area, and skin due to widespread peripheral vasodilation. The region can be influenced by afferents from many sources. Thus weak stimulation of the sciatic nerve may produce a reflex fall of blood pressure while strong stimulation elicits a pressor effect. It is clear that a close relationship exists between the two regions and that under normal circumstances the vascular response patterns are the result of interactions occurring between them.

Certain vasodilator neurons may be sensitive to a rise in brain temperature. The blood pressure recording illustrated in Figure 144 is from the femoral artery of the decerebrate cat. A heating electrode is inserted into the medullary reticular formation. It can be seen that a dramatic fall of blood pressure occurs when the temperature of the brain is raised to about

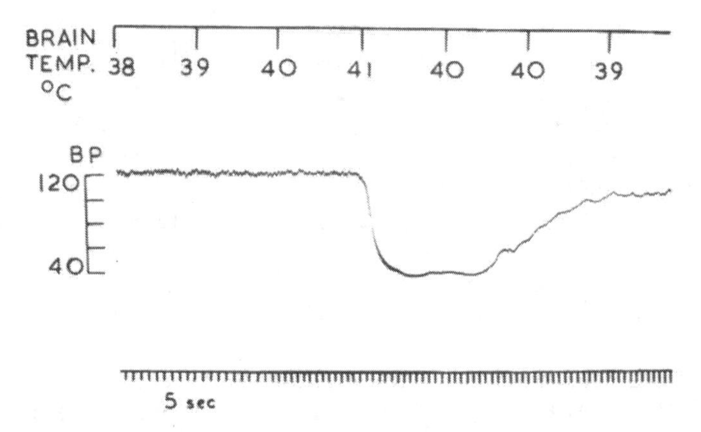

Fig. 144. Effect of heating the brain stem reticular formation on arterial blood pressure in a cat under sodium pentobarbital anesthesia. Femoral blood pressure is at 120 mm Hg. When the brain temperature reaches 41°C the blood pressure rapidly falls and rises again when the brain is cooled.

41°C. The reaction is due to vasodilatation and recovers when the brain is cooled. It may help to explain the circulatory collapse that sometimes occurs in heat exhaustion.

Respiratory reflexes

The respiratory center. The reflex centers controlling respiratory movements are located bilaterally in the medulla and pons. They are divided into three functional areas–inspiratory, expiratory, and pneumotaxic.

1. Inspiratory neurons are found in the medial reticular formation rostral to the obex; they can be identified by their discharge patterns being locked to the inspiratory phase of the respiratory cycle. These neurons are extremely sensitive to changes in carbon dioxide tension, hydrogen ion concentration, and to a lesser extent, the oxygen tension of the blood. Descending axons make contact with the motoneurons of the spinal cord from which emerge the motor nerves supplying the muscles of respiration. The phrenic nerve supplies the diaphragm; it arises from the third, fourth and fifth cervical segments. The upper thoracic nerves supply the intercostal muscles and give some branches to the diaphragm.

2. Expiratory neurons are found in the dorso-lateral reticular formation caudal to the obex; they can be identified by their discharge patterns occurring at the end of the inspiratory phase and again at the end of the respiratory cycle. There are, however, other respiratory neurons that discharge

continuously throughout both phases of the cycle. They probably serve as modulators of respiratory function during forced expiration.

3. Pneumotaxic neurons are found in the dorso-lateral region of the pons. They are connected to the medullary respiratory centers but appear to have little influence on the respiratory cycle during quiet breathing. They may, however, be activated from the hypothalamus, and this suggests that they can modify the rate and depth of breathing to meet the demands of the body in exercise and in other conditions of stress.

Reflex mechanisms. Respiratory neurons possess a limited degree of spontaneous activity but are not by themselves capable of giving a normal pattern of rhythmic breathing. The latter is maintained by reflex action arising from stretch receptors in the lungs, while impulses from many other sources contribute to the changing patterns of respiration that accompany different physiological conditions.

1. Vagal stretch reflexes. Expansion of the lungs during quiet inspiration stimulates inflation receptors in the bronchi and bronchioles. Impulses are carried by the vagal nerves to the medulla and, when the discharge reaches a peak frequency, the inspiratory center is inhibited and expiration follows. As the lungs collapse, the frequency of discharge falls, allowing the inspiratory neurons to become active again, and a new cycle commences. The effect is called the Hering-Breuer reflex. Deep expiration, forcing air out of the lungs, excites deflation receptors in the walls of the alveoli and these in turn inhibit the expiratory center. Thus when one center is active, the other center becomes inactive resulting in alternating movements of inspiration and expiration.

The mechanism of rhythmic breathing has been studied by making various sections in experimental animals:

- If both vagal nerves are cut, the Hering-Breuer reflex is lost, and breathing assumes an apneustic pattern with prolonged, slow inspiration and short expiration. Rhythmic breathing is not abolished, despite the loss of vagal stretch afferents, because impulses from the pneumotaxic center operate a second inhibitory mechanism.
- If a section is made between the medulla and pons to cut off the pneumotaxic center, rhythmic breathing is maintained, providing the vagi remain intact.
- If the pneumotaxic center is eliminated and both vagi cut, the inspiratory center is isolated and rhythmic breathing abolished. The animal breathes in gasps until breathing ceases.

An interpretation of these results is depicted in Figure 145. From what has been said above, it is clear that inspiration is the active phase of respiration and that expiration is passive, aided by the elastic recoil of the lungs. By itself, the inspiratory center merely produces a sustained discharge of

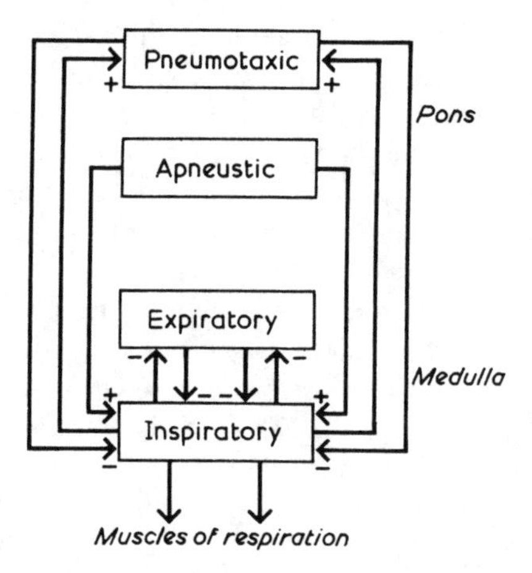

Fig. 145. Schematic diagram of respiratory centers showing possible interconnections and mode of action. At the level of the medulla, inspiratory and expiratory neurons interact to provide periodic discharge of impulses to the respiratory muscles. This accounts for the respiratory cycle during quiet breathing. At the level of the pons, the apneustic and pneumotaxic centers modify the rhythm by exerting facilitatory or inhibitory influences. The apneustic center is capable of driving the inspiratory neurons, which relay impulses to the pneumotaxic center. This center in turn exerts an inhibitory action on the inspiratory neurons.

impulses which, of course, can serve no useful purpose. The discharge is converted into a rhythmic action by means of two inhibitory mechanisms—vagal and pneumotaxic—that regulate the rate and depth of breathing in proportion to the existing requirements of the body.

2. Sino-aortic reflexes. Moderate degrees of oxygen lack reflexively stimulate breathing. The sensitive areas are chemoreceptors in the carotid and aortic bodies. The carotid body is situated at the bifurcation of the common carotid artery. It consists of a mass of nucleated cells richly supplied with blood and a dense sensory innervation provided by terminals of the sinus nerve. The aortic body is found in the concavity of the aortic arch near the origin of the pulmonary trunk. Its structure is similar to that of the carotid body but it is innervated from branches of the vagus nerve. A fall in arterial oxygen concentration causes a marked increase of impulse discharge in the afferent nerves, whereupon the pulmonary ventilation is considerably increased.

The sino-aortic reflexes have been studied experimentally in a number of ways:

- Electrical recordings from single fiber preparations show irregular, spontaneous firing when the animal is breathing air. On switching over to a low-oxygen mixture, an intense impulse barrage is provoked. Similar effects on the chemoreceptors are obtained by stimulating the cervical sympathetic, causing vasoconstriction of the blood vessels supplying the bodies and a consequent fall of arterial pO_2.

• Perfusion of the carotid bifurcation with blood at low oxygen tension reflexively excites the respiratory center. The hyperpnea observed can be abolished by section of the chemoreceptor nerves.

• The chemoreceptors can be stimulated by drugs such as lobeline or by hemorrhage, which lowers the oxygen tension and reflexively stimulates breathing. If the sino-aortic nerves are cut, anoxia causes only respiratory depression.

• The chemoreceptors are also sensitive to excess CO_2 and acidemia, but in this case section of the sino-aortic nerves has little effect. Respiration is significantly increased by direct action of these factors on the respiratory center.

The physiological importance of the sino-aortic reflexes is in their response to anoxia. This response is the only mechanism available for increasing pulmonary ventilation when the oxygen tension is lowered, as, for example, when the body is exposed to high altitudes. Breathlessness occurs when climbing to heights over 10,000 feet, and cyanosis develops above this level unless oxygen masks are worn. If the oxygen tension of the blood falls too low from any cause, breathing may become slow and shallow or depressed due to failure of the respiratory center.

Alimentary reflexes

The brain stem is a center of two important alimentary reflexes—vomiting and salivation.

1. The act of vomiting is a reflex effect by which the stomach empties its contents upward through the esophagus. It is often preceded by nausea and salivation. The stimuli that cause vomiting may come from the stomach or other parts of the alimentary tract or they may be central in origin, including unpleasant psychological factors. The motor reactions are very complex and involve the accessory muscles of respiration. Squeezing of the gastric contents occurs by simultaneous contraction of the diaphragm and muscles of the abdominal wall.

The center for the vomiting reflex has two functional parts, a reflex zone and a chemoreceptor trigger zone.

The *reflex zone* is situated on the dorso-lateral border of the lateral reticular formation. It receives afferents from the stomach and duodenum via the vagus and sympathetic nerves. Electrical stimulation induces immediate vomiting while adjacent points are ineffective.

The *chemoreceptor trigger zone* is located in the area postrema in the floor of the fourth ventricle. It contains clusters of large cells surrounded by a prominent capillary network and bundles of nerve fibers arising from smaller cells. The large cells are sensitive to emetics in the circulating blood, and the small cells serve as a conducting pathway for impulses initiating the reflex.

The dual mechanism can be demonstrated experimentally in the fasting dog. When a solution of copper sulphate is introduced into the stomach, vomiting occurs within a few minutes, but if the nerves to the stomach are cut, vomiting is delayed for about two hours–the time required for the drug to reach an adequate concentration in the circulating blood. On the other hand, if the drug is given intravenously, even after denervation of the stomach, vomiting results almost immediately. The peripheral mechanism for vomiting is initiated by excessive irritation of the gastro-intestinal tract. The central mechanism can be influenced directly by the administration of certain drugs–e.g., apomorphine and digitalis–or by impulses reaching the trigger zone from the higher centers. In motion sickness, impulses from the vestibular nuclei reach the trigger zone after passing through the flocculonodular lobe of the cerebellum.

2. Salivation. Saliva is a watery secretion produced reflexively from the parotid, submaxillary, and sublingual glands. It contains ptyalin, an enzyme for digesting starches, mucous for lubricating purposes, and a high concentration of potassium and bicarbonate ions. Under normal conditions the reflex is a response to the introduction of food into the mouth in order to aid mastication and swallowing; but the sight and smell of food or merely anticipating a good meal can also initiate a reflexive flow of saliva.

Reflex centers are scattered through the reticular substance of the brain stem. The superior salivary nucleus controls the submaxillary and sublingual glands, while the inferior salivary nucleus controls the parotid gland. Afferent fibers from the mouth, tongue, and palate are carried in the glossopharyngeal and lingual nerves to these centers. Efferent fibers belong mainly to the parasympathetic outflow accompanying the branches of the fifth, seventh, and ninth cranial nerves. The details are as follows:

- From the superior salivary nucleus, secretory fibers run in the chorda tympani branch of the facial nerve to join the lingual nerve on the deep surface of the mandible where they enter the submaxillary ganglion. The postganglionic fibers arising from here supply the submaxillary and sublingual glands.
- From the inferior salivary nucleus, secretory fibers travel in the lesser superficial petrosal nerve (a branch of the glossopharyngeal) to the otic ganglion. The postganglionic fibers join the auriculo-temporal nerve (the largest division of the trigeminal) to reach the parotid gland.

In addition to the parasympathetic secretory fibers, the salivary glands have a sympathetic supply derived from cells in the superior cervical ganglion. Stimulation of the sympathetic constricts the blood vessels to the glands, causing a diminished blood flow and inhibition of the salivary reflex. This may in part explain the dryness of the mouth that occurs at times of acute emotional distress. Some authors believe that the sympathetic can influence the chemical composition of the secretion.

Mid-brain

The oculomotor reflexes control the diameter of the pupil. Dilatation is brought about by the radial fibers and constriction by the sphincter of the iris. The response of the pupil to light entering the eye and the changes due to accommodation were described in Chapter VIII. It will be recalled that the oculomotor nucleus is the common efferent pathway for both light and accommodation reflexes and that the afferent influences come from the eye itself. However, the mid-brain is a region where somatic and visceral afferents converge from all over the body, many of them terminating on the oculomotor nucleus itself. Therefore, it is not surprising that impulses of extraocular origin also play an important role in controlling the diameter of the pupil.

Pupillary dilatation

Reference has already been made to the use of pupillary dilatation as an index of response in tracing splanchnic afferent pathways in the spinal cord. Stimulation of the central cut end of the splanchnic nerve causes dilatation of the pupil in the chloralosed animal. The diameter can be measured on a graticule inserted into the eye-piece of a low power microscope. In their course through the medulla and pons the afferent fibers occupy the lateral part of the reticular formation and are then directed forward to the central gray of the aqueduct to end in the oculomotor nucleus. Further evidence on the oculomotor mechanism is as follows:

1. The diameter of the pupil is maintained by a tonic parasympathetic discharge derived from cells in the oculomotor nucleus. The pupillo-dilator pathway has an inhibitory action on these cells, as shown by the fact that the reflex is present after section of the cervical sympathetic, but abolished if the oculomotor nerve is cut.
2. Stimulation of the sympathetic or general emotional excitement causes dilatation of the pupil. This is due in part to activation of the radial fibers of the iris and in part to inhibition of the oculomotor nucleus.
3. Pupillary dilatation occurs in response to noxious stimulation. Many spinal afferents carrying pain nerve impulses send inhibitory collaterals to

the oculomotor nucleus. The pathways for the pupillary and pain mechanisms are not the same, but have many features in common—e.g., bilaterality, diffuse projections, and high-threshold fiber components.

Pupillary constriction

The circular fibers of the iris form the sphincter that surrounds the margin of the pupil. It is a powerful muscle supplied by the short ciliary postganglionic fibers of the ciliary ganglion. The preganglionic fibers constitute the parasympathetic root of the oculomotor nerve.

1. Pupillo-constrictor neurons are found in the small-celled component of the oculomotor nucleus, comprising parts of the Edinger-Wesphal and anteromedian nuclei. Increased activity of these neurons causes constriction of the pupil and decreased activity causes dilation.

2. The discharges may be facilitated or inhibited by influences from elsewhere. Thus the light reflex is an excitatory drive resulting in constriction of the pupil, whereas impulses descending from the hypothalamus generally exert an inhibitory influence and dilate the pupil.

3. Stimulation of the limbic cortex has a profound effect on most autonomic functions including the oculomotor reflexes. For this reason the mid-brain may be regarded as part of a neuronal network integrating the reactions for emergency and defense.

VISCERAL AFFERENT FUNCTIONS IN THE CEREBELLUM

In previous chapters we have seen that the cerebellum is vitally concerned with the control of movements and posture, a fact that is fully supported by studies of human deficiencies. The cerebellum achieves its purpose by making corrective adjustments to the motor systems during the time that the movements are being carried out. As expected, much of the information required for these adjustments comes from the muscles and joints themselves. More recently it has been shown that the cerebellum receives information continously from receptors all over the body, including auditory, visual, cutaneous, and visceral signals. The contributions made by impulses of visceral origin will be considered here.

Methods of study

Many workers favor the use of sodium pentobarbital for mapping the distribution of localized responses as this anesthetic tends to reduce the background activity; others prefer the decerebrate unanesthetized animal. Single unit discharges are readily recorded from Purkinje cells or from the deep cerebellar nuclei by means of microelectrodes inserted into the substance at selected sites. Small balloons tied into the stomach, gallbladder, or loop of intestine provide a suitable form of stimulus by distension. For electrical stimulation, a distal pair of electrodes is applied to the central cut end of the splanchnic nerve, while a proximal pair records the nerve action potential. In this way the intensity of the stimulus can be evaluated in terms of the fiber spectrum excited.

The majority of electrode points responding to stimulation of the abdominal viscera are found in the posterior culmen of the anterior lobe, lateral to the paravermian vein (Fig. 146). The culmen corresponds to the anatomic subdivision respresenting the trunk of the animal and is designated the "splanchnic receiving area," which, however, also receives cutaneous afferents from the body wall.

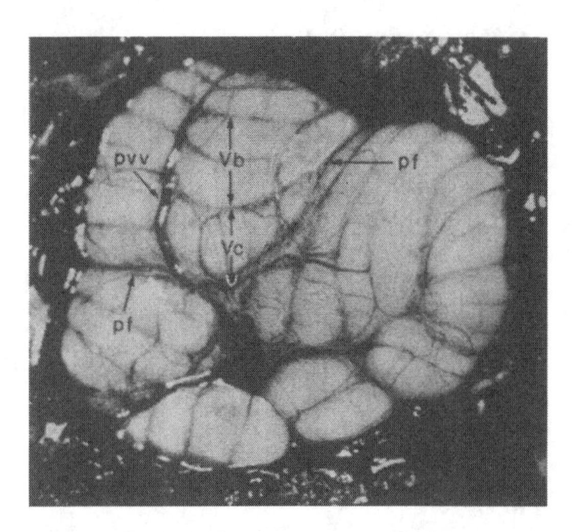

Fig. 146. A photograph showing the dorsal surface of the right side of a cat's cerebellum; pf is the primary fissure between culmen and simplex; pvv, the paravermian vein. Lobules Vb, Vc indicate sites of maximal evoked responses to stimulation of abdominal viscera.

Stimulation of abdominal viscera

The activity of single units in the culmen of the anterior lobe can be studied during distension of a hollow viscus by means of a balloon. Some units are silent in the absence of stimulation and respond only during the period of distension by discharging a burst of spikes. Other units fire spontaneously at low frequency and respond to stimulation by a greatly increased discharge (Fig. 147). A third type of activity recorded is a spontaneous discharge at high frequency. Figure 148 shows complete cessation of firing during the period of distension. These experiments suggest that impulses set up by distension of abdominal viscera can influence cerebellar activity in several ways: The silent unit is excited and the spontaneously discharging unit is either facilitated or inhibited.

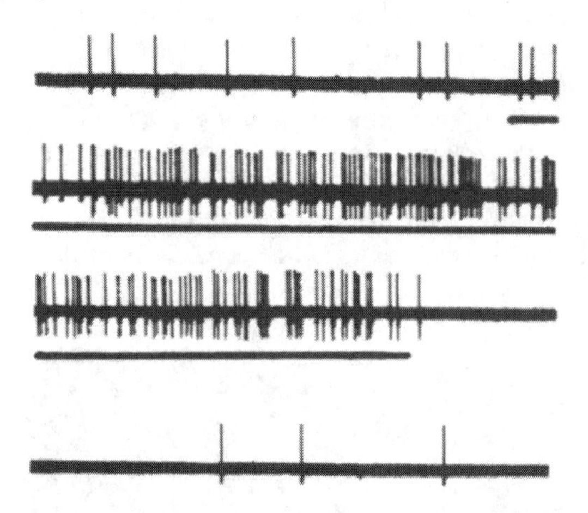

Fig. 147. Effect of distending the gall bladder on a spontaneously discharging cerebellar unit. Continuous record shows increased frequency of discharge during the period of distension, which is indicated by the signal line. Note post-excitatory depression after stimulus ceases. The cat used was under sodium thiopental anesthesia. The gall bladder was distended by means of a small balloon.

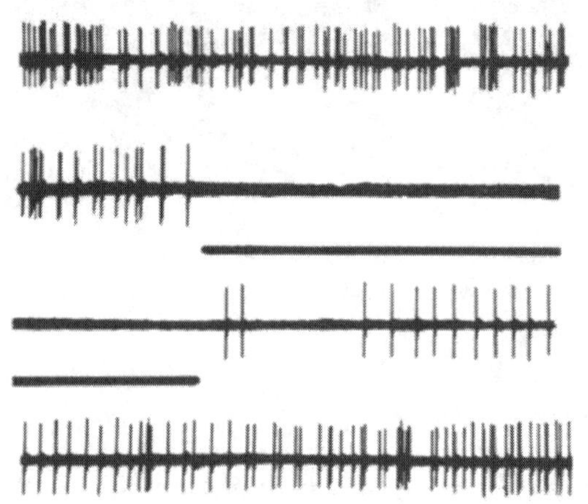

Fig. 148. Inihibition of single cerebellar unit discharging spontaneously at high frequency. Continuous record shows complete cessation of firing during period of stimulation. Note gradual return to original frequency on deflation of the balloon.

Stimulation of the splanchnic nerve

The majority of impulses originating in abdominal viscera are conveyed to the spinal cord by the splanchnic nerves. A single shock delivered to the central cut end of the nerve evokes a primary wave in the culmen of the cerebellum, When the recording electrode is on the pial surface, the primary wave is initially positive. When the electrode penetrates the substance of the cerebellum, the primary wave is initially negative. With further advancement of the electrode, a second positive wave is recorded (Fig. 149). The primary waves represent localized potential changes set up by activity in the terminals of the splanchnic afferent projections. They constitute physiological evidence of a nervous pathway between the point of stimulation and the recording site. The successive reversals of polarity are characteristic of recordings from the cerebellum, where the cortex is thrown into folds and several layers are encountered as the electrode is advanced toward the central core of white matter. In the spinal cord, the A beta fibers of the splanchnic nerve pass into the ipsilateral posterior column, close to

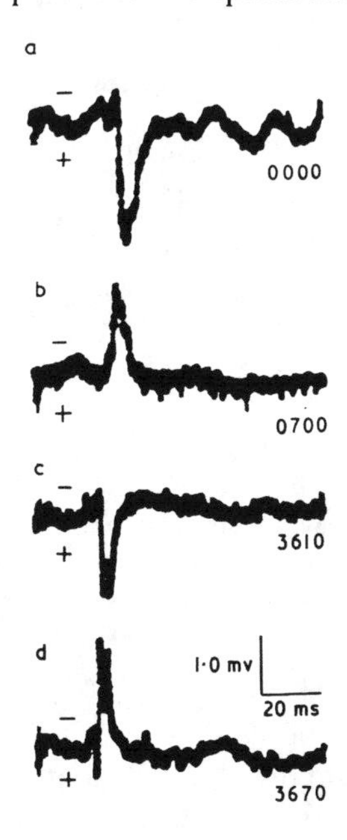

Fig. 149. Evoked primary responses recorded from the culmen of the cerebellum following electrical stimulation of ipsilateral splanchnic nerve. Note a. initially positive primary wave at pial surface: b. reversed primary wave 700 μ below pial surface: c. deep positive wave: and d. deep negative wave. The successive reversals of polarity are characteristic of recordings from the cerebellum where several layers of the cortex are encountered as the microelectrode is advanced.

the mid-line; the A delta fibers ascend bilaterally in the anterolateral columns.

Cerebellar units responding to splanchnic nerve stimulation fall into two categories–low threshold and high threshold. Low-threshold units are usually silent in the absence of stimulation. They are activated by the A beta group of splanchnic nerve afferents and have a latency of about 15 msec. High-threshold units are activated by the A delta group of fibers and have latencies between 20 msec and 25 msec. The traces shown in Figure 150 are simultaneous recordings from the cerebellum and splanchnic nerve in the cat.

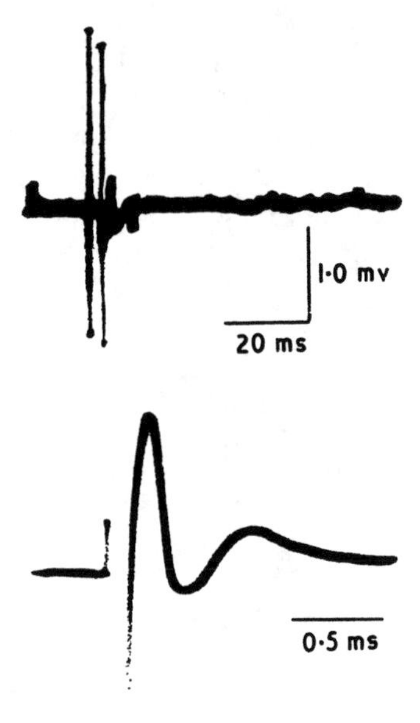

Fig. 150. Simultaneous recordings from the cerebellar cortex and the splanchnic nerve of the cat. Upper trace shows high-threshold unit discharge; latency of the first spike is 20 msec. Lower trace represents splanchnic nerve action potential showing A beta and A delta components. (Calibrations: upper trace, 20 ms and 1 mv; lower trace, 0.5 msec.)

A useful way of studying the discharge properties of single units is by observing the results of changing the stimulus intensity. It is found that progressive increase of intensity has effects on the latency, number of spikes and probability of firing.

1. Reduction in the mean latency. With weak intensities of stimulation the latency of the evoked spike is variable. As the stimulus intensity is increased the mean latency is reduced; longer latencies drop out, and a greater incidence of the shorter latencies is observed.

2. Increase in number of spikes. The number of spikes in an evoked discharge is also variable, yet a small increase of stimulus intensity will convert a one-spike discharge into two or more spikes. This is shown in Figure 151 in which multiple bursts of spikes are recorded with stronger stimulation.

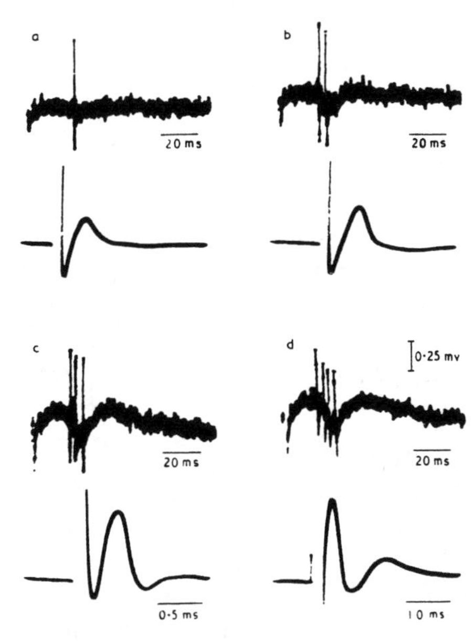

Fig. 151. Relationship between number of spikes and intensity of stimulus. The splanchnic nerve action potential is recorded below each cerebellar response. Note correlation between number of evoked spikes and size of the nerve action potential.

3. Increase in the probability of response. An evoked discharge does not always follow each successive stimulus. However, increasing the intensity of the stimulus usually increases the probability that the unit will discharge at least one spike. The relationship between stimulus intensity and probability of firing is illustrated in the graph of Figure 152. It can be seen that probability is low with weak intensities and rises rapidly as the stimulus gets stronger.

Interaction between cutaneous and splanchnic afferents

To demonstrate the possibility of interaction between the afferent pathways sharing the same pupulation of neurons, the technique of conditioning and testing is employed. A stimulus to one pathway is followed by a

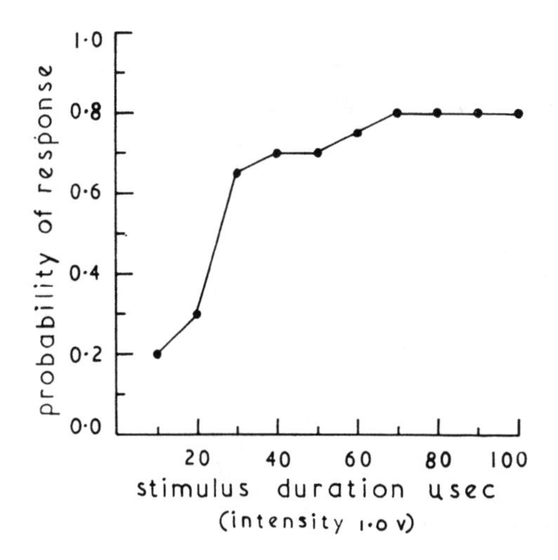

Fig. 152. Relationship between probability of response and strength of stimulus. As the strength is increased, the probability of the unit firing at least one spike is also increased. The ordinate represents probability factor, the abscissa, stimulus duration in μ sec.

stimulus to the other over a range of time intervals. The splanchnic receiving area in the posterior culmen overlaps to some extent the representation of the abdominal wall; this provides the structural basis for synaptic convergence and opportunities for sensory interaction. The experimental data described below show that when the separation between conditioning and testing stimuli reaches a critical value, the most prominent change observed in the anesthetized animal is inhibition of the test response.

1. Effects on primary wave responses. In the experiment illustrated in Figure 153, a conditioning stimulus is delivered to the skin of the abdominal wall, and a testing stimulus delivered to the splanchic nerves. The conditioning-testing (C-T) interval is 70 msec and a primary wave is recorded from the cerebellum in response to each stimulus. It can be seen that when the C-T interval is shortened to 60 msec, the amplitude of the test primary wave is decreased and, when the C-T interval is 40 msec, the response to the testing stimulus is absent, but reappears when the conditioning stimulus is switched off.

2. Effects on single unit responses. An example of cutaneous-splanchnic interaction is illustrated in Figure 154, in which the upper trace shows that the unit responds to both stimuli when the interval of separation is about 70 msec. In the lower trace, the C-T interval is reduced to 40 msec and the test response is absent. Comparable results are obtained when the sequence of stimuli is reversed.

3. Effects on response probability. Interaction effects are best demonstrated by computing the probability that a unit will respond to stimulation

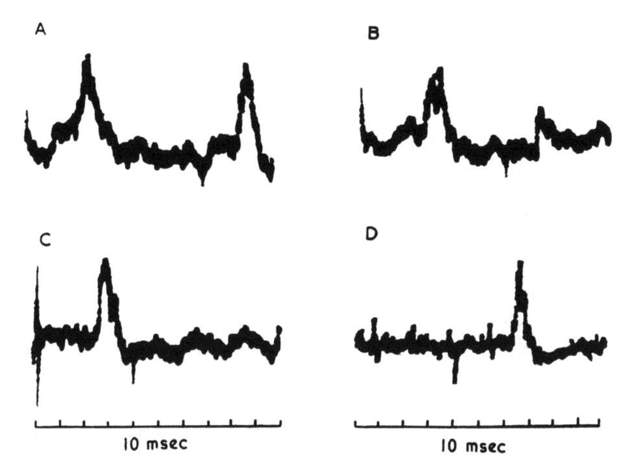

Fig. 153. Interaction between cutaneous and splanchnic afferents revealed by effect on test primary wave. Conditioning stimulus is applied to skin, testing stimulus to splanchnic nerve. In A, where the interval of separation between the two stimuli is 70 msec, two primary waves are recorded. In B, where the interval of separation is reduced to 60 msec, the test primary wave is decreased in amplitude. The lower traces show the effect of reducing the C-T interval to 40 msec. The test primary wave is now absent but reappears when conditioning stimulus is switched off.

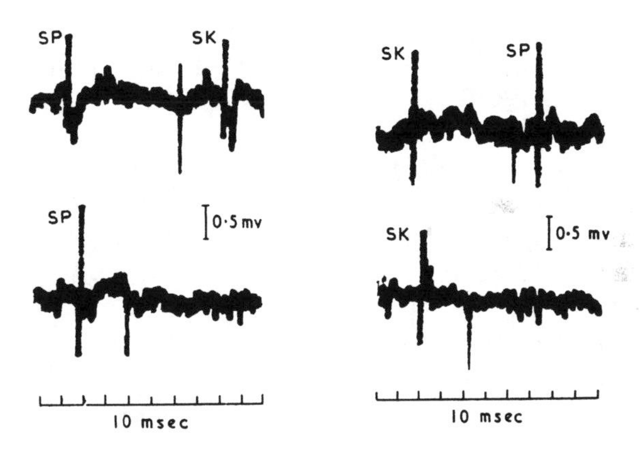

Fig. 154. Effect of interaction between cutaneous and splanchnic afferents on single cerebellar unit. On the left, upper trace shows that the unit responds to both splanchnic and skin stimulation when interval of separation is 70 msec; lower trace illustrates complete blocking of test response when interval is reduced to 40 msec. Traces on right show comparable results when sequence of stimuli is reversed. SP represents response to splanchnic nerve, SK, reponse to skin.

on any given trial. If the presence or absence of a response to a testing stimulus is determined for each interval of time, the results can be plotted in the form of a graph, as shown in Figure 155. From this graph it can be

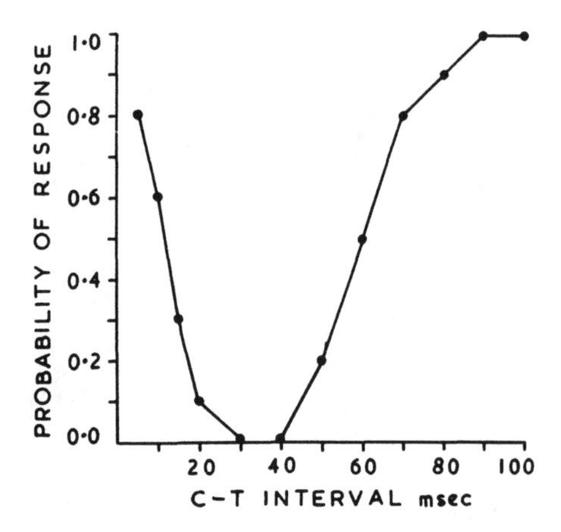

Fig. 155. Graph showing relationship between probability of response to a testing stimulus and the conditioning-testing (C-T) interval. Probability = 1.0 when C-T interval is 90 msec or longer. Note diminishing probability as the C-T interval is reduced and complete cessation of the test response (probability = 0) with intervals between 25 msec and 40 msec. With intervals shorter than 25 msec, probability rises again—i.e., inhibition is less intense at the briefer intervals.

seen that the probability of response is progressively lowered as the C-T intervals are shortened until, at an interval of 40 msec between the two stimuli, the unit ceases to respond (P = O). Inhibition of the test response is maintained for intervals as short as 25 msec, but the effect becomes less intense at briefer intervals.

All these experiments emphasize an important aspect of neuronal function: When two pathways converge on the same neuron, activity in one pathway can block the influence of the other pathway, the effect depending on the time interval between the two sets of impulses. This feature is common to many other parts of the nervous system, but is obviously of great significance to the normal functioning of the cerebellum, owing to the extremely wide range of its input from almost every kind of sense organ.

VISCERAL AFFERENT FUNCTIONS IN THE CEREBRUM

It is now established that impulses of visceral origin ascend in the central nervous system to reach localized areas in the thalamus and cerebal cortex. In their ascent they contribute to synaptic mechanisms at various levels of integration—spinal, cerebellar, and reticular—before being distributed to their thalamo-cortical destinations. Visceral afferents form a fairly complex projection system having many features in common with the somatosensory system. Under normal circumstances they do not play a major role in consciousness, being more concerned with reflex mechanisms controlling the internal environment. Yet, when stimulated excessively, they can dominate the entire nervous system involving all the diverse reactions associated with visceral pain.

Thalamus

Despite the complexity caused by the inflow of impulses from all over the body, the thalamus maintains a high degree of topographical preference for different sensory modalities. Splanchnic nerve afferents are relayed in the lateral part of the ventrobasal nucleus where arm and trunk areas overlap; vagal nerve afferents are relayed in the posteromedial nucleus. Thus at this level there is no opportunity for convergence and interaction between these two major inputs from the viscera. Much of the experimental work on visceral responses in the thalamus has been concentrated on splanchnic relays to the sensory cortex, possibly because these have assumed some clinical importance, but vagal afferent relays to the orbital cortex are well documented and will be considered later in this chapter.

Evoked activity

Electrical stimulation of the central cut end of a splanchnic nerve evokes primary wave and single unit responses: The *primary waves* are initially positive with a latency of 5 msec to 6 msec when only the A beta group of fibers is excited. It is possible to trace evoked splanchnic potentials in the ipsilateral dorsal column, gracile nucleus, contralateral lemniscus, and contralateral thalamus. Thus there exists a fast-conducting fiber system carrying impulses from the abdominal viscera to the contralateral thalamus. When stronger stimulation is used to excite the A delta group of fibers, evoked responses are obtained in the ipsilateral as well as the contralateral thalamus. The impulses travel by the slower-conducting route in the anterolateral columns on both sides of the cord. The dual pathway for splanchnic afferent impulses resembles the lemniscal and anterolateral projections of the somatosensory system.

Single unit responses evoked by splanchnic nerve stimulation usually take the form of a repetitive discharge of spikes. A beta responses have a mean latency of about 12 msec and probably signal states of distension in the internal organs. A delta responses have much longer latencies and are thought to be implicated in visceral pain. The majority of units responding to splanchnic stimulation also respond to cutaneous stimulation and sensory interaction can be readily demonstrated. Some units, however, are identified with splanchnic activity alone. The latter represent only a small number out of the total population samples, so it must be concluded that the discriminative capacity of the thalamus is limited for visceral sensation.

Sensory cortex

There are two different areas in the sensory cortex that are known to receive splanchnic afferent projections from the relay nuclei of the thalamus; these are found in the primary and secondary sensory areas. In the cat, the primary sensory area is located just anterior to the angle of the ansate sulcus; the second sensory area lies in the rostral strip of the anterior ectosylvian gyrus (Fig. 156). The methods used for mapping the cortical responses are similar to those described for the cerebellum and the results reveal that the functional differences between the low and high threshold fiber groups are faithfully maintained at the cortical level.

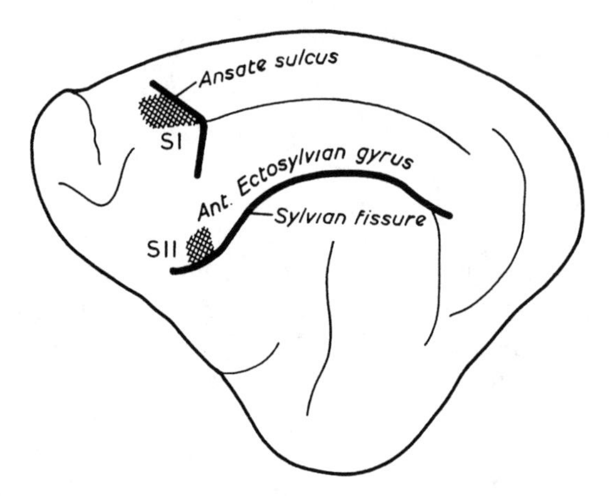

Fig. 156. Diagram of left cerebral hemisphere of the cat showing splanchnic afferent projection areas. The primary area (SI) is located anterior to the ansate sulcus. The secondary area (SII) is located in the rostral part of the anterior ectosylvian gyrus.

Evoked activity

Potential changes may be recorded from the surface of the cortex during distension of a viscus by means of a balloon. Cortical activity evoked by peripheral nerve stimulation may be displayed simultaneously with the nerve action potential.

1. Mechanical stimulation. In lightly anesthetized animals, the potential changes are superimposed on the electrocorticogram. Deep anesthesia reduces the spontaneous activity of the cortex when the evoked potentials can be more easily recognized. They consist of burst discharges varying in duration and amplitude according to the degree of distension and the site of the recording electrode. Figure 157 illustrates the distribution of potentials evoked by distension of the stomach. Maximal responses are recorded at points close to the angle of the ansate sulcus, while points further away show decreased responses in both rostral and caudal directions. The experiment gives ample proof of the discrete localization of the sensory projections to the cerebral cortex.

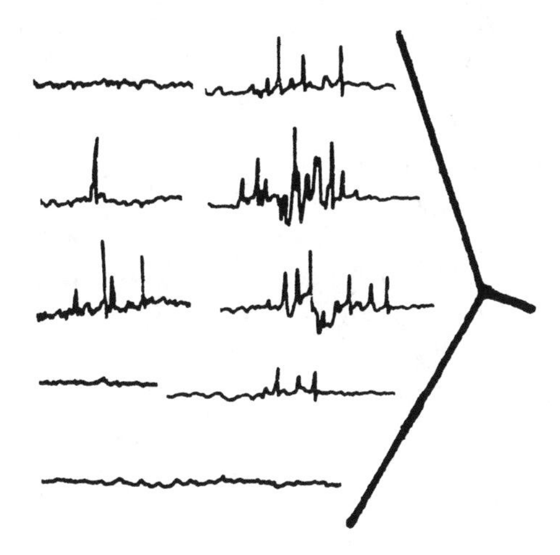

Fig. 157. Surface potential changes evoked in cat cerebral cortex during distension of the stomach by means of a balloon. Position of the ansate culcus is seen on the right. Note maximal responses are recorded from points close to the sulcus, while points further away have decreased responses.

2. Electrical stimulation. A single shock delivered to the central cut end of the splanchnic nerve evokes a primary wave and single unit discharges in the corresponding area of the contralateral cortex.

The *primary response* may be recorded from each of the two sensory areas. When recorded from the pial surface, it takes the form of an initially positive wave, often followed by three or four secondary waves suggesting a spread of activity from the primary afferent focus. The response is due to current flowing from the surface toward the deeper layers of the cortex where the thalamic afferent fibers terminate. An evoked primary wave is produced only when impulses reach the recording electrode and therefore it indicates the presence of a link between the stimulated nerve and the recording site.

Single unit responses may occur as initially negative spikes firing either singly or in short bursts and rarely more than 1 mv in amplitude. Or, more typically, the response takes the form of a positive-negative spike, 3 mv to 5 mv in amplitude, occurring about the peak of the primary wave (Fig. 158). Spontaneously discharging units may be influenced in one of two ways. Either the unit increases its firing rate up to 400/sec or responds by giving an evoked multiple spike discharge (Fig. 159). The effects of increasing the intensity of the stimulus are comparable to the patterns observed in units of the somatosensory cortex as described in Chapter XV. These effects may be summarized as follows:

- The mean latency for the first spike of the discharge is reduced and there is a greater incidence of the shorter latencies. The synchronous arrival of impulses due to the stronger stimulus may possibly speed up transmission processes.
- The number of spikes per discharge is increased, i.e., single spikes are converted into multiple spikes.
- The probability of the unit firing at least one spike is increased. At low stimulus intensities, exciting only a small fraction of the A beta fibers, prob-

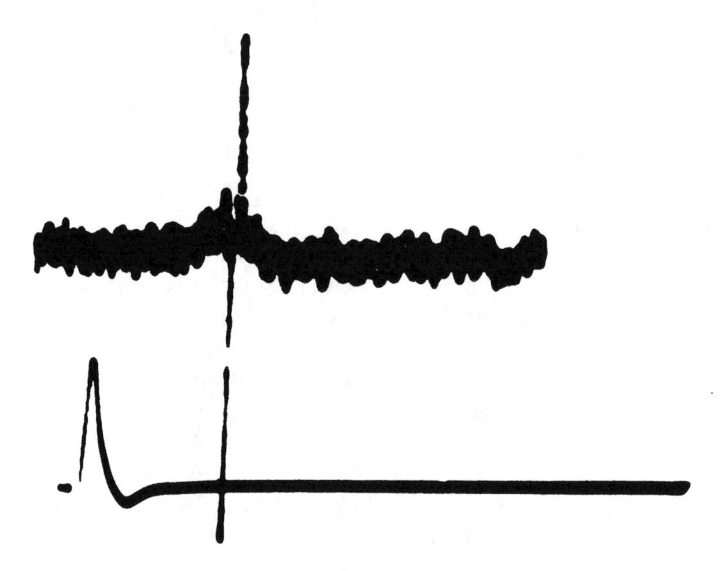

Fig. 158. Single cortical unit response to stimulation of contralateral splanchnic nerve. Upper trace illustrates typical positive-negative spike superimposed on evoked primary wave. In lower trace, note splanchnic nerve action potential is simultaneously recorded.

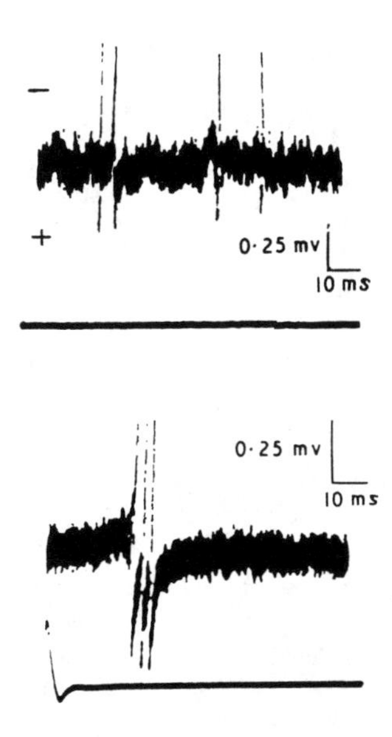

Fig. 159. Effect of splanchnic afferent stimulation on spontaneously active units. Upper trace, before stimulation, demonstrates the unit firing irregularly. Lower trace shows how this unit responds to a single shock delivered to the central cut end of the splanchnic nerve (action potential simultaneously recorded). Calibrations are 0.25 mv and 10 msec.

ability is low; as the stimulus intensity is increased, exciting a larger fraction of fibers, the probability curve becomes steeper until a high level of probability is reached with maximum excitation of the A beta group.

3. Interaction. The cortical discharge evoked by stimulation of the splanchnic nerve can be prevented by prior stimulation of a cutaneous nerve. This phenomenon depends upon afferent convergence and neuronal sharing. In the experiment illustrated in Figure 160, a cortical unit responds to stimulation of the abdominal skin by discharging a spike; the same unit responds again to stimulation of the splanchnic nerve. The interval between the two stimuli is 70 msec. In the lower trace of the figure the interval is reduced to 30 msec, with the result that the unit fails to respond to the splanchnic stimulus. Interaction plays an important physiological role. It is a means by which cortical neurons create a barrier to unlimited bombardment from impulses arriving synchronously at the thalamo-cortical terminals. The duration of this inhibitory effect is variable, but, in general, depression of the spike-generating mechanism lasts about 25 msec to 50 msec and is therefore related to the temporal separation of the conditioning and testing stimuli. During this period of time other excitatory influences are unlikely to succeed unless they are extremely powerful.

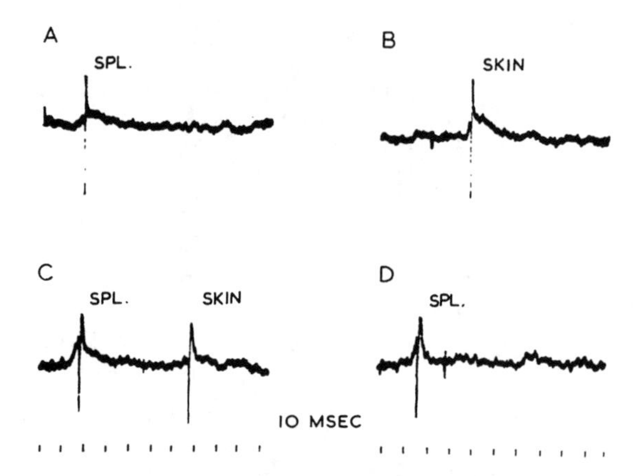

Fig. 160. Interaction between cutaneous and splanchnic nerve afferents. Upper traces show, respectively, response of a cortical unit to cutaneous stimulation alone and response of same unit to stimulation of the splanchnic nerve alone. Lower traces show effect of a conditioning-testing sequence: When the C-T interval is 70 msec, the unit responds to both stimuli; when the C-T interval is reduced to 30 msec, the test response is blocked. Time calibration is 10 msec.

It can be concluded from the above evidence that there is a remarkable parallel of organization between the visceral and somatic afferent systems. Splanchnic afferent impulses ascend in the central nervous system to the thalamus and are relayed to both primary and secondary areas of the sensory cortex. The initial discharges of the cortical neurons arise in the deeper layers and spread to the surface along intracortical pathways, yet retain a discrete functional localization in which A beta and A delta responses can be separately identified. The variability of a cortical response to peripheral stimuli becomes less variable as the stimulus intensity is increased, suggesting that the amount and type of information conveyed to the cortex are reflected in the patterns and timing of the cortical unit discharges. Finally, indiscriminate transmission of impulses from thalamus to cortex is subject to modification by interaction effects enforcing prolonged periods of depression.

Orbital cortex

There are good reasons for believing that the orbital cortex is a visceral receiving area and that it also serves as an integrating center for viscerosomatic reactions. Anatomically, it is one of the components of the limbic complex, contributing to the circuitry that ultimately feeds into the hypothalamus. In addition, pathways have been traced from the orbital cortex to the brain stem reticular formation and cerebellum. Physiologically, it has wide-ranging influences on many functions of the body, including the cardiovascular and respiratory systems, gastric functions, and temperature regulation. Clinically, it may be involved in the autonomic disturbances manifested in all kinds of emotional stress.

The afferent pathways to the orbital cortex originate from all over the body, but topographical patterns which are so characteristic of the sensory areas are almost absent. In fact, there is considerable overlapping of representation for vagal, splanchnic, and cutaneous afferent projections with no obvious subdivisions into leg, arm, and face areas. In the cat, the responsive cortical points are limited to the posterior orbital gyrus, a region lying between the orbital sulcus and the olfactory tract (Fig. 161).

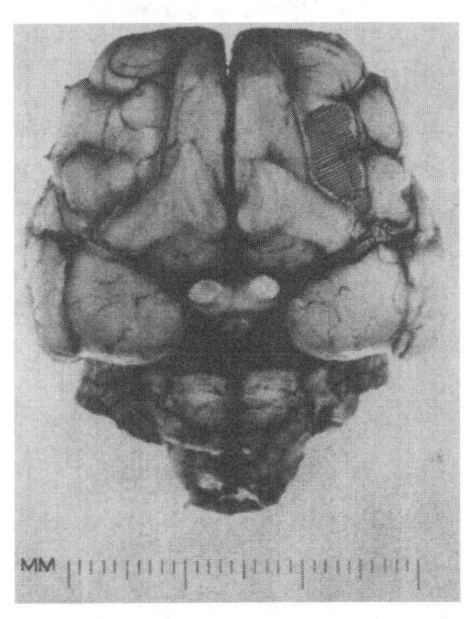

Fig. 161. Photograph showing ventral surface of cat brain. Shaded area over posterior orbital gyrus indicates site of visceral afferent projections. The olfactory tract is to the left and the cut optic nerves can be seen centrally.

Vagal afferent responses

In 1938 Bailey and Bremer described the "isolated encephalon" preparation in which the spinal cord and both vagi were cut and the animal kept alive by artificial respiration. They observed that induction shocks delivered to the central end of either vagus produced potentials changes on the orbital surface of the frontal lobe.

Under chloralose anesthesia, primary waves can be obtained from all parts of the posterior orbital gyrus following single shocks to the central cut end of the cervical vagus. Ipsilateral and contralateral stimuli are equally effective. Single unit responses appear as a single diphasic spike at threshold intensity of stimulation, but occurring in bursts of three or four spikes with stronger stimulation (Fig. 162). The same cortical points nearly always respond to stimulation of the body wall and inhibitory actions can be demonstrated in conditioning and testing experiments.

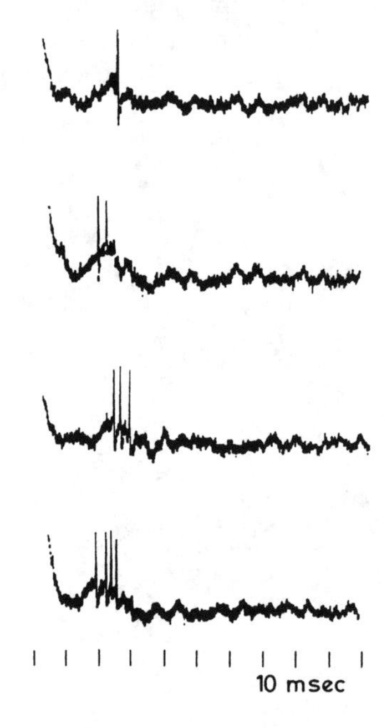

Fig. 162. Single unit responses recorded from orbital cortex. Cat was under chloralose anesthesia. From above downward, note increase in number of spikes per discharge with increasing intensities of stimuli delivered to central cut end of cervical vagus. Time calibration is 10 msec.

Splanchnic afferent responses

Evoked responses following electrical stimulation of one splanchnic nerve may be recorded from the orbital cortex in both cerebral hemispheres. Surface-positive primary waves are obtained over an area that virtually overlaps the representation of the lower intercostal nerves. Unit responses evoked by splanchnic nerve stimulation are of two types—low threshold and high threshold.

1. A few orbital units are capable of being driven by stimulation of the A beta group of fibers. Such units are usually silent or fire irregularly in the absence of stimulation and respond to the stimulus by firing in a periodic burst lasting about 70 msec. The latency for the first spike of the evoked discharge is about 15 msec (Fig. 163). Low-threshold units cannot be driven by vagal or cutaneous stimuli and therefore, despite their relative scarcity, they supply evidence for a discrete projection pathway to the orbital cortex.

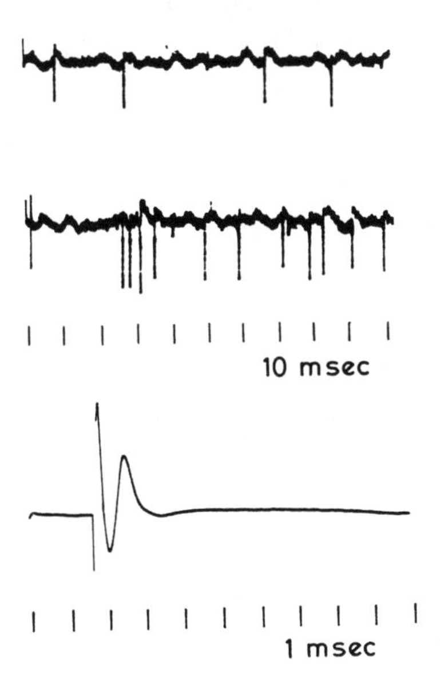

Fig. 163. Periodic firing of orbital unit evoked by electrical stimulation of the splanchnic nerve. The upper trace shows irregular firing of the unit in the absence of stimulation. The middle trace demonstrates the effect on firing frequency of the unit following single shock stimulation to the splanchnic nerve. The lower trace illustrates simultaneous recording of the splanchnic nerve action potential. Note that only the A beta group of afferents is excited.

2. The vast majority of orbital units respond to stimulation of the A delta group of fibers. The same units can also be driven by vagal and cutaneous stimuli suggesting that they belong to overlapping fields of representation shared by many peripheral sources. Accordingly, interaction between these various afferent projections can easily be demonstrated. An example of splanchnic-vagal interaction is illustrated in Figure 164. The upper trace shows the response of an orbital unit to vagal stimulation alone; the middle trace shows how the vagal response is blocked by a preceding conditioned

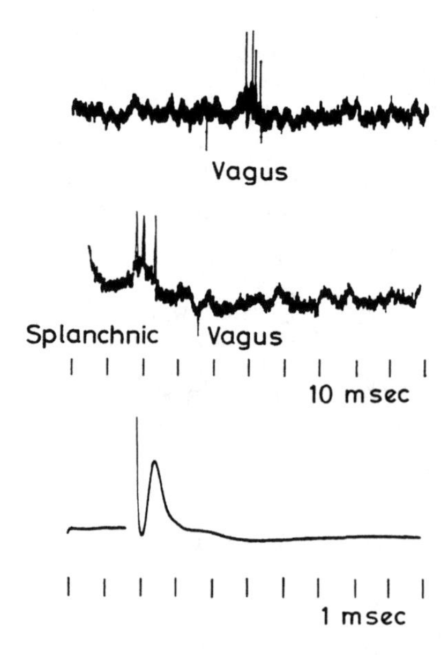

Fig. 164. Splanchnic-vagal interaction. Cat was under chloralose anesthesia. Upper trace demonstrates response of orbital unit to vagal stimulation alone. Note delay of stimulus artifact on sweep. Middle trace illustrates response of same unit to splanchnic nerve stimulation. When this is presented in a conditioning-testing sequence, the vagal response is inhibited. C-T interval was about 30 msec. In lower trace, simultaneous recording of splanchnic nerve action potential shows activation of A delta fibers.

stimulus to the splanchnic nerve; the lower trace is a simultaneous recording of the splanchnic nerve action potential showing activation of the A delta fibers. Comparable forms of interaction may be observed between splanchnic and cutaneous afferents.

Functional interpretation

It may be concluded from the experimental evidence that the orbital cortex is concerned in some way with the higher control of visceral functions. Vagal and splanchnic afferents converge on this structure and are capable of changing the patterns of its neuronal discharges. The influence exerted by the orbital cortex on other structures in the brain, particularly the hypothalamus, will in turn depend upon the information received from the viscera themselves. From a physiological point of view the orbital cortex can be distinguished from the sensory cortex in a number of important ways:

1. There is no definite topographical organization to indicate a discrete body representation like that found in the sensory areas.

2. The extent of convergence between visceral and cutaneous afferents shows almost complete overlap of responsive points.

3. The same unit can often be influenced by several different sensory modalities.

4. Afferent projections to the orbital cortex are bilateral in character and are equally effective.

5. The orbital cortex is susceptible to barbiturate depression to a degree that is not found in the sensory areas.

All these observations suggest that the two afferent projection systems operate so differently that the term "visceral brain" may be an apt description. By this is meant the parts of the cortex together with their subcortical connections which collectively control the visceral functions of the body. The orbital cortex has an essential role to play in this organization. It is thought to act as an integrating center for many channels of information, where the necessary adjustments can be made to meet the requirements of environmental change.

SELECTED BIBLIOGRAPHY

Bailey, P., and Bremer, F. A sensory cortical representation of the vagus nerve. *J. Neurophysiol.* 1: 405-412, 1938.

Downman, C.B.B., and McSwiney, B.A. Reflexes elicited by visceral stimulation in the acute spinal animal. *J. Physiol.* 105:80-94, 1946.

Franz, D.N., Evans, M.H., and Perl, E.R. Characteristics of viscero-sympathetic reflexes in the spinal cat. *Amer. J. Physiol.* 211:1292-1298, 1966.

Gernandt, B., and Zotterman, Y. Intestinal pain: An electrophysiological investigation on mesenteric nerves. *Acta Physiol. Scand.* 12:56-72, 1946.

Korn, H. Splanchnic projection to the orbital cortex of the cat. *Brain Res.* 16:23-38, 1969.

Maclean, P.D. Psychosomatic disease and the "visceral brain." *Psychosom. Med.* 11:338-353, 1949.

McLeod, J.G. The representation of the splanchnic afferent pathways in the thalamus of the cat. *J. Physiol.* 140:462-478, 1958.

Newman, P.P. Single unit activity in the viscero-sensory areas of the cerebral cortex. *J. Physiol.* 160:284-297, 1962.

Newman, P.P. *Visceral Afferent Functions of the Nervous System.* London: Edward Arnold, 1974.

Newman, P.P., and Paul, D.H. The representation of some visceral afferents in the anterior lobe of the cerebellum. *J. Physiol.* 182:195-208, 1966.

Sillito, A.M., and Zbrożyna, A.W. The localization of pupilloconstrictor function within the mid-brain of the cat. *J. Physiol.* 211:461-477, 1970.

CHAPTER XXI

Nervous Control of the Viscera—Efferent System

The hypothalamus may be regarded as the origin of the neuro-endocrine outflow. The *neural* part of the outflow is also known as the autonomic nervous system. It extends downward into the brain stem and spinal cord and its peripheral fibers carry motor impulses to the viscera, blood vessels, and secretory glands. As in the somatic nervous system, the transmission of impulses is mediated by chemical substances liberated at the nerve terminals. However, many of the synaptic connections of the autonomic nerves are made outside the spinal cord in ganglion cells and numerous nerve plexuses that accompany the blood vessels to the organs they supply. The *endocrine* part of the outflow produces its effects through hormones secreted into the blood stream. These substances are transported by the circulation to their target organs which in turn regulate the internal secretions by reacting back on the hypothalamus. A reciprocal relationship therefore exists between the central nervous system and the endocrine glands. If the activity of the hypothalamus is abnormally increased or decreased, corresponding derangements occur in endocrine function, producing symptoms of excessive secretion or insufficiency.

AUTONOMIC OUTFLOW

The autonomic nerves are classified into two major subdivisions called the sympathetic and the parasympathetic. Some physiological systems are controlled predominantly by one or the other, but in many instances the organs are innervated by both. In such cases, the two divisions act reciprocally, excitation of one causing inhibition of the other. Reciprocal action is obviously important to the function of the hollow organs. For example, emptying of the urinary bladder is accomplished by contraction of the muscular wall and relaxation of the sphincter. The detrusor muscle is supplied by parasympathetic fibers from the sacral segments of the cord, while the internal sphincter is supplied by sympathetic fibers from the lumbar segments of the cord.

Sympathetic division

Functional anatomy

The sympathetic nerves arise from columns of cells in the gray matter of the lateral horn of the spinal cord. The columns extend from the first thoracic segment (T1) to the second lumbar segment (L2). The axons pass out of the cord through the ventral roots as preganglionic fibers or white rami in company with the spinal motor nerves supplying the skeletal muscles. After emerging from the cord, they leave the spinal nerves and proceed to the ganglia of the sympathetic chain, where they make synaptic contacts. The white rami are all myelinated and belong to the B group of fibers. From the sympathetic chain, postganglionic fibers take devious routes to their destinations in one of the organs. Many of them, however, pass back as the grey rami to join a spinal nerve through which they are distributed to the blood vessels and skin. The grey rami are mostly unmyelinated and belong to the C group of fibers.

The sympathetic chain extends down the whole length of the vertebral column; it is divided on each side into four parts:

1. The cervical part consists of three ganglia–superior, middle, and inferior–from which postganglionic fibers are distributed along the arteries to structures in the head, neck, and thorax. The superior cervical ganglion gives rise to the internal carotid plexus which sends branches to the dilator muscle of the pupil. The inferior cervical or stellate ganglion is often fused with the first thoracic ganglion and contributes to the deep part of the cardiac plexus.

2. The thoracic part comprises a series of ganglia from each thoracic spinal segment. Branches from the upper five ganglia are distributed to the aortic, cardiac, and pulmonary plexuses. Branches from the lower seven ganglia form the greater and lesser splanchnic nerves. The lowest splanchnic nerve arises from the last thoracic ganglion and ends in the renal plexus.

3. The lumbar part is situated in front of the vertebral column. It is composed of a dense network of fibers, some of which enter the celiac ganglion, while others form numerous secondary plexuses. Most of the branches run with the arteries to supply the abdominal viscera; others descend into the pelvis to join the hypogastric plexus.

4. The pelvic part lies in front of the sacrum. It comprises the sacral ganglia which, together with the presacral nerves, contribute to the hypogastric and pelvic plexuses. The branches are distributed to the pelvic viscera and arteries of the lower limbs.

Effects of sympathetic stimulation

These may be described as local and general. Local effects include the following:

1. The eye. Dilatation of the pupil and retraction of the eye lids produce a staring gaze. In Horner's syndrome, in which the cervical sympathetic fibers are damaged, the pupil on the affected side is constricted owing to unopposed action of the parasympathetic supplying the sphincter muscle of the iris.

2. The thoracic viscera. Effects are acceleration of the heart, increase in the force of contraction, and dilatation of the coronary arteries. The pulmonary arteries are constricted by vasomotor fibers, but the bronchi are dilated by inhibition of the smooth muscle in their walls.

3. The abdominal viscera. Local effects in this area are as follows:

• Gastrointestinal tract. Inhibition of peristalsis and increased tone of the sphincters. The net result is relaxation of the gut and slowed propulsion of its contents.

• Adrenal medulla. This structure may be regarded as a specialized ganglion of the sympathetic outflow supplied by preganglionic splanchnic nerve fibers. Its chromaffin cells manufacture the catecholamines noradrenaline and andrenaline (epinephrine) which are released into the circulation. The action of these hormones will be described later in this chapter.

• Splanchnic arterioles. Vasomotor fibers in the splanchnic nerves constrict the blood vessels, increase the peripheral resistance, and cause a considerable rise in systemic blood pressure.

4. The pelvic viscera. Relaxation of the wall of the bladder and rectum and closure of the sphincters occur. Contraction of the muscle coat around the seminal vesicles and prostate result in ejaculation.

5. The skin. Local effects include erection of the hairs, constriction of cutaneous blood vessels, and the secretion of perspiration. The sympathetic fibers to the sweat glands are cholinergic, but the apocrine sweat glands are sensitive to circulating adrenaline.

6. The limbs. The main blood vessels and arterioles to the skin constrict. Active dilatation of the vessels supplying the skeletal muscles leads to increased blood flow.

General effects

Sympathetic activity is responsible for the reflex adjustments of the body under normal environmental conditions and also for mobilizing the resources of the body in times of stress. The former may be described as a stability mechanism and the latter as an emergency mechanism.

1. Stability mechanism. The idea of homeostasis was mentioned briefly in the previous chapter. Essentially, the various tissues of the body perform their functions in a way that helps to maintain constant conditions in the internal environment. The sympathetic participates in these homeostatic mechanisms partly by direct action on the organs concerned and partly by releasing adrenaline into the blood stream. For instance, the body temperature is maintained within normal limits by regulating the balance between heat loss and heat conservation. In a warm environment heat is lost from the body through cutaneous vasodilatation and perspiration. In a cold environment heat is conserved by vasomotor constriction in the skin and the secretion of adrenaline to stimulate metabolic activity and raise heat production. Again, the arterial blood pressure is maintained within normal limits by means of the sino-aortic reflexes. A rise of blood pressure tends to slow the heart and inhibit the vasomotor center, causing reflex vasodilatation and lowering of the peripheral resistance. A fall of blood pressure, on the other hand, reflexively accelerates the heart and increases the sympathetic discharge to the blood vessels.

2. Emergency mechanism. When the body is called upon to deal with a sudden emergency such as an increased demand for oxygen in muscular exercise or when confronted by any situation of stress, all the resources of the body may be activated. A good example is the so-called "fight or flight" reaction of animals exposed to attack. Arousal of the defence mechanisms initiates physiological responses in which all the signs of sympathetic overactivity may be widely displayed. These responses include acceleration of the heart, rise in blood pressure, pupillary dilatation, pilo-erection, and perspiring. The reactions are maintained and enhanced by the accompanying secretion of adrenaline.

Actions of adrenaline

This hormone affects all sympathetically innervated structures throughout the body; its principal effects are as follows: (1) It increases the force of the heart, coronary blood flow, and cardiac output; (2) it dilates the pupils and bronchi; (3) it inhibits the peristaltic activity of the gastrointestinal tract; (4) it diverts the blood flow from the skin and splanchnic area to the skeletal muscles; and (5) it converts liver glycogen into glucose for the extra needs of the tissues.

Sympathectomy

This operation is most commonly performed to relieve the symptoms of peripheral vascular disease and causalgia. Nerve block by the injection of local anesthetic is useful as a preliminary procedure. Successful treatment depends upon a detailed knowledge of the vasoconstrictor outflow to ensure that total denervation of the affected area is achieved. Unfortunately, tissues deprived of their sympathetic supply become remarkably sensitive to circulating adrenaline, so that attacks of vascular spasm may recur.

1. Peripheral vascular disease. In certain disorders of the vascular system, e.g., Raynaud's disease, intermittent spasm of the terminal arteries causes numbness, pallor of the skin, and perspiring. The condition usually affects the hands and feet, which are particularly influenced by exposure to cold; the digital arteries become closed and cyanosis develops in the skin. If the main arteries of the limbs are involved, the patient has difficulty in walking and complains of pain which compels him to stop, owing to ischemia of the limb muscles. Sympathectomy helps by causing immediate vasodilatation and increased blood flow through the tissues; the skin becomes flushed and warm again although reflex sweating is abolished. Denervation is accomplished in the arm by dividing the preganglionic fibers from the second and third thoracic ganglia. The appropriate operation in the leg is removal of the second, third, and fourth lumbar ganglia.

2. Causalgia. This condition is characterized by burning pain associated with nutritional disturbances of the skin mostly affecting the upper limb. The pain usually follows injury to a peripheral nerve, but it is not relieved by interruption of pain tracts in the spinal cord. The mechanism of causalgia has not been satisfactorily explained, although sympathetic overactivity is said to have a modulating influence of the central transmission of pain. The condition is permanently relieved by sympathetomy to block efferent autonomic impulses.

The parasympathetic division

The parasympathetic nerves emerge as two major outflows, cranial and sacral. They have both preganglionic and postganglionic neurons, but the arrangement differs from the sympathetic in that the ganglia are for the most part located in the walls of the organs supplied.

Cranial outflow. The fibers follow the distribution of the third, seventh, ninth, and tenth cranial nerves.

Third nerve (oculomotor). Preganglionic fibers arise from the small-celled component of the oculomotor nucleus and pass through the orbit with the nerve to the inferior oblique muscle to end in the ciliary ganglion. Postganglionic fibers run in the short ciliary nerves to the ciliary muscle and sphincter of the iris. Stimulation results in constriction of the pupil.

Seventh nerve (facial). Preganglionic fibers arise from the superior salivary nucleus and pass through the tympanic cavity as the chorda tympani nerve to reach the submaxillary ganglion by way of the lingual nerve. Postganglionic fibers are distributed to the submaxillary and sublingual salivary glands. Stimulation results in vasodilatation and salivary secretion.

Ninth nerve (glossopharyngeal). Preganglionic fibers arise from the inferior salivary nucleus and pass through the tympanic cavity as the lesser superficial petrosal nerve to reach the otic ganglion. Postganglionic fibers are distributed by way of the auriculo-temporal nerve to the parotid gland. Stimulation results in vasodilatation and salivary secretion.

Tenth nerve (vagus). This is the major outflow of the cranial parasympathetic. It has a more extensive course and distribution than any of the other cranial nerves since its branches pass downward into the thorax and through the diaphragm into the abdomen. Preganglionic fibers arise from the dorsal nucleus of the vagus in the medulla and terminate in the ganglia of plexuses and in the walls of the viscera.

1. Cardiac branches. Postganglionic fibers supply the coronary arteries, atria, and junctional tissue including the pacemaker of the heart. Stimulation results in depression of all the activities of the heart–i.e., excitability, conductivity, heart rate, and force of contraction. The heart is therefore slowed, and the blood pressure falls. The vagus also supplies constrictor fibers to the coronary arteries.

2. Pulmonary branches. Postganglionic fibers from the pulmonary plexuses serve as a motor to the circular muscles of the bronchi. Stimulation results in bronchoconstriction.

3. Gastric branches. The right vagus nerve supplies mainly the posterior surface of the stomach; the left vagus supplies the anterior surface, lesser curvature, and pylorus. Both nerves contribute to the gastric plexus from which branches pass to the liver, pancreas, and spleen. The postganglionic fibers are both motor and secretory. Motor fibers supply the muscular walls of the stomach, promoting gastric contractions, but their action on the pyloric sphincter is inhibitory. Secretory fibers pass to the digestive glands and oxyntic cells of the mucosa.

Stimulation results in an increase of gastric movements and flow of gastric juice. The secretion induced by psychic stimuli–sight, smell, and taste of food–is entirely eliminated after bilateral vagotomy.

4. Intestinal branches. The vagus supplies the small and large intenstine down to the transverse colon. Postganglionic fibers are derived from local nerve plexuses situated in the wall of the bowel. Auerbach's plexus lies between the circular and longitudinal muscular layers of the small intestine; Meissner's plexus lies in the submucosa. The action of the parasympathetic is secretory to the glands of the mucosa and motor to the muscular wall, but the ileocolic sphincter is inhibited.

Stimulation results in increased peristalsis, which carries the contents of the bowel forward in a series of waves. Local nervous reflexes initiate contractions of the intestinal villi, but the latter are influenced more directly by the products of digestion.

Sacral outflow. The pelvic viscera are innervated from a parasympathetic outflow arising in the second, third, and fourth sacral segments of the spinal cord. The fibers pass into the hypogastric plexus on the front of the sacrum between the two common iliac arteries, and secondary plexuses accompany the arterial branches to the descending colon, rectum, bladder, and uterus.

1. Large intestine. The main function of the large intestine is to act as a terminal reservoir and to propel the contents to the exterior. The rectum is normally empty but responds to distension by reflex evacuation of the contents which pass into it from the colon. The parasympathetic supplies motor fibers to the muscular wall and inhibits the internal sphincter to cause it to relax. Emptying of the bowel is aided by voluntary contraction of the diaphragm and abdominal muscles which are, of course, supplied by somatic nerves.

2. Urinary bladder. Voluntary micturition is accomplished by reciprocal action between the sympathetic and parasympathetic nerves to the bladder. During filling, the muscular wall adjusts to the increasing volume so as to lower the internal pressure. Tonic sympathetic activity to the sphincter keeps the bladder closed. On emptying, impulses initiated by the cerebral cortex pass in the sacral outflow to the bladder wall. The detrusor muscle contracts while at the same time the sphincter of the bladder relaxes. Emptying is also assisted by a rise of intra-abdominal pressure.

ROLE OF CHEMICAL MEDIATORS

The chemical substances released by the autonomic nervous system are classified into two groups–adrenergic and cholinergic. *Adrenergic* substances are found in the adrenal glands and at the postganglionic sympathetic nerve terminals. *Cholinergic* substances are released by preganglionic fibers in the autonomic ganglia of the sympathetic and parasympathetic and at the postganglionic parasympathetic nerve terminals. In addition, the sympathetic postganglionic fibers innervating the sweat glands are also cholinergic. The principal adrenergic mediators are the catecholamines, noradrenaline, and adrenaline; the principal cholinergic mediator is acetylcholine.

The catecholamines

These substances are derivations of the amino-acid tyrosine. Their chemical structures are shown beow. It is seen that adrenaline is the methylation product of noradrenaline.

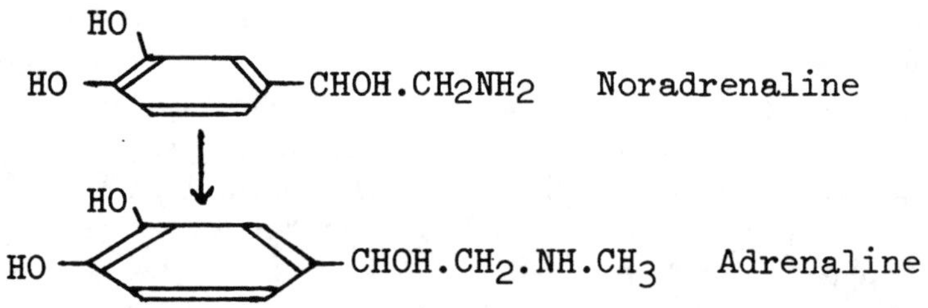

HO, HO — C_6H_3 — $CHOH.CH_2NH_2$ Noradrenaline

↓

HO, HO — C_6H_3 — $CHOH.CH_2.NH.CH_3$ Adrenaline

1. Noradrenaline (norepinephrine). The release of noradrenaline from the adrenal medulla is brought about by impulses in the preganglionic fibers of the splanchnic nerve. Acetylcholine is the transmitter agent at this site. Most of the noradrenaline escaping into the circulation is inactivated before it reaches the tissues. The release of noradrenaline in the peripheral organs is brought about by impulses reaching the postganglionic sympathetic terminals. The transmitter is stored as granules in a large number of vesicles. The nerve impulse promotes the entry of calcium ions into the vesicles, causing them to empty their contents across the synaptic gap; the transmitter then combines with specific receptor sites on the postsynaptic membrane. About 85 percent of the noradrenaline is rapidly removed from the receptor sites into the adrenergic nerve terminals where it is once again bound to the storage granules within the vesicles. The remainder is either inactivated or destroyed. Inactivation is brought about by the enzyme catechol-o-methyl transferase present on the postsynaptic membrane and by the enzyme monoamine oxidase in the mitochondria of the axoplasm.

2. Adrenaline (epinephrine). The hormone is stored as granules in the chromaffin cells of the adrenal medulla. Unlike noradrenaline, it is not

present in the postganglionic nerve terminals and still produces its effects after these nerve endings have degenerated. Splanchnic nerve stimulation releases the hormone into the blood stream. Its actions on the viscera are similar to those following direct stimulation of the sympathetic nerves except that the actions are much more prolonged. Of greater significance is the fact that the two catecholamines often produce different effects on the same tissues owing to their affinity for a specific type of receptor.

Alpha and beta receptors. Although Dale postulated many years ago that different receptor mechanisms are probably present in tissues innervated by sympathetic nerves, his theory was not substantiated until Ahlquist suggested the existence of two types of receptor, arbitrarily named "alpha" and "beta." Ahlquist observed that certain catecholamines, including some synthetic drugs, produced various effects in cardiac and smooth muscle. For example, in the peripheral arterioles both noradrenaline and adrenaline produce vasoconstriction with a consequent rise of blood pressure. This is due to stimulation of alpha receptors. On the other hand, adrenaline and the synthetic drug, isoprenaline, cause vasodilatation in the blood vessels of skeletal muscle due to stimulation of beta receptors.

The situation is different in the heart. Noradrenaline has no direct effect on cardiac muscle, whereas the administration of adrenaline or isoprenaline causes increased heart rate and increased force of contraction. The receptors in the heart are mainly of the beta type. Both types of receptor are present in the wall of the gut where all the catecholamines produce relaxation of the smooth muscle and inhibit motility. It can be concluded, therefore, that sympathetic stimulation causes excitatory effects in some organs and inhibitory effects in others. The distribution of alpha and beta receptors determines the differences in responses to the various catecholamines. Noradrenaline possesses predominantly alpha-receptor stimulating properties; isoprenaline stimulates only the beta receptors while both types of receptor respond to adrenaline.

Adrenergic blockade

Certain drugs block the effects of adrenergic transmission by competing at the receptor sites on the postsynaptic membrane. Since they prevent the undesirable consequences of excessive sympathetic drives, they are widely used in the management of clinical disorders. In small doses, these drugs are specific for one type of receptor, but in excessive amounts they will depress both types of receptor. Alpha-receptor blocking agents such as dibenamine and phentolamine antagonize the vasoconstrictor effects of noradrenaline but have no direct effect on the heart. They are used to produce peripheral vasodilatation. Beta-receptor blocking agents such as propranolol protect the heart from over-stimulation, reduce the heart rate, and lower the blood pressure. The tachycardia associated with anxiety states and the pain of angina due to emotion or strain are effectively controlled by the administration of these drugs.

Cholinergic transmission

Acetylcholine has a wide distribution in the autonomic nervous system. It is the transmitting agent for nerve impulses at the following sites—(1) all preganglionic sympathetic nerve terminals, including those of the splanchnic nerves supplying the adrenal glands; (2) the postganglionic sympathetic fibers supplying the sweat glands; and (3) all preganglionic and postganglionic terminals of the parasympathetic nerves.

The effects of this substance on the autonomic ganglia are described as a "nicotinic" action due to a specific type of receptor with nicotine-like properties. In small doses, nicotine and similar drugs cause autonomic effects by stimulating the ganglion receptors. Impulse transmission from the ganglia may be blocked by drugs such as hexamethonium. The effects of acetylcholine on postganglionic terminals are described as a "muscarinic" action due to a different type of receptor found in the cells of the peripheral organs. These effects are blocked by atropine.

Action on sympathetic ganglia

1. Evidence produced by Feldberg and Gaddum that acetylcholine is the transmitter at sympathetic ganglia was originally based on perfusion experiments. In these investigations, eserine was added to the perfusion fluid to prevent hydrolysis of acetylcholine. The venous effluent from the superior cervical ganglion was collected after repetitive stimulation of the preganglionic nerve and the substance released identified as acetylcholine by a series of tests. For example, it lowers the blood pressure and inhibits the heart on intravenous injection.
2. Very low concentrations of acetylcholine are effective in depolarizing the ganglion cells and promoting their discharge. Nicotine receptors in the postsynaptic membrane are stimulated, leading to firing along the postganglionic fibers, while the preganglionic nerve remains quiescent.
3. Acetylcholine is rapidly destroyed by the enzyme cholinesterase to

form choline and acetate, thus limiting the accumulation of the transmitter around the ganglion cells.

4. Ganglion blockade is produced by hexamethonium which combines with the nicotine receptors. Acetylcholine is released by preganglionic stimulation, but transmission of impulses through the ganglion is interrupted.

Action on sweat glands

The cholinergic action of the sympathetic nerves to the sweat glands was first demonstrated by Dale and Feldberg in the cat. If acetylcholine is injected directly into the skin, there is local vasodilatation and secretion of sweat. Cholinergic drugs, e.g., pilocarpine, enhance the action of acetylcholine, causing profuse sweating even after division of the nerves to the glands. The action is abolished by the injection of atropine.

Action on parasympathetic

Acetylcholine is the chemical transmitter for preganglionic and postganglionic synapses of parasympathetic nerves. The mechanism of its release from the nerve terminals and its action on receptors in the postsynaptic membrane resemble the mechanisms at the neuromuscular junction described in Chapter IV. Briefly, acetylcholine is formed from acetyl-coenzyme A and choline under the influence of choline acetylase and it is broken down by the enzyme cholinesterase. Accordingly, the action of the transmitter is prolonged by the administration of anti-cholinesterase drugs and depressed by blocking agents which compete with it for the receptor sites.

The changes produced by acetylcholine on the postsynaptic membrane may be either excitatory or inhibitory depending upon the particular organ involved. The wall of the gut and the detrusor muscle of the bladder are examples of excitatory action; the heart is an example of inhibitory action. During resting conditions acetylcholine is constantly being released at the nerve endings due to the steady arrival of impulses discharging at a low frequency. This accounts for parasympathetic tone, which promotes mild peristalsis in the gut but restrains the heart. During activity the nerve endings release larger amounts of acetylcholine; gastrointestinal motility is increased, while the heart is slowed and the force of contraction reduced.

HYPOTHALAMUS

The hypothalamus provides the most important output pathway controlling the internal organs of the body. Different hypothalamic centers are associated with specific functional systems, but the region is so compact that it is difficult to investigate them separately. Many of the effects of lesions are often due to involvement of more than one system or else result from damage to fibers passing through the hypothalamus from other parts of the brain. Also, the diffuse mingling of tracts and cells between one center and another implies that experimental findings must be interpreted with caution. These difficulties are sometimes reflected in complex neuro-endocrine disturbances that accompany a clinical disorder.

Functional anatomy

The hypothalamus extends from the lamina terminalis anteriorly to the mammillary bodies posteriorly. On the medial side it adjoins the third ventricle, while the internal capsule and optic tract form the lateral boundary. The inferior relations are defined by the optic chiasm, median eminence, and hypophysis. The hypothalamus contains many nuclear masses which may be conveniently divided into four regions:

1. The *anterior* region contains the supraoptic and paraventricular nuclei. They give rise to the hypothalamo-hypophysial tract which runs to the posterior lobe of the pituitary.
2. The *medial* region contains the ventromedial and dorsomedial nuclei.
3. The *lateral* region contains the tuberal nuclei and a collection of fibers called the medial forebrain bundle through which efferent pathways run from the hypothalamus into the brain stem.
4. The *posterior* region contains the mammillary body, composed of several nuclear groups lying close to the posterior perforated substance.

Organization of activities

The following scheme may be useful in classifying the more important activities of the hypothalamus despite the fact that the functions ascribed to the various regions are not nearly so discrete as they appear to be in Figure 165. Nevertheless, when these functions are taken together, it is abundantly clear that the hypothalamus plays a key role in the regulation of the internal environment, especially in its control of the entire autonomic outflow.

1. The anterior region is traditionally the site controlling the parasym-

HYPOTHALAMIC FUNCTIONS

ANTERIOR	MEDIAL	POSTERIOR
Parasympathetic	*Energy balance*	*Sympathetic*
Heat loss	*Thirst. hunger*	*Heat conservation*
Antidiuretic	*Sexual behavior*	*Shivering*
Oxytocin	LATERAL	*Lactation*
Sleep	*Emotion*	*Arousal*
	Defence reactions	

Fig. 165. Scheme of hypothalamic functions. The anterior group includes the supraoptic and paraventricular nuclei; the medial group contains the ventromedial and dorsomedial nuclei; the lateral group includes the tuber cinereum and the posterior group includes the mammillary bodies.

pathetic system and heat loss mechanisms. Parasympathetic effects produced by electrical stimulation include increased peristalsis, bladder contraction, slowing of the heart, and peripheral vasodilatation. Localized heating by implanted electrodes or a rise in the temperature of the blood bathing this region activates heat-sensitive neurons. Sympathetic tone is inhibited, allowing vasodilatation and loss of heat from the skin.

Stimulation of the supraoptic nucleus causes the release of anti-diuretic hormone from the posterior pituitary gland; the hormone promotes reabsorption of water from the distal tubules of the kidney. The cells of the supraoptic nucleus are believed to have osmoreceptor properties since they are affected by the electrolytic content of the blood and respond to an injection of hypertonic saline by an increased discharge. Stimulation of the paraventricular nucleus causes the release of the hormone oxytocin, which in turn acts on the pregnant uterus and mammary glands. These functions are described in more detail at the end of the chapter. There is also evidence that a mechanism inducing sleep may operate through a hypothalamic center close to this region.

2. The medial region is concerned mainly with food intake and energy balance. The ventromedial nucleus, in particular, exerts a stabilizing control on the fat stores and inhibits the appetite. This region also contributes neurosecretory substances that are carried through the portal system of blood vessels to cause the release of anterior pituitary hormones. For example, electrical stimulation promotes the secretion of gonadotrophic hormones on which depend the intrinsic sexual rhythm and mating behavior of most mammals.

3. The lateral region is generally associated with mechanisms of emotional expression. Its intricate mass of neural elements elaborate the pat-

terned responses of integrated defence reactions. Electrical stimulation in conscious animals evokes an outburst of rage manifested by typical autonomic signs and alterations in posture. The region is also associated with mechanisms controlling hunger and thirst since excitation causes an animal to search for food or to drink large quantities of water.

4. The posterior region is the principal site of the sympathetic outflow and heat conservation mechanisms. The sympathetic effects are transmitted mainly through the cardiovascular centers of the brain stem and include acceleration of the heart, increased cardiac output, and rise of blood pressure.

A fall of body temperature removes the inhibitory influence of the anterior hypothalamus and allows sympathetic tone to increase. As a result, intense vasoconstriction occurs in the skin vessels preventing loss of heat to the exterior. Shivering increases heat production when the body is exposed to cold. The hypothalamus behaves as a central thermo-detector sending impulses down the cord to the motoneurons innervating the skeletal muscles. After prolonged exposure to cold, the metabolic rate is stepped up by an increased output of thyrotrophic hormone which in turn stimulates the thyroid gland. The posterior hypothalamic nuclei may also serve as a waking center. There is good evidence to believe that the ascending reticular activating system excites this region to maintain the waking state and to trigger alerting mechanisms, as explained in Chapter XXIV.

It must be emphasized that the hypothalamus is a compact network of neural, vascular, and secretory elements which cannot be considered in isolation. There is little value in relating different cell groups to specific functions. However, the above scheme may help to piece together individual functional patterns and to give an overall picture of a region of the brain that appears to influence almost every aspect of bodily activity.

Lesions of the hypothalamus

Human hypothalamic lesions may arise from malformations, degenerative changes, or from the presence of tumors or cysts. The resultant disturbances often give a complicated clinical picture involving several functional systems.

1. Diabetes insipidus. In this condition, patients suffer from excessive thirst and pass large quantities of dilute urine. It is due to a deficiency of antidiuretic hormone and consequent fall in the amount of water reabsorbed by the kidney tubules. Deficiency of antidiuretic hormone (ADH) may result from reflex inhibition of the hypothalamus caused by emotional disturbances, although these are fortunately amenable to correction. More significantly, diabetes insipidus is caused by a lesion in the supraoptic-

hypophyseal tract which normally controls the release of the hormone or by damage to the posterior pituitary gland where the hormone is stored.

2. Adiposo-genital syndrome. Fröhlich described a clinical disturbance affecting mainly the functions of the medial group of hypothalamic nuclei. It sometimes occurs in children after encephalitis. The principal features are adiposity and immature sexual development.

A number of factors may be involved in the development of adiposity. One factor is simply overeating. After bilateral damage to the ventromedial nuclei there is loss of the ability to regulate food intake in relation to energy balance. Another factor is a deficiency of hormone stimulated lipolysis. Obese subjects have low rates of lypolitic activity and thus have difficulty in utilizing their fat stores. Sexual immaturity is associated with underdevelopment of the gonads and of the secondary sexual characteristics. Males become impotent, the skin remains smooth, and the onset of puberty is arrested; in females the reproductive organs cease to function and the menstrual cycles are absent.

3. Autonomic epilepsy. Tumors of the third ventricle may cause attacks of sweating, tachycardia, pupillary dilatation, and other signs of sympathetic overactivity. Lesions affecting the anterior region of the hypothalamus may cause a parasympathetic seizure characterized by slowing of the heart, fall of blood pressure, and increased peristalsis.

4. Abnormal body temperature. Failure of the temperature regulating mechanism may have serious consequences if the temperature of the body falls below 25°C or rises above 43°C. Hypothermia is always a potential risk in newborn babies before adaptation processes have become established and also in elderly individuals who are naturally prone to cerebrovascular changes. Prolonged exposure to excessive cold may put the hypothalamic thermostat out of action by depression of its vasomotor and shivering mechanisms.

Hyperthermia develops in hot environments with high humidity. Loss of salt through sweating, dehydration, and hard muscular work all aggravate the condition until the heat regulating mechanism breaks down and the individual collapses from heat stroke. Many infections also give rise to pyrexia. Heat loss is then relatively inadequate to compensate for the increased heat production and increased metabolic rate produced by the circulating pyrogens.

HIGHER CONTROL OF THE HYPOTHALAMUS

Our knowledge of the cerebral influences on hypothalamic activity is still largely at an exploratory stage. This is partly due to the difficulty of sorting out the profuse network of fiber connections passing into and out of the hypothalamus and partly to the fact that some of the neural mechanisms are regulated by hormones in the blood stream. The brain structures that appear to exercise overall control of the hypothalamus are collectively known as the limbic complex which, as we have already seen, acts as an integrating center for many viscero-somatic channels. These structures comprise the anterior cingulate, posterior orbital and hippocampal gyri, and the subcortical group of amygdaloid nuclei.

Input-output relations

The pathways relating the hypothalamus to other parts of the brain are indicated in Figure 142 (see previous chapter). Further details are as follows:

1. The anterior cingulate is the cortical destination of a tract that runs from the mammillary bodies through the anterior nuclei of the thalamus. Since the mammillary bodies receive a major projection from the fornix, this tract is in reality a communication channel between the hippocampus, hypothalamus, and cingulate gyrus.
2. The posterior orbital gyrus is the cortical destination of a complex system of fibers passing from the hypothalamus through the dorsomedial nucleus of the thalamus. Since this nucleus also receives a heavy projection from the orbital cortex, it may be regarded as a neural center playing an active part in autonomic adjustments. In addition, the orbital cortex and the hypothalamus are strongly linked by direct tracts which have been traced histologically to individual hypothalamic nuclei.
3. The hippocampus consists of two divisions, Ammon's horn and the dentate gyrus. The latter is a narrow strip of cortex lying under cover of the fimbria. *Efferent* fibers emerging from the fimbria pass above the dentate gyrus to form the posterior column of the fornix. The two posterior columns come together and follow the under surface of the corpus callosum to form the anterior columns which bend backward to reach the mammillary bodies. *Afferent* pathways in the fornix come from widely scattered points in the hypothalamus; they terminate on granule cells in the dentate gyrus whose axons relay to the hippocampal pyramidal cells.
4. The amygdaloid nuclei lie in the depth of the temporal lobe in the roof of the inferior horn of the lateral ventricle. They form two major groups of subcortical nuclei–basolateral and corticomedial. Fibers from the basolateral nuclei pass to the hypothalamus along two efferent pathways: One pathway proceeds through the anterior commissure to the preoptic and anterior nuclei of the hypothalamus; the other pathway contributes to the stria terminalis, the best known pathway connecting the amygdaloid with

the hypothalamus. Fibers from the corticomedial nucleus converge on the same region via the stria terminalis. The amygdaloid projections are represented schematically in Figure 166.

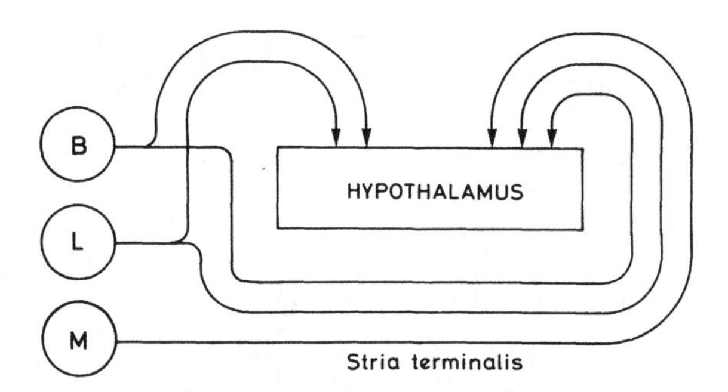

Fig. 166. Diagram of principal pathways between amygdaloid and hypothalamus. Fibers from the basal (B) nuclei and lateral (L) nuclei pass via the anterior commissure and stria terminalis. Fibers from the medial (M) nuclei reach the hypothalamus via the stria terminalis.

Electrical stimulation

The results of experimental work on animals show that the limbic regions are capable of modifying reactions in many systems of the body. The results, however, are not always consistent and sometimes contradictory. This can be explained to some extent by the fact that excitatory and inhibitory points exist side by side in the cortex. Furthermore, the level of cortical excitability is altered by anesthesia, while responses may vary according to the parameters of stimulation used. It is also apparent that each limbic component has certain affinities for a particular functional system and that the amygdaloid structure possesses complete flexibility of action.

1. Anterior cingulate gyrus. The cellular structure of the anterior cingulate is agranular, resembling that of the motor cortex. Through its connections with the caudate nucleus it is regarded as a powerful suppressor area for somatic effects can be elicited from the same stimulated site to produce the kind of organized behavioral patterns seen in defense reactions. In regard to the visceral effects the most prominent changes observed are slowing or arrest of respiration, fall of blood pressure, inhibition of gastric contractions, dilatation of the pupil, and pilo-erection. These are all mixed parasympathetic-sympathetic effects presumably mediated by the hypothalamus. However, different response patterns may occur in unanesthetized animals with chron-

ically implanted electrodes, suggesting that excitation and inhibition are both essential to produce a meaningful physiological reaction.

2. Posterior orbital gyrus. Electrical stimulation of the orbital cortex is known to have a marked effect on the autonomic nervous system. Some of these effects are excitatory, others inhibitory.

- Alimentary. Usually, the stomach is relaxed and peristalsis reduced or suppressed. On the other hand, the volume and concentration of the gastric secretions are increased, in particular, the concentration of HCl and pepsin. These findings have been confirmed in the unanesthetized animal and in man. It has long been known that increased acidity of the stomach is caused by vagal stimulation of the gastric glands and that the so-called "psychic" secretion is eliminated after bilateral vagotomy. The orbital cortex therefore exercises a dual influence on the vagal outflow, causing increased or decreased responses according to the specific hypothalamic region activated.
- Cardiovascular. Stimulation of the orbital cortex generally causes acceleration of the heart beat and a rise of blood pressure when the stimulus parameters are large enough to excite the orbito-hypothalamic projections. The effects are presumably the result of vasomotor sympathetic discharges. Weaker stimuli do not produce any significant change in arterial blood pressure, but they may prevent a fall in blood pressure resulting from some other cause. This effect may be mediated by orbito-medullary pathways inhibiting vasodilator mechanisms in the reticular formation.
- Respiratory. Stimulation of the orbital cortex inhibits respiration with slowing or arrest in the expiratory phase. Escape from inhibition with shallow breathing occurs after about 30 seconds, and the normal rate is resumed when the stimulus ceases. The respiratory and cardiovascular changes appear to be independent.
- Thermal. Stimulation of the orbital cortex has a powerful inhibitory effect on vasodilatation induced by a rise in brain temperature. It will be recalled from the previous chapter that heat-sensitive neurons are located in the brain stem reticular formation. When the temperature of the brain is artificially raised to 41.5°C the blood pressure falls rapidly, but recovers as soon as the brain is cooled. In Figure 167, the orbital cortex is stimulated during the period of heating when the temperature is allowed to reach 43°C. The blood pressure recorded from the femoral artery is maintained at 120 mm Hg. When the stimulus is switched off the blood pressure immediately falls to 40 mm Hg, but rises again as the brain is cooled.

Under physiological conditions it is doubtful whether the orbital cortex has any role to play in temperature regulation since this is controlled effectively by the hypothalamus. Heat is removed from the body by cutaneous vasodilatation without altering the level of the systemic blood pressure. Excessive heat stimulates vasodilator neurons, which in turn modify vas-

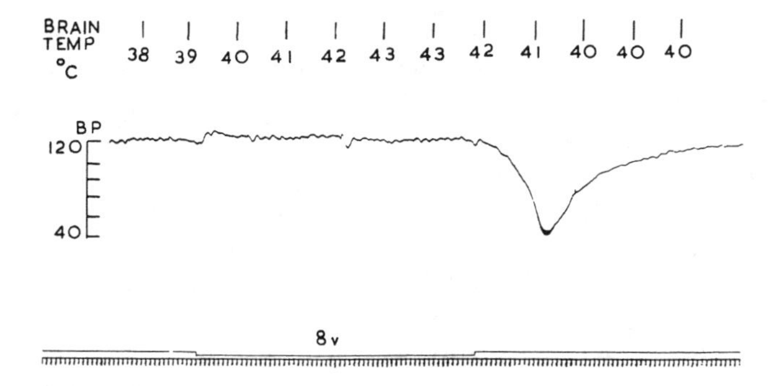

Fig. 167. Effect of stimulating the orbital cortex. Blood pressure was recorded from the femoral artery. During period of stimulation (signal 8 volts) localized heating causes brain temperature to rise to 43°C. Blood pressure is maintained at 120 mm Hg. When stimulation ceases, blood pressure falls to 40 mm Hg and rises again as brain is cooled. Time intervals were 5 sec. (Compare with Fig. 144.)

omotor tone in the peripheral blood vessels. Cortical influences may raise the threshold of the vasodilator mechanism and thus prevent the fall of blood pressure that would ensue from loss of vasomotor tone. Compensation fails only if a critical temperature is reached when massive vasodilatation occurs accompanied by circulatory collapse. Comparable changes in the human individual working in a hot environment may be the basis of heat exhaustion.

3. Hippocampus. The effects of hippocampal stimulation on the various functions of the body are somewhat confusing. Stimulation generally results in a fall of blood pressure, slowing of respiration, dilatation of the pupils, and a reduction in concentration of the gastric secretions. It is unlikely that the hippocampus exerts a direct control on any physiological system, and the changes suggest interference with mechanisms rather than specific actions. Other effects of hippocampal stimulation include movements of the body, especially those involved in defence and attack. It is not thought that the hippocampus initiates the movements, but it is quite capable of inhibiting cortically induced movements. There is, in fact, a definite tendency toward inhibition of most structures involved, whether somatic or visceral. Eccles has drawn attention to the long-lasting inhibitory potentials derived through basket cell synapses on the soma of the hippocampal pyramidal cells. A powerful recurrent inhibitory mechanism is thereby established.

The predominance of hippocampal inhibition is seen only under experimental conditions and is not exclusive of excitatory influences that may occur in the conscious animal. There is good reason for believing that the hippocampus is a link in a circuit, which also includes the thalamus and anterior cingulate gyrus, and that the controlled output of this circuit discharges directly to the hypothalamus and along more diffuse paths to the brain stem and cerebellum. The output influences the firing patterns of all these structures so that action can switch over to favor facilitation or inhibition according to circumstances. Thus under conditions of intense excitement, such as occurs in emotional stress, inhibition is overcome and the result is a powerful driving force on the hypothalamic outflow.

4. Amygdaloid. A wide variety of visceral and somatic responses may be elicited by electrical stimulation of the amygdaloid nuclei. Often the stimulus results in opposite effects on the same functional system. For example, respiration may be slowed or increased, the heart may accelerate or decrease in rate, or the blood pressure may rise or fall. More consistent responses are obtained from the alimentary system where the effects of amygdaloid stimulation are usually an increase in gastric motility, an increase in the volume of gastric juice, and a rise in blood sugar. Dilatation of the pupil is also commonly reported, whatever the conditions or techniques used. In addition, organized movement patterns may be observed with changes in posture, often developing into aggressiveness and other signs of increasing muscular tension. Evidently, the amygdaloid nuclei contribute to the overall excitability of the hypothalamus and provide the necessary background for the reactions of the animal in defence or flight.

Summary

The regions of the brain collectively described as the limbic complex have at least one function in common—the regulation of the internal environment. They appear to achieve this end by modulating the discharges of the hypothalamus, the most important output pathway to the viscera. There are still many uncertainties about the mechanisms of higher control, details of circuitry, and mode of operating in conscious states. It seems that each limbic component can operate independently on the same input signals and produce a fairly wide range of response patterns. But it is unlikely that the fragmented responses induced by electrical stimulation have any real physiological significance. Unfortunately, we know very little about the integration of cerebral control systems. All that can be said at present is that the limbic regions as a whole exercise a profound influence on hypothalamic activities and show a considerable degree of flexibility in achieving their purpose.

ENDOCRINE OUTFLOW

The hypothalamus is responsible for regulating the activity of the entire endocrine system. This is achieved by direct nervous effects on two endocrine glands, the posterior pituitary and the adrenal medulla. The posterior pituitary receives its nerve supply via the hypothalamic-hypophyseal tract, while the adrenal medulla is connected to the hypothalamus by descending tracts in the brain stem and spinal cord and by the preganglionic fibers which pass out to it in the sympathetic nerves. All the other endocrine glands are controlled by hormones secreted into the circulation from the anterior pituitary gland. The secretions of the anterior pituitary are in turn regulated by neurovascular releasing factors manufactured in the hypothalamus.

The pituitary gland. This gland is situated in the sella turcica at the base of the brain between the optic chiasm and posterior perforated substance. It is divisible into two distinct portions, the anterior pituitary and posterior pituitary. The anterior lobe is developed from the pouch of Rathke, which is derived from the pharyngeal epithelium, and consists of three parts—tuberalis, intermedia, and distalis. The tuberalis is a thin collar of highly vascular tissue encircling the median eminence. The intermedia lies adjacent to the posterior lobe but is poorly developed in the human being. The pars distalis contains epithelial cells of various size and shape; these are classified into two main kinds—chromophobe cells, which have little affinity for dyes, and chromophil cells, which are responsible for secreting the hormones. The posterior lobe, also known as the neurohypophysis, is developed as an outgrowth of the hypothalamus with which it is directly connected by nerve fibers. It consists of two parts, the infundibular stem and the infundibular process. Hormones stored in the posterior lobe are derived from cells in the anterior region of the hypothalamus (Fig. 168).

Anterior lobe

Hypophysial portal vessels

Very few nerve fibers pass from the hypothalamus to the anterior lobe and none are secreto-motor in function. The anterior lobe hormones are brought under hypothalamic control by neurovascular agents transported down a fine network of capillaries. Popa was the first to report on extensive capillary sinuses in the median eminence, which ended in a vascular bed surrounding the glandular cells. This important discovery was confirmed by Harris, who also indicated the direction of blood flow along portal vessels

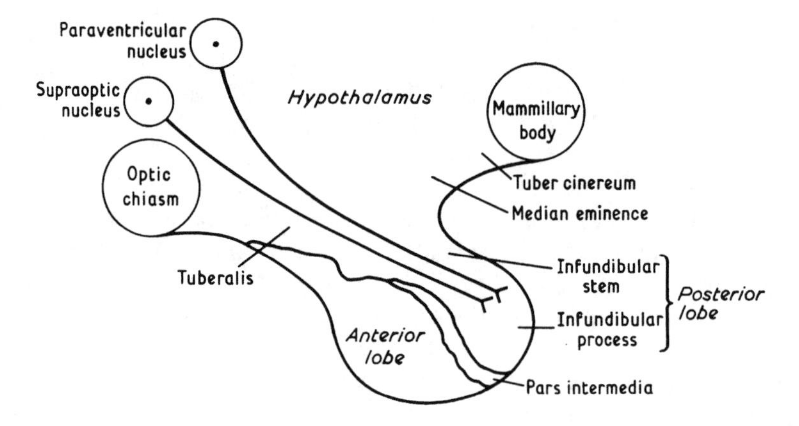

Fig. 168. Sagittal section through the hypothalamus and pituitary gland. The anterior lobe arises from the tuberalis. The neurohypophysis or posterior lobe receives the terminals of the hypothalamo-hypophyseal tract.

connecting the two capillary networks–i.e., from the median eminence to the pituitary gland. It is now firmly established that chemical substances released from the neural elements of the hypothalamus are carried by the portal vessels to control the secretion of the anterior pituitary hormones. The neurovascular agents synthesized in the hypothalamus are known as "releasing factors."

Evidence for neurosecretion

Evidence for believing that the hypothalamus regulates anterior pituitary activity is based on the following experiments:

1. Changes in the external envionment. Adequate feeding, temperature, humidity, and light have all been shown to play a part in the breeding habits of most animals. Such factors appear to influence the secretion of gonadotrophic hormones, which stimulate the reproductive organs and determine seasonal mating. For example, the estrous cycle of the female ferret is greatly prolonged by exposure to extra light.

2. Electrical stimulation. Harris introduced a remote control method of stimulating the hypothalamus in which a secondary coil was placed under the scalp and a primary coil encircled the cage. Induction shocks applied to various regions of the hypothalamus caused the release of gonadtrophic hormone and ovulation in the rabbit. Ovulation did not occur if the pituitary was stimulated directly.

3. Transplantation. Most endocrine glands will resume their function

when transplanted to a different site in the body. The anterior pituitary is an exception. If the pituitary is removed in the rat and grafted elsewhere in the same animal, the reproductive organs soon atrophy. If it is regrafted to its natural site in the sella turcica, normal function is resumed as soon as vascularization is established. It is clear that the anterior lobe is dependent on its blood supply via the hypophysial portal vessels.

Hypothalamic releasing factors

The median eminence is the major source of releasing factors carried by the portal vessels to the anterior pituitary. These chemical substances are mostly peptides secreted at the nerve endings of special neurons in the hypothalamus. Despite the multiplicity of their physiological actions, most of them have been identified and their chemical structure determined. The hormones on which they exert their influence are thyrotropin, gonadotropin, adrenocorticotropin, prolactin, and growth hormone.

1. Thyrotropin rleasing factor (TRF). Guillemin was one of the first to separate TRF from other releasing factors in the hypothalamus by employing chromatograph techniques. TRF contains the amino acids histidine, proline, and glutamic acid. When injected into the circulation it increases all the known activities of the thyroid cells by causing the formation in the cells of cyclic AMP. The result is increased proteolysis of the thyroglobulin in the follicles of the gland and secretion of the thyroid hormones, thyroxine and triiodothyronine. Calcium ions may be a necessary part of the stimulating mechanism. The production of TRF is itself controlled by the level of circulating thyroxine, representing a form of feedback control.

2. Gonadotrophin releasing factors. The onset of puberty and the regulation of the reproductive cycles in adult life are dependent upon two pituitary gonadotrophins, follicle stimulating and luteinizing hormones. These in turn respectively control the output of the ovarian hormones, estrogen and progesterone.

- Follicle stimulating releasing factor (FRF). Evidence for the existence of a separate follicle stimulating factor is scanty. Some workers have identified a polypeptide which appears to be distinct from the luteinizing factor but causes the release of both gonadotrophic hormones.
- Luteinizing hormone releasing factor (LRF). This substance was extracted from the median eminence and shown to be a decapeptide. It can be prepared synthetically and has a potent effect on the anterior pituitary gland. LRF liberates the luteinizing hormone, which acts on the ovary to promote ovulation and the development of the corpus luteum. If pregnancy supervenes, the level of luteinizing hormone is significantly increased.

3. Corticotropin releasing factor (CRF). This peptide can be extracted from the median eminence after prolonged electrical stimulation. Its synthesis can be enhanced after adrenalectomy and by the addition of acetylcholine and calcium to the medium. CRF causes the release of adrenocorticotrophine from the anterior pituitary, which in turn promotes the secretion of the cortico-steroids, aldosterone and cortisol, from the adrenal cortex. The close relationship between the anterior pituitary and the adrenal cortex is seen in a number of clinical syndromes–e.g., Simmonds' disease, in which there is a deficiency of ACTH and a grave disturbance to essential metabolic processes. Cortico-steroids have a direct feedback effect on the hypothalamus, whereby the release of CRF is automatically controlled.

4. Prolactin inhibitory factor (PIF). Prolactin is unique among the hormones secreted by the anterior pituitary in that its regulation by the hypothalamus is through inhibition. If the anterior pituitary is separated from the hypothalamus by making a section through the pituitary stalk, the rate of prolactin secretion is increased. Under normal conditions, PIF is continuously transported to the gland from the median eminence so that prolactin secretion is suppressed. However, during lactation the releasing factor itself is suppressed, especially by the act of suckling, and large amounts of prolactin are made available for the secretion of milk.

5. Growth hormone releasing factor (GRF). In contrast to other endocrines growth hormone does not have a target organ but instead exerts an effect on most tissues of the body. It increases the rate of protein synthesis, conserves carbohydrates and uses up the fat stores for energy. The size and weight of the body increas proportionately. As a consequence of such generalized metabolic processes there are numerous ways in which the secretion of GRF can be influenced. The most important factor is the state of nutrition of the cells, but any alteration in energy balance can stimulate hormone production. Other factors influencing secretion are exercise, starvation, hypoglycemia, and fevers.

Posterior lobe

The hormones secreted by the neurohypophysis are vasopressin and oxytocin. Vasopressin is a polypeptide containing eight amino acids. When injected into the blood stream, it raises the arterial blood pressure and increases the glomerular filtration rate. However, its main action is on the kidney tubules where it promotes the reabsorption of water and reduces the volume of the urine. Oxytocin has no pressor or antidiuretic activity, yet its chemical structure is similar to that of vasopressin except that in vasopressin phenylalanine and arginine replace leucine and isoleucine of the oxytocin molecule. The main actions of oxytocin are on the pregnant uterus and mammary glands.

The hormones are not secreted in the posterior lobe. They are formed in the supraoptic and paraventricular nuclei and transported by axoplasmic flow from the hypothalamus to the pituitary gland, where they are stored as granules pending release into the circulation.

Release of antidiuretic hormone (ADH)

The antidiuretic action of the posterior pituitary gland was first noted in the clinical treatment of diabetes insipidus when the administration of extracts resulted in a reduced urinary flow and alleviation of thirst. Experimental proof was provided by Verney, who used dogs prepared with carotid loops tied under the skin. Injection of hypertonic saline into the carotid artery caused inhibition of diuresis and a concentrated urine. Depriving the animal of water had the same effect since dehydration promoted the release of antidiuretic hormone. Conversely, injection of hypotonic saline or drinking pure water resulted in a large volume of dilute urine. In other experiments, removal of the posterior lobe caused diuresis and thirst, which, however, could be prevented by the administration of vasopressin. Verney concluded that the secretion of antidiuretic hormone was controlled by the concentration of electrolytes in the plasma and that the sensitive region was located in the anterior part of the hypothalamus.

Mechanisms of ADH release

Several factors are believed to contribute to the control of ADH secretion; they include the plasma osmotic pressure, the volume of circulating blood, cardiac receptors, and emotional stress. All of these factors exert their influence on neurons in the hypothalamus. When impulses are transmitted down their axons, the hormone is released from the nerve endings.

1. Plasma osmotic pressure. It has been postulated that neurons in the supraoptic nuclei function as osmoreceptors. The total osmotic pressure of the extracellular fluid is largely due to its content of Na^+ ions and other dissolved electrolytes. At the normal osmotic pressure of the blood, a small quantity of ADH is released. If the osmotic pressure rises, the cells of the supraoptic nuclei are stimulated and more ADH is released; reabsorption of water is increased and the urine becomes concentrated. If the osmotic pressure falls, water enters the cells and the release of ADH ceases; less

reabsorption occurs in the kidney and the urine becomes dilute. In this way the hypothalamus provides a mechanism for controlling the sodium ion concentration and the osmotic pressure of the body fluids.

2. Blood volume. An increase of blood volume causes an increase of urine flow. This may be due to stimulation of volume receptors causing reflex inhibition of ADH secretion. A fall of blood volume, as occurs following hemorrhage, increases the secretion of ADH to promote retention of water in the body. The increased secretion is believed to have a potent pressor effect on the arterioles, thereby helping to maintain the arterial blood pressure.

3. Cardiac receptors. Stimulation of cardiac receptors by means of small balloons inserted into the atria of anesthetized dogs results in an increase in urine flow. It was originally believed that the antidiuretic hormone was involved in this response on the theory that impulses in the cardiac vagal nerves reflexively inhibited the release of ADH. The theory has been challenged on the grounds that increased urine flow occurs when the atria are distended during intravenous infusion of vasopressin. It might be argued that injecting vasopressin merely alters the blood flow through the kidney, increasing sodium excretion and thereby causing diuresis. However, water diuresis following atrial distension results after complete destruction of the pituitary gland. This would dispute the idea that the antidiuretic hormone has any role to play in this diuretic response. The matter is still unresolved.

4. Emotional stress. Afferent impulses from many sources—physical exercise, anxiety, and emotional excitement—all result in a decreased urine flow. The effect is brought about by increasing hypothalamic activity and hence ADH secretion. Retention of water in the body is a frequent concomitant of stressful situations.

Release of oxytocin

The principal actions of this hormone are on the pregnant uterus and mammary glands. Since the early observations of Dale that extracts of the posterior lobe cause contractions of the uterus both in vivo and in isolated preparations, attention has been given to its functional role in parturition and to its clinical value in obstetric practice. Oxytocin also plays an essential role in the process of lactation by promoting milk ejection.

1. Parturition. The onset of labor is attributed to the release of oxytocin from the neurohypophysis. There is little contractile activity in the uterus until near the end of gestation, when increasing secretions of the hormone culminate in powerful contractions to complete the process of childbirth.

The stimulus arises from stretching of the birth canal. Impulses are transmitted through sensory nerves to the spinal cord and pass upward through the brain stem finally reaching the paraventricular nuclei in the anterior hypothalamus. This is supported by the following facts: Uterine contractions are recorded by inserting a balloon into the uterus of the postpartum rabbit. Stimulation of the pituitary stalk results in increased frequency and amplitude of the contractions associated with a rise in the plasma content of oxytocin. Dilatation of the cervix elicits nervous reflexes. which are lost after sectioning the spinal cord. The uterine contractions are abolished in the hypophysectomized animal.

The use of oxytocin by obstetricians is now well established. It helps to promote stronger contractions in the third stage of labor and diminishes bleeding after delivery of the placenta. In women who have a deficiency of the hormone, labor may be prolonged, and surgical intervention then becomes necessary. On the other hand, in cases of premature labor and threatened miscarriage, the release of oxytocin can be inhibited by drugs such as ethanol. Inhibition of oxytocin release may be brought on by strong emotional factors, an important cause of uterine inertia.

2. Milk ejection. The act of suckling initiates a neurogenic reflex in the nursing mother to bring about the ejection of milk already present in the mammary glands. Impulses reaching the hypothalamus enhance the secretion of oxytocin into the blood stream. The hormone acts on the mammary glands to cause contraction of the myoepithelial cells surrounding the alveoli and ducts, thereby expressing the milk. Lactation may be inhibited by a number of factors:

- Noradrenaline depresses milk flow owing to vasoconstriction of the mammary arteries and thus prevents access of circulating oxytocin to its site of action. It may also constrict the mammary ducts.
- Adrenaline blocks the release of oxytocin from the neurohypophysis.
- Emotional disturbances are a common cause of poor milk yield. Many psychological factors as well as generalized sympathetic stimulation can inhibit oxytocin secretion and consequently depress milk ejection. Under such circumstances the administration of oxytocin has a beneficial effect.

SELECTED BIBLIOGRAPHY

Babkin, B.P., and Speakman, T.J. Cortical inhibition of gastric motility. *J. Neurophysiol.* 13:55-63, 1950.

Dreifuss, J.J., and Murphy, J.T. Convergence of impulses upon single hypothalamic neurones. *Brain Res.* 8:167-176, 1968.

Green, J.D. The hippocampus. *Physiol. Rev.* 44:561-608, 1964.

Harris, G.W. *Neural Control of the Pituitary Gland.* London: Edward Arnold, 1955.
Hess, W.R., Akert, K, and McDonald, D.A. Functions of the orbital gyri of cats. *Brain* 75:144-258, 1952.
Hoff, E.C., Kell, J.F. Jr., and Carroll, M.N., Jr. Effects of cortical stimulation and lesions on cardiovascular function. *Physiol. Rev.* 43:68-114, 1963.
Isaacson, R.L. (Ed.) *The Limbic System.* New York: Plenum Press, 1974.
Isaacson, R.L., and Pribram, K.H. (Eds.). *The Hippocampus.* London: Plenum Press, 1975.
Martini, L., Motta, M., and Fraschini, F. (Eds.). *The Hypothalamus.* New York: Academic Press, 1970.
Newman, P.P., and Wolstencroft, J.H. Influence of orbital cortex on blood pressure responses in cat. *J. Neurophysiol.* 23:211-217, 1960.
Pribram, K.H. The limbic systems, efferent control of neural inhibition and behavior. In: Adey, W.R., and Tokizane, T. (Eds.). *Structure and Function of the Limbic System.* Amsterdam: Elsevier, 1967.
Schally, A.V., Arimura, A., and Kastin, A.J. Hypothalamic regulatory hormones. *Science* 179:341-350, 1973.
Ward, A.A. Jr., and McCulloch, W.S. The projection of the frontal lobe on the hypothalamus. *J. Neurophysiol.* 10: 309-314, 1947.

CHAPTER XXII

Physiology of Emotion

Emotion is a sensory experience through which we express our feelings. We recognize emotions in others and communicate our own feelings in a language of gestures, postures, and facial expressions. Yet it is not easy to define an emotion or to account for the way a person will react to a particular situation. These difficulties arise from a number of factors influencing our upbringing such as age, sex, race, and social conventions. Young children, for example, often swing from one extreme to another; tears are quickly replaced by laughter in circumstances that appear to be trivial. Women are far more attuned to their emotions than men and tend to react more expressively. Certain races are described as warm-blooded and others as reserved or withdrawn. Emotional behavior varies considerably according to the conventions of society and obedience to its rules. Most adults learn to exercise a degree of control over their reactions, however intense their feelings, except perhaps for occasional outbursts of temper or mirth.

Emotional stimuli may give rise to pleasant or unpleasant sensations. Pleasant sensations like joy, gladness, love, and delight produce signs of satisfaction and reward. Unpleasant sensations like fear, grief, and hate may lead to nervous tension or depression. Changes aroused by an emotional stimulus are generally expressed in the viscera and skeletal muscles. The sympathetic outflow is predominantly affected; thus fear is accompanied by tachycardia, pupillary dilatation, and perspiration. If these reactions are appropriate to the stimulus, we accept them as normal. For example, a person addressing an audience for the first time will probably feel apprehensive. If, however, the fears are unwarranted or deemed inappropriate, the reactions might be considered symptoms of an anxiety state. Clinical disorders of the emotions range from mild psychological states to extremes of psychotic behavior.

NEUROPHYSIOLOGICAL BASIS

The neural mechanisms responsible for emotional behavior have been studied from many different aspects including ablation experiments, electrical stimulation and recording, clinico-pathological findings, and neurosurgical techniques. The earliest laboratory studies were made simply by removing a portion of the brain and noting any changes in the animal's responses to handling. Later came detailed studies of the peripheral changes in the autonomic nervous system following selective cerebral ablations. With the advent of stereotaxic techniques, it became possible to observe the effects of discrete lesions in the unrestrained, unanesthetized animal. More recently, the effects evoked by electrical stimulation of the brain have been analyzed in animals equipped with intracerebral electrodes. A natural development of this technique is the employment of radio controlled methods in which selected areas of the brain can be influenced by the operator sending electrical signals for stimulating, recording, or injecting drugs.

Animal experiments

Effects of ablation

1. Goltz noted signs of anger in the form of growling or snarling in the decerebrate preparation, but the responses were brief and poorly organized. More typical patterns of aggressive behavior developed when sections were made above the level of the hypothalamus. Decorticate dogs exhibited reactions of attack like snapping of the jaws and lowering of the head accompanied by a rise of blood pressure.
2. Cannon described a condition of "sham rage" in the thalamic cat. A relatively mild stimulus provoked marked offensive reactions like raising the back, erection of the hair, baring the claws, dilatation of the pupils, hissing, and biting. These effects resulted from removal of inhibitory cortical control.
3. Klüver and Bucy emphasized the importance of forebrain structures in producing alterations of emotional behavior following extensive cortical lesions in the monkey. Their findings drew attention to another component of cortical control in which the operated animals showed loss of fear and undue docility.
4. Bard demonstrated that sham-rage behavior was dependent upon the integrity of the hypothalamus. Subsequent work with Mountcastle suggested the existence of opposing cortical influences affecting the patterns of

emotional behavior. In their studies, neocortical structures (frontal and temporal poles) were described as driving influences lowering the rage threshold, while paleocortical structures (limbic complex) operated as restraining influences raising the rage threshold. Bilateral removal of the neocortex produced a placid animal with an apparent lack of emotional responsiveness to stimulation. In other animals, bilateral ablation of limbic structures elicited marked rage reactions, attacking movements, and autonomic manifestations. It was concluded that the hypothalamic centers for emotion were largely influenced by the opposing mechanisms of neocortical facilitation and limbic inhibition.

Effects of stimulation

1. Ranson, Olds, and many other workers have reported the effects of direct hypothalamic stimulation in cats and monkeys. Electrical stimulation of the lateral hypothalamic region evokes aggressive acts and postures of attack. Pain or punishment centers have been located in the hypothalamus by training the animal to press a lever on the side of the cage in order to switch the stimulus off. The technique has also been used for locating centers giving satisfaction or reward; the animal continued to press the lever again and again in a robot fashion to obtain rewards.
2. Hess demonstrated a whole range of emotional response patterns by discrete electrical stimulation of animals equipped with chronically implanted electrodes. On stimulation, the animal immediately became aggressive, showing its claws and looking for a fight, then returning to its peaceful activities as soon as the stimulus ceased. There were no sharp limits between one electrode point and another to determine precisely which areas controlled each type of behavior.
3. Delgado developed the "chemitrode" system for electrical and chemical studies of the brain by remote control in unrestrained animals. The chemitrode consists of a fine double tube implanted in the brain; it permits electrical excitation or the injection of drugs by means of radio signals from a transmitter. Brain stimulation was found to modify behavior by increasing or reducing the aggressive tendencies of the animal. This could have a profound effect on all the other animals in the cage. For example, if stimulation made the monkey more docile it would cease to be feared by the others, while increased aggressiveness often disturbed the social relations of the whole colony.

Influence of the limbic system

In 1937, Papez published his theory of central emotion in which he proposed that the hypothalamus may be involved in a circuit coursing between forebrain structures and the temporal lobe. He claimed that impulses from the hippocampus passed by way of the fornix to the mammillary bodies and were then transmitted through the anterior thalamic nuclei to the cingulate gyrus. The downward course of the impulses was presumed to follow pathways between the hypothalamus and the brain stem (Fig. 169).

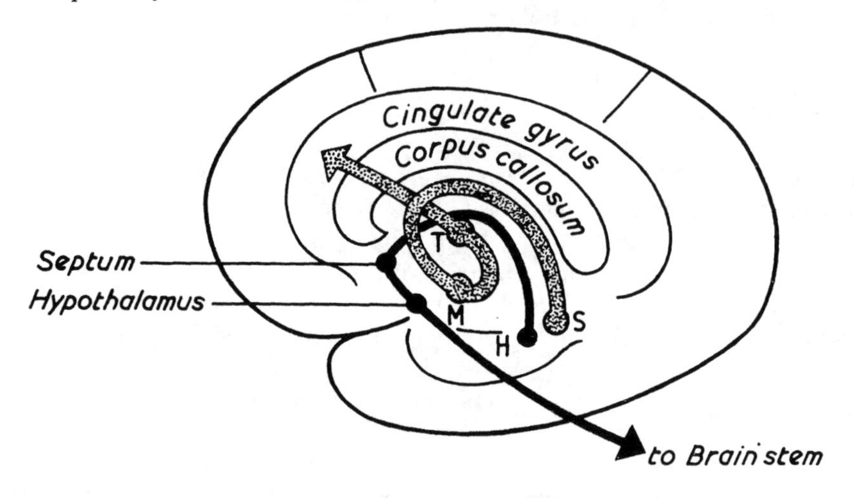

Fig. 169. Medial aspect of right cerebral hemisphere showing components of limbic structures and their principal relations. The hippocampal outflow (H) passes via the fornix to the septal nuclei and hypothalamus. The outflow from the subiculum (S) is directed toward the mammillary body (M) and continued through the anterior nucleus of the thalamus (T) to the cingulate gyrus.

The influence of the limbic system on emotional behavior has been the subject of numerous investigations in recent years. They include lesion experiments in various species, electrical stimulation studies of the freely moving animal, work on the effects of pharmacological agents, and a host of clinical studies. Despite all these efforts, however, much more information is required before we can be certain about the mechanism of emotional behavior. Our knowledge of the anatomical relationships between the limbic structures and the hypothalamus is far from complete. Moreover, it is difficult to evaluate the role of these structures since many of their activities are shared or else they give responses that are either inconsistent or are meaningless in isolation. Unfortunately, we know little about the integration of cerebral activity, especially between limbic and non-limbic systems, and for the present we must rely upon the analysis of the effects known to occur in the four principal limbic components.

Anterior cingulate

1. Stimulation. The results of experimental work on unanesthetized animals show that the anterior cingulate is capable of modifying reactions in many systems of the body. One of the most characteristic features is inhibition of spontaneous movements such as walking or grooming, which are replaced by fixed attention and alertness; the animal raises its head and pricks up its ears. Increased tension in the facial mucles of expression may also be recorded as part of the arousal response pattern.

2. Ablation. Bilateral removal of the anterior cingulate gyri results in behavioral changes in the direction of increased tameness. Monkeys show no attempt to escape from the cage and lose their fear of being handled; they do not fight or exhibit any hostility toward their companions. Beneficial effects of anterior cingulectomy have been reported in patients suffering from agitated and obsessional states. However, any conclusion concerning the functional significance of this portion of the brain must be qualified by the fact that the operation is performed only when severe behavorial disturbances are already present.

Posterior orbital

1. Stimulation. The changes induced by electrical stimulation of the orbital surface are expressed largely in various autonomic reactions, principally those affecting the cardiovascular and respiratory systems and the gastrointestinal tract. Arousal responses elicited in the unanesthetized animal are similar to the anterior cingulate effects with the adoption of specific attitudes associated with feeding, defence, or attack.

2. Ablation. Bilateral removal of the orbital cortex results in motor hyperactivity and restlessness manifested by increased pacing of the animal in its cage. Other behavioral changes are not so marked or are only transient since presumably the undamaged parts of the limbic system are sufficient for maintaining the basic processes of control. In the human being, selective orbital ablations are performed for the relief of abnormal behavior. The types of operation and the indications for treatment will be discussed in the following chapter.

Hippocampal

1. Stimulation. The effects of hippocampal stimulation on behavioral reactions are somewhat inconclusive. All that can be said is that a broad range of somato-visceral response patterns emerges, including defensive movements of the body, pupillary dilatation, salivation, and perspiring—activities conventionally associated with emotion and arousal. Single hippocampal units show characteristic bursts of activity when the animal is rewarded with food, while long silent periods are observed if the brain is stimulated and food is not given as reinforcement. The evidence suggests that the hippocampus may have a built-in discriminative mechanism for integrated actions and that it is not primarily concerned with controlling specific functions.

2. Ablation. It is doubtful whether the hippocampus can be effectively removed without damaging the overlying neocortical structures. Electrolytic lesions can be more precise but may not result in complete destruction. Secondary damage to the hypothalamus by interruption of its blood supply is a further complication. It is therefore not surprising that the results are extremely varied, or negative, or productive of little permanent change. In general, a bilateral lesion in the anterior part of the hippocampus lowers the rage threshold, increases fear, and makes the animal unsuitable for training. Hippocampal ablation also produces a deficit in learning ability. These findings suggest that the hippocampus may form part of a neuronal system that exerts an inhibitory control over behavior.

Amygdaloid

1. Stimulation. In freely moving, unanesthetized animals, electrical stimulation can cause a variety of involuntary movements such as raising the head, turning the body, stretching the limbs, licking, sniffing, and swallowing. Most of these movements are related to searching for food or to preparations for attack or sudden flight. The movements are accompanied by unmistakable signs of neuro-endocrine involvement. The behavioral effects elicited from the amygdaloid depend upon the intensity of the stimulus; low intensities induce fear and growling, while higher stimulus intensities lead to rage reactions. A crude form of functional organization of the stimulated regions may be present: The corticomedial nuclei are associated with positive reactions to fear with increased motor excitement; the basolateral nuclei, on the other hand, elicit crouching and flexing postures, arrest of movement, and decreased social activities.

2. Ablation. The existence of functional localization within the amyg-

daloid nuclei is supported by ablation experiments since contradictory results are obtained when lesions are placed at different sites. Following bilateral amygdalectomy in cats and monkeys, some investigators have observed increased docility, some have observed increased aggression and rage reactions, and others have failed to detect any changes in emotional behavior. The commonest change observed after complete removal of the amygdaloid nuclei is placidity. The animals become unusually tame and display no reactions of fear or aggression even when attacked by others in the colony. Increased sexual excitement is a prominent feature of amygdaloid lesions. This may be due to release of the ventromedial hypothalamic nuclei from inhibitory control and consequent increased production of sex hormones. Comparable findings have been reported after bilateral amygdalectomy in man. It is apparent that the amygdaloid nuclei can exert both excitatory and inhibitory influences as the occasion demands, and they must therefore help to control the overall pattern of conscious behavior.

In summary, the various components of the limbic system have each been shown to play an important role in the emotional life of the animal. A close functional association exists between these regions and the hypothalamus, which may be regarded as the target or final common path for emotional expression. The results of experimental stimulation indicate that normal behavior is expressed as the mean of excitatory and inhibitory influences producing integrated and reproducible response patterns. The emotional changes attributed to bilateral lesions of the limbic system emphasize its modulatory functions and flexibility in the direction of activation or suppression. Hence the underlying cause of apparently contradictory disturbances can be explained on the basis of differences in the location and extent of the damage.

EMOTIONAL REACTIONS

The extent to which the body will react to an emotional situation is not easy to predict. Sometimes the outward signs of an emotion are not readily apparent, while at other times an emotional stimulus may cause an explosion of uncontrollable behavior. The character of the stimulus often determines the nature of the response—e.g., tachycardia in anxiety, or perspiring in embarrassment. Since many physiological systems participate in the expression of emotion, it may be helpful to place them in the following categories:

1. Sympathetic system. The responses are due to sympathetic activation and the discharge of adrenaline. The principal effects are acceleration of

the heart, rise in blood pressure, dilatation of the pupils, dryness of the mouth, pilo-erection, and perspiring. Adrenaline mobilizes liver glycogen, raises the blood sugar, constricts the capillaries in the skin, and aids the redistribution of blood to the heart and skeletal muscles.

2. Parasympathetic system. Slowing of the heart, peripheral vasodilatation, and fall of blood pressure may reduce the blood flow to the brain and cause fainting. Loss of consciousness due to an intense emotional experience is called vaso-vagal syncopy. Other parasympathetic effects are increased salivation, hyperacidity, diarrhea, and loss of bladder control.

3. Pituitary-adrenal system. The release of ACTH from the anterior pituitary gland stimulates the activity of the adrenal cortex. As a result there is an immediate rise in the blood level of corticosteroids for use in the general metabolic processes of the tissues and for rapid mobilization of energy stores. The release of ADH from the posterior pituitary gland promotes reabsorption of water from the distal tubules of the kidney, resulting in a marked fall in the volume of urine excreted.

4. Muscular system. There is usually increased tension in all skeletal muscles, as revealed by electromyographic records, although the subject may show no other visible signs of emotion. In extreme cases, tremors of the arms and facial twitching are plainly evident. Opposite effects may also occur when the muscles become limp or fatigue rapidly with use. Or the individual may become temporarily paralyzed.

While all categories of bodily changes contribute to the physical side of emotion, certain types of stimuli have preferences for one or other form of expression. Fear and anxiety, for example, usually elicit reactions in the sympathetic nervous system, while shock excites the parasympathetic nerves to the vascular system and may cause circulatory collapse. Long-term emotional stress has an adverse effect on many systems of the body. Patients develop all kinds of symptoms like digestive disturbances, nervous headache. and insomnia, which may tax the skill and understanding of the clinician far more than symptoms do in the treatment of organic disease.

DIAGNOSTIC AIDS

Attempts have been made to measure emotional reactions in patients suffering from psychological stress. Some of the methods employed will now be described. None of the methods help to identify the underlying cause of the abnormality, but they can provide a means of assessing the response patterns evoked by a particular emotional situation.

Autonomic tests

1. General analysis of bodily changes includes measurements of heart rate, arterial blood pressure, respiratory rate, alveolar CO_2 tension, metabolic rate, pupillary diameter, salivary secretion, gastric secretion, and skin temperature.

2. Laboratory analysis of blood, urine, and cerebrospinal fluid includes quantitative estimations of electrolytes, neuro-humoral agents, biogenic amines, and hormones. Although many of these tests are difficult to perform and require highly skilled techniques, they are probably a more reliable index of change than the measurements of peripheral autonomic mechanisms.

Sudo-motor tests

These tests are based on the fact that perspiring is very commonly induced in emotional states. Selected sites are the forehead, neck, axilla, palms of the hands, and the soles of the feet.

1. In the starch-iodine method, a weak solution of iodine is applied to the skin and starch-impregnated paper is then pressed against the test area. Activity of the sweat glands causes a local staining reaction, the intensity of which is proportional to the number of functioning glands.

2. In the skin galvanometer method, the subject is connected to a Wheatstone Bridge by means of electrodes wrapped around two fingers of his hand. The electrical resistance of the skin is measured during a control period of rest and again during a period of emotional excitement. The fall in skin resistance due to sweating is read directly from a meter. A resistance curve is then constructed from readings taken at set intervals over a period of about 20 minutes. Although the method is familiar to psychologists, it is often difficult to obtain suitably controlled records owing to variable skin resistances in normal subjects and failure to attain complete relaxation in highly emotional subjects.

Depression of alpha rhythm

The correlation between an emotional stimulus and the depression of alpha rhythm can often be demonstrated on the electroencephalogram. The emotional content of different words or phrases can be deduced, since neutral words have no effect on the trace, whereas stimulus words may cause a significant depression of the rhythm. Examples are given in Figure 170, in which the subjects had their eyes closed. One of the subjects reacted in a very definite way to the word "speech"; depression of alpha rhythm persisted for about 12 seconds. Previously, he had volunteered the information that he disliked the idea of giving a speech owing to a past embarrassing experience.

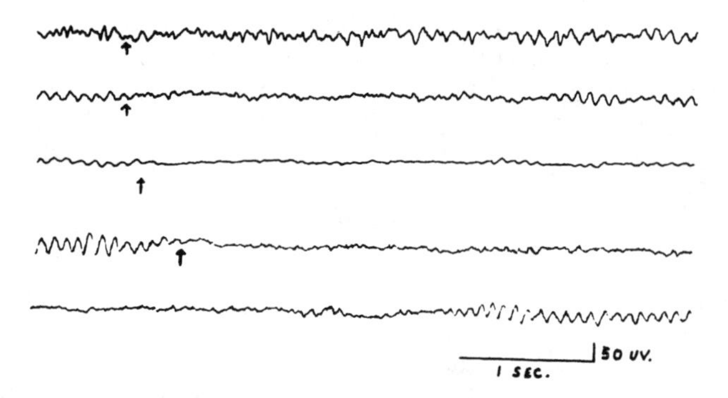

Fig. 170. Depression of alpha rhythm. Examples are from four individuals' responses. Top trace shows no change after a neutral word. Second trace shows slight depression following the stimulus word 'quarrel.' Third trace shows depression of alpha rhythm following the stimulus word 'hospital.' Continuous record of lower two traces is from a subject who reacted to the word 'speech.' Stimuli are indicated by arrows. Calibrations are 1 sec and 50 μv.

Increased muscle tension

The relationship between emotion and skeletal muscle tension was first reported from the study of pressure points in handwriting. The idea was applied to electromyographic recordings from limb and trunk muscles in which bursts of muscle action potentials were observed when the subject became excited. Almost any muscle can be used for these tests. One technique employs standard EEG electrodes placed over the temporal muscles of the scalp. When the subject is completely relaxed with the eyes closed, no muscle artifacts appear in the tracings. The individual is not required to answer to test words, but only to listen. Stimulating words tend to bring out muscle spikes superimposed on the tracings owing to action currents in the scalp muscles. If the word happens to arouse a strong emotional feeling, the action potentials build up in amplitude and frequency until the operator decides to withdraw the stimulus (Fig. 171). The subject rarely shows any visible signs of emotion during the interview except on the electromyographic record. The method is considered extremely useful for detecting emotional changes and for providing a means of measuring them in an accurate way. It also underlines the practical association of psychological stimuli with muscle tension.

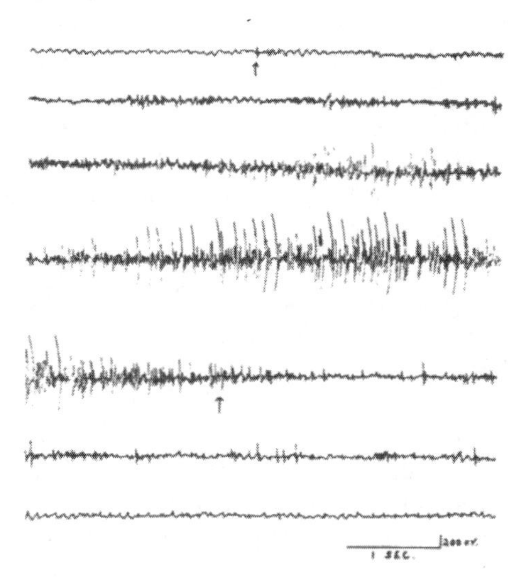

Fig. 171. Electromyographic response to suggestion. Continuous record from scalp muscles of female subject with eyes closed. At first arrow, subject 'sees' a spider; it 'jumps' on her leg; 'crawls' over her body; a big, hairy spider appears; she is unable to brush it off. At second arrow, spider has gone out of sight. Note disappearance of muscle action potentials when stimulus is withdrawn.

HUMAN EMOTIONAL DISTURBANCES

The stresses to which man is exposed only too frequently result in abnormal patterns of behavior. These abnormalities may range from mild anxiety states to a total lack of restraint as is seen in the psychopath. To account for these changes from a neurophysiological standpoint, it must be supposed that there is some loss of inhibitory control or that control is ineffective under certain tense situations. Lack of control may result from inadequate training in childhood or it may develop after long-standing stresses or when the particular stress is unduly strong. Because of the difficulties in defining human emotions, except in terms of what is conventionally acceptable to society, it is equally difficult to assign a realistic classification. For example, an emotional outburst such as a person showing anger when insulted can be accepted as perfectly normal. But if the reaction is inappropriate to the circumstances, his behavior is interpreted as unreasonable.

Fear and other symbols of danger set in motion a pattern of protective reactions; these in turn cause profound changes in bodily functions. Thus headache, vasomotor disturbances, vomiting, and other gastrointestinal disturbances are common experiences following an unpleasant episode but tend to disappear when the crisis has passed. An individual who tends to overact or one who fails to make suitable adjustments to resolve his conflicts may develop troublesome symptoms which then become a permanent feature of his life. Accordingly, it may be useful to distinguish between emotional disorders due to psychological stress and those that arise from a pathological disturbance of the control mechanisms.

Psychological causes of emotional problems represent a failure to cope with stress situations. Almost any tissue of the body may be involved. Some patients direct their fears toward one particular organ, while others present a much wider complex of apparently unrelated symptoms. The components that make up the expression of an emotional disturbance belong to the immense field of psychosomatic medicine. They include such diverse disorders as migraine, peptic ulcer, asthma, hypertension, ulcerative colitis, and skin rashes. As the nervous system is particularly sensitive to psychological stress, a distinction can be made between predominantly excitatory and predominantly inhibitory states, because different physiological mechanisms are in operation.

1. Excitatory states. These are indicated by a display of anxiety, aggressiveness, restlessness, insomnia, tension, and increased sexual drives.

2. Inhibitory states. All forms of depression come under this heading. The symptoms include apathy, lack of concentration, suppressed tensions, feelings of guilt, and diminished sexual potency.

Pathological causes of emotional problems represent changes resulting from disease. Certain lesions interfere with the peripheral mechanisms of emotion and are not primarily concerned with the affective processes. A good example is seen in Parkinson's disease, in which the facial expression becomes mask-like and emotional movements of the facial muscles seldom occur. The condition, of course, is due to the stiffness of the musculature resulting from loss of inhibitory signals from the substantia nigra. Lesions more directly concerned with the central mechanisms of emotion are located in the hypothalamus, limbic regions, and prefrontal lobes.

1. Hypothalamus. Tumors, infections, and vascular lesions involving the hypothalamus may cause a variety of behavioral changes in addition to the more usual features of neuro-endocrine dysfunction described in the previous chapter. The clinical picture may be complicated by mixed irritative and destructive influences on closely related anatomical structures. In general, anterior lesions tend to cause loss of restraint and aggressive behavior, while posterior lesions tend to produce states of depression.

2. Limbic regions. The hippocampus and temporal cortex are common

sites of seizure discharges in epileptic patients. Many such patients suffer from feelings of inadequacy, introspection, loss of sexual drive, or impotence. These disturbances may represent a repressive function of limbic activity on normal motivations and drives. Behavioral disturbances due to destructive lesions are seen in patients with tumors or abscesses of the temporal lobe. Damage to the limbic circuitry leads to loss of control manifested by unreasonable outbursts of temper, hostile urges, and other attitudes against the conventions of society.

3. Prefrontal lobes. Functional disturbances of the prefrontal lobes may account for subtle changes of general behavior and may complicate various life situations. Often no other symptoms are present. Fantasies or sudden outbreaks of inexplicable anxiety may develop. Bilateral injury to the prefrontal lobes usually results in marked emotional and personality changes. Patients lose interest in social relationships, become careless about their appearance and clothes, and often develop compulsive obsessions. The implications are obviously of great clinical importance and will be discussed in detail in the next chapter.

SELECTED BIBLIOGRAPHY

Bard, P., and Mountcastle. V.B. Some forebrain mechanisms involved in the expression of rage. *Res. Publs. Ass. Nerv. Ment. Dis.* 26:362-404, 1948.

Hilton, S.M., and Zbrozyna, A.W. Amygdaloid region for defence reactions and its efferent pathway to the brain stem. *J. Physiol.* 165:160-173, 1963.

Fernandez de Molina, A., and Hunsperger, R.W. Organization of the subcortical system governing defence and flight reactions in the cat. *J. Physiol.* 160:200-213, 1962.

Kaada, B.R., Andersen, P., and Jansen, J., Jr. Stimulation of the amygdaloid nuclear complex in unanesthetized cats. *Neurology* 4:48-64, 1954.

Maclean, P.D. The hypothalamus and emotional behavior. In Haymaker, W., Anderson, E., and Nauta, W.J.H. (Eds.): *The Hypothalamus.* Springfield, Ill. Thomas, 1969.

Maclean, P.D., and Delgado, J.M.R. Electrical and chemical stimulation of fronto-temporal portion of limbic system in the waking animal. *Electroen. Clin. Neurophysiol.* 5:91-100, 1953.

Newman, P.P. Electromyographic studies of emotional states in normal subjects. *J. Neurol. Neurosurg. Psychiat.* 16:200-208, 1953.

Papez, J.W. A proposed mechanism of emotion. *Archs. Neurol. Psychiat.* 38:725-743, 1937.

Parmeggiani, P.L., Azzaroni, A., and Lenzi, P. On the functional significance of the circuit of Papez. *Brain Res.* 30:357-374, 1971.

Schreiner, L., and Kling, A. Behavioral changes following rhinecephalic injury in cat. *J. Neurophysiol.* 16:643-659, 1953.

Ward, J.W., and Back, J.B. Responses elicited from orbital cortex of cats: overt activity and related pathways. *Amer, J. Physiol.* 207:740-749, 1964.

CHAPTER XXIII

The Prefrontal Lobes

For many years the prefrontal lobes were considered to be "silent areas" since electrical stimulation elicited no visible responses and removal in experimental animals failed to produce any obvious deficits. Nevertheless the literature of the nineteenth century contains some extremely important observations on the possible role of this part of the brain, which have particular relevance to man's study of his own mind.

In 1875, Ferrier published his now historic account of experiments on the brains of monkeys from which he concluded that animals deprived of their prefrontal lobes exhibited marked changes in character and disposition. To this evidence must be added the famous case of Phineas Gage who suffered a change of personality after recovering from a severe, penetrating wound through the skull. A classical account of the behavioral changes observed in two tame chimpanzees was reported by Fulton and Jacobsen who removed first one frontal area and three months later the second frontal area from each animal. It appeared that the second operation led to the elimination of frustrational behavior, a finding that encouraged Moniz and Lima to perform bilateral prefrontal leucotomy for relieving anxiety states in man.

The initial success of surgical intervention in patients resistant to other forms of treatment stimulated the development of better techniques and more accurate location of lesions. The operation became very popular during the last few decades in the treatment of personality disorders and is still useful in selected individuals. However, the indications for psychosurgery are now very limited owing to the increasing effectiveness and availability of modern drugs.

Functional anatomy

The prefrontal lobe occupies the frontal pole anterior to the excitomotor cortex (areas 4, 6, and 8) and also extends on to the orbital surface and adjacent medial surface of the hemisphere. The cortical neurons receive a heavy projection of fibers from the dorsomedial nucleus of the thalamus and in turn project back to this nucleus, as shown in the diagram of Figure 172. Since two-way projections are known to exist between the orbital surface and the hypothalamus, it follows that the prefrontal lobe has extensive roundabout connections with the anterior cingulate gyrus, temporal cortex, and brain stem. The complex circuitry between these cortical and subcortical structures provides the anatomical basis for correlating some of the higher functional activities of the brain such as the intellect, personality, motivations, emotional drives, and social conscience.

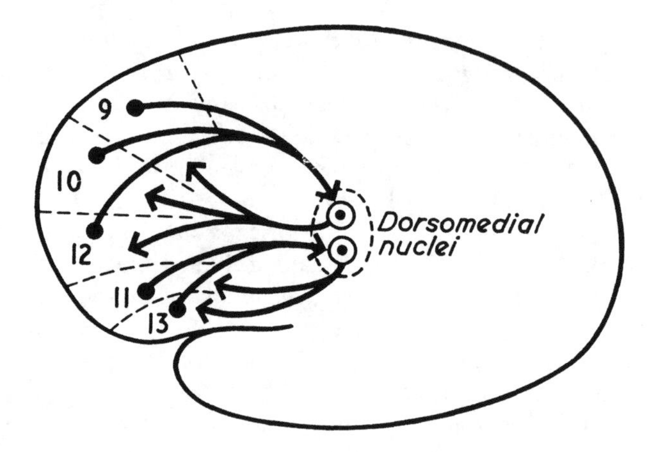

Fig. 172. Diagram of reciprocal connections between the dorsomedial nuclei of the thalamus and the prefrontal cortex. The small-celled portion of the nucleus projects to the frontal pole; the large-celled portion projects to the orbital surface.

Experimental studies in subhuman primates

Intelligent animals are first submitted to a period of pre-operative training and then tested for performance and reactions after bilateral ablation of their prefrontal lobes.

1. Behavioral tests. Training procedures include problem solving, discriminative abilities, and intelligence tests. A careful assessment is made of the animal's reactions to reward and punishment, emotional behaviour, temper tantrums, ease of handling by the operator, and relationship to other members of the colony. One classic training procedure is known as the "stick and platform" test. The cage lies between two platforms on which sticks of different sizes are placed. The monkey must grasp a stick within its reach to obtain longer sticks in order to secure food it can see at the end of one platform.

2. Ferrier's observations. "An animal deprived of its frontal lobes retains all its powers of voluntary motion and sensory impressions, yet appears to lose the faculty of intelligent and attentive observation. It ceases to exhibit any interest in its surroundings, looking indifferently at what formerly would have excited intense curiosity or indulging in restless and purposeless wanderings."

3. Becky and Lucy operations. In 1934, Fulton and Jacobsen performed the operation of lobectomy on two trained chimpanzees. After removal of one frontal area in each instance (areas 9, 10, 11, and 12 in the Brodmann map) no sign of behavioral change could be detected during three months of post-operative testing. When the second frontal area was removed from each animal, it was evident that a profound change had occurred in the disposition and character of the animals; they seemed devoid of emotional expression and also failed the stick and platform test.

4. Types of disturbance. The effects of prefrontal lobectomy depend upon the extent of the damage inflicted by the operation on the cortico-thalamic and hypothalamic projections.

- Impairment of intelligence results from interruption of the connections between the frontal poles and the dorsomedial nuclei. The animals are slow to learn simple discrimination tests and perform badly with many errors before making the correct choices.
- Restlessness results from lesions encroaching upon the basal ganglia. The animals assume various postures and pace about the cage in a stereotyped fashion.
- Lack of emotional responsiveness and indifference to training failures are commonly caused by lesions of the orbital cortex interrupting projections to the hypothalamus.

Experimental studies in man

Changes in the intellect and personality have been described in persons who have recovered from severe head wounds affecting the prefrontal lobes. More subtle changes of character are documented in the clinical literature of patients with slowly growing frontal lobe tumors.

1. The case of Phineas Gage. The case report on Phineas Gage provides a classic description of personality changes as a consequence of brain damage. The injury sustained by Gage was caused by an iron crowbar blown by a stick of dynamite though the front part of his skull. Prior to the accident he had been an efficient and capable foreman, but subsequently his behavior became so outrageous that his employers could not give him back his place, and his friends hardly recognized him as the same person. He was

fitful, irreverent, impatient, and vacillating to the extent that his mind was radically altered.

2. Head injuries. The effects of damage to the brain from head injuries can be very serious at any age, but the consequences are far more tragic for babies and young children who have not yet developed their full faculties. Adults tend to become distractible, turning from one idea to another without any sense of purpose and lack tenacity in pursuing a course of action. Quickly changing moods, euphoric episodes, or sudden depression are frequent occurrences. Prefrontal lesions also generally result in loss of motivation and ambition, although the intellect and memory may not be obviously impaired.

3. Frontal lobe tumors. Large space occupying tumors produce features of raised intracranial pressure—headache, vomiting, and papilledema. Smaller frontal lobe lesions may be silent clinically but usually produce subtle changes in personality. At first there may be little evidence of cerebral disorder apart from an increased irritability or minor deviation from judicious and responsible behavior. Later, more definite signs are apparent as the patient becomes more and more disorganized. Intellectual efficiency deteriorates, memory suffers, wrong decisions are made, and employment may be at risk. Finally, the patient loses pride in his personal appearance and habits, often remaining insensible to repercussions to the extent of becoming a social embarrassment.

Prefrontal leucotomy

The practice of opening the skull to relieve mental derangement existed in primitive times, but the first planned operation to sever the frontothalamic projections was not introduced until 1935 when Moniz reported successful results on a small number of cases. In the original operation a burr hole was made on each side of the skull and a blunt knife or leucotome was inserted into the brain. The sweep of the knife was made blindly for a measured distance in the plane of the section. During the next two decades a modified form of leucotomy was performed with increasing popularity in many countries and, despite misgivings on ethical grounds, the newer methods gave extremely impressive results with minimal side effects.

Types of operation

While the standard method of leucotomy met with some initial success, the operation failed to give clinical improvement in a percentage of patients and some were made worse. Accordingly, attempts have been made to control the lesion more accurately and to limit its extent.

1. Prefrontal lobotomy, as practiced by Freeman and Watts, involved section of the fronto-thalamic projections in different planes.
2. Orbital lobotomy was just as radical except that the approach was made through the roof of the orbit.
3. Gyrectomy was performed by open operation in which the projections were severed under direct vision.
4. Bilateral anterior cingulectomy was performed with beneficial results as an alternative to lobectomy.
5. Stereotaxis allowed the location of lesions with a high degree of accuracy and with virtually no operative risk. In one kind of operation, selected cortical sites were permanently destroyed by electrocoagulation. In another kind of operation, the hypothalamus was the target. A guiding probe was first inserted in the conscious patient and a small stimulating current applied to the tip. The patient showed signs of anxiety, indicating the correct placement of the probe. Heating electrodes were then applied, with the patient under anesthesia, to destroy the site on each side by coagulation.

Indications for psychosurgery

Despite great advances in modern drug therapy, surgery is justified when all other forms of treatment, including electroshock, have been tried without success. Some clinicians, however, believe that leucotomy should not be postponed until the patient's condition is desperate and that if the patient has suicidal tendencies, the operation may have a life-saving value. The general indications for psychosurgery are as follows:

1. Antisocial behavior. Persons who cannot fit into family or social life may be confined to mental institutions. Their unwillingness to cooperate or to listen to reason makes them unsuitable for useful employment.
2. Criminal behavior. Persons who suffer from chronic obsessional states may come into conflict with the law. Habitual criminals manifest their lack of restraint by exhibitionism, fire-raising, or irrational destruction of property. Others are driven to irresponsible actions of violence, sexual assaults, or murder.
3. Intractable depression. Persons with a long-standing psychiatric history may subside into a morbid state bordering on melancholia. The mood may swing to extremes of euphoria often associated with paranoid delusions.
4. Intractable pain. Persons who are not generally considered to be abnormal may suffer from a deep anxiety in which physical pain is the domi-

nant symptom. The disability may be sufficiently severe to prevent the normal pursuit of hobbies and occupation.

Results

Only a general statement will be made here on the evaluation of the results of leucotomy, because so much depends upon the selection of patients, the duration of the illness, and the type of operation performed. Furthermore, different surgeons employ different diagnostic criteria, different methods of post-operative assessment and different standards of classifying recovery. Patients deemed to be improved by the operation are those who are able to leave the hospital and return to their families and eventually to work. The term "recovery" should be applied only to those patients who require no further psychiatric help. On the whole, the best results are obtained in cases of chronic anxiety and obsessional states, whereas poor results can be expected in patients with aggressive or psychopathic personalities. Reports from close relatives are most helpful in evaluating the beneficial effects or disadvantages of the operation. The important features to investigate in the follow-up are the patient's ability to cope with day-to-day problems and to occupy his time in useful activities.

Functions of the prefrontal lobes

It is difficult to make any clear-cut statements on the functions of the prefrontal lobes since most studies are based on clinical observations of disturbances associated with tumors of the frontal lobe or of the disturbances following injury or surgery. While the extent of a lesion can be verified by postmortem findings, the functional damage inflicted on intact neural tissue cannot be accurately estimated. The removal of one prefrontal lobe may have surprisingly little obvious effect on the behavior and abilities of the patient, and even bilateral ablation may leave the individual without any striking neurological deficit. There is no doubt that the prefrontal lobes contribute in a general way to the processes of cerebration and therefore reach their greatest prominence in man. From a practical point of view, the important functions are those that deal with personality, intellect, emotional behavior, social conscience, and the tolerance to pain.

1. Personality. The prefrontal lobes take part in developing the personality and maintaining originality of thought. The structure of character is molded by reactions to events and willingness to accept responsibility

according to circumstances. Personality changes are the commonest sign of disturbed function, as evidenced by a lack of interest and initiative. The individual becomes careless of his appearance and is unable to concentrate or plan effectively.

2. Intellect. The prefrontal lobes are not essential to the intellectual faculty but are needed for the more critical aspects of strategy and its application. Normal persons are capable of assessing the odds and making logical deductions before committing themselves to action. The patient, on the other hand, lacks insight, makes errors when faced with an alternative choice, and generally has a dull intellect, while holding an exalted opinion of his abilities.

Emotional behavior. The prefrontal lobes help to promote a proper balance between drives and restraints. Emotional drives provide an incentive for ambition, creative thought, the development of new activities, and above all the will to find solutions to problems. Restraints are the basis of self-discipline and sacrifice; to impose restraint is to enable the individual to be strong and withstand disappointments and frustrations. If the balance is disturbed for any reason, the outcome may lead to excesses in either direction, resulting in total unreasonable behavior. Loss of restraint causes the patient to overact emotionally, language may become abusive, and temper easily aroused without provocation. Alternatively, loss of emotional drive leads to introspection; the patient becomes unnaturally quiet and insecure. In the end, routine activites become a burden and severe depression sets in.

4. Social conscience. It is not possible to define human relationships in terms of structural localization in the brain. Nevertheless, it is clear that human beings have concern for the welfare of their fellows and that very strong mechanisms must exist for arousing feelings of compassion. In prefrontal lesions marked changes occur in a patient's relationship with other people. He may remain completely indifferent to the sufferings of others and show no regard or affection even to those who were formerly close to him.

5. Tolerance to pain. The perception of pain is not, strictly speaking, a function of the prefrontal lobes. But pain can enter so much into the emotional life of a person that his whole attention is concentrated on it and then his work, hobbies, and family are adversely affected. As medication rarely succeeds in removing the pain, the patient's demands for relief increase with the attendant risk of addiction. Prefrontal leucotomy is recommended in selected cases. The procedure does not in fact abolish the pain but only the reactions to it. Increased tolerance allows the patient to return to a useful and contented life without further demands for relief by drugs.

Thalamo-cortical relationships

The functions ascribed to the prefrontal lobes demonstrate how the brain uses information from different sources and draws upon experience in preparing for a course of action. Personality is developed by the way an individual performs intellectual tasks and makes decisions, which sometimes may be distasteful but are deemed necessary and fall within bounds acceptable to society. Contributions to the study of personality and behavior come from many different disciplines, yet little is known about the mechanisms by which the brain performs its complex operations. Unfortunately, neurophysiological concepts relating cerebral processes to their neural organization are still poorly understood.

It may be useful to look at the organization of the thalamus as the key to the role of the prefrontal lobes. The anterior nuclei of the thalamus are part of the neuronal circuitry between limbic and hypothalamic structures that appears to operate a restraining influence on behavior. Damage to any part of the circuitry interferes with the delicate balance that normally directs our thoughts and actions along well-defined paths. Therefore, if the anterior thalamic nuclei were destroyed, the functions of the corresponding cortical areas would become grossly deranged. Loss of restraint deprives an individual of the power to exercise finer judgments and leaves him at the mercy of his passions and innate desires.

The dorsomedial nuclei of the thalamus are essential to the driving influences that are believed to emanate from the orbitofrontal cortex. Such influences are indispensable for developing profitable activities and for pursuing ambitions and plans for the future. Accordingly, destruction of the dorsomedial nuclei causes loss of motivation and marked deficits of emotional and intellectual capacity. It must be recognized, however, that the thalamic nuclei are only part of a profuse network of cerebral interconnections. Many other aspects of brain function, such as learning and memory, remain to be considered, and the danger of oversimplification is only too obvious.

SELECTED BIBLIOGRAPHY

Clemente, C.D., and Chase, M.H. The neurological substrates of aggression behavior. *Ann. Rev. Physiol.* 35:329-356, 1973.

Falconer, M.A. Relief of intractable pain of organic origin by frontal lobotomy. *Res. Publ. Ass. Nerv. Ment. Dis.* 27:706-714, 1948.

Fulton, J.F. *Functional Localization in the Frontal Lobes and Cerebellum.* Oxford: Clarendon Press, 1949.

Heath, R.G. Electrical self-stimulation of the brain in man. *Amer. J. Psychiat.* 120:571-577, 1963.

Hitchock, E., Laitinen, L.V., and Vaernet, K. (Eds.). *Psychosurgery.* Springfield, Ill. Thomas, 1972.

Knight, G.C. Stereotaxic tractotomy in the surgical treatment of mental illness. *J. Neurol. Neurosurg. Psychiat.* 28:304-310, 1964.

Lewin, W. Observations on selective leucotomy. *J. Neurol. Neurosurg. Psychiat.* 24:37-44, 1961.

Livingston, K.E. Surgical contributions to psychiatric treatment. In Freedman, D.X., and Dyrud, J.E. (Eds.). *Amer. Handbook of Psychiatry* 5:548-563, 2nd ed., 1975.

Meyer, A., Beck, E., and McLardy, T. Prefrontal leucotomy: A neuro-anatomical report. *Brain* 70:18-49, 1947.

Moniz, E. Prefrontal leucotomy in the treatment of mental disorders. *Amer, J. Psychiat.* 93:1379-1385, 1937.

Rylander, G. Personality analysis before and after frontal lobotomy. *Res. Publ. Ass. Nerv. Ment. Dis.* 27:691-705, 1948.

Ward, A.A. Jr. The anterior cingulate gyrus and personality. *Res. Publ. Ass. Nerv. Ment. Dis.* 27:438-445, 1948.

Watts, J.W., and Freeman, W. Frontal lobotomy in the treatment of unbearable pain. *Res. Publ. Ass. Nerv. Ment. Dis.* 27:715-722, 1948.

CHAPTER XXIV

The Reticular System

The central core of the brain stem between the medulla and the thalamus has been described as a third neural system that coordinates the motor and sensory functions of the body and grades the activity of most other nervous mechanisms. Its lower portion is a center for many visceral reflexes controlling respiratory, cardiovascular, and gastrointestinal functions. From here, fibers descend into the spinal cord to influence muscle tone, posture, and limb movements; other influences modify inputs from the sense organs, acting as a filter of all sensory signals.

There are also important two-way links with the cerebellum. Its upper portion is involved in the general level of neuronal excitability affecting all cerebral processes, especially those maintaining the waking state. It is the neural mechanism of consciousness and also a storage system for the subconscious. The degree of alertness and comprehension is determined by ascending influences relayed from the reticular core to widespread areas of the cerebral cortex. Finally, the multi-synaptic character of the reticular structure accounts for its susceptibility to anesthetic agents and many other types of drugs.

Functional anatomy

The organization of the reticular system comprises three major structural components—the brain stem reticular formation, the diffuse thalamic system, and the cortico-reticular projections.

1. Brain stem reticular formation. According to Cajal, the reticular formation consists of a chain of interneurons occupying the mid-line between the medulla and thalamus. It begins a little above the decussation of the pyramids, where its main input is from the anterolateral columns of the cord. Throughout its ascent it receives numerous collaterals from the medical lemniscus, lateral lemniscus, cerebellum, and the principal motor and sensory nuclei in the brain stem. Attempts to divide the reticular formation

into distinct anatomical regions are not very convincing, although certain specific nuclei are described. These include the nucleus reticularis gigantocellularis, nucleus reticularis pontis caudalis, and nucleus reticularis tegmenti pontis.

2. Diffuse thalamic system. The rostral portion of the reticular formation ends in the mid-line and intralaminar nuclei of the thalamus, subthalamus, and hypothalamus. Some of the axons from these nuclei project only to subcortical structures, but the majority pass from the thalamus to the cerebral cortex in a diffuse projection extending into both cerebral hemispheres. All parts of the cortex are reached by these unspecific reticular fibers.

3. Cortico-reticular projections. Descending pathways from the cerebral cortex terminate at various levels in the reticular formation. Many of them relay as mossy fibers to the cerebellum. According to Brodal, dense collections of cortico-reticular fibers are to be found in the cortico-spinal tracts of the same side. On the other hand, descending pathways from the orbital cortex and temporal lobe exert their influence bilaterally.

Visceral functions

At all levels of the brain stem reticular formation, there are reflex arcs serving as centers for visceral functions. These functions have been described in detail in Chapter XX and only a brief summary will be given here.

1. Respiratory. There are three respiratory centers–inspiratory, expiratory, and pneumotaxic. Rhythmic breathing is the result of the Hering-Breuer reflex, whereby the inspiratory phase of respiration is brought to an end. The pneumotaxic center modulates the periodicity of breathing, for example, increasing the rate of respiration during exercise.

2. Cardiovascular. There are three cardiovascular centers–cardiac inhibitory, cardiac acceleratory, and vasomotor. The heart rate and force of contraction are reduced by vagal inhibition and increased by sympathetic action. The vasomotor center controls the arterial blood pressure by reflex influences on the peripheral arterioles.

3. Alimentary. There are three alimentary centers–gastric, vomiting, and salivary. Vagal reflexes control the volume and concentration of the gastric juice and regulate movements of the stomach. The vomiting center is influenced by afferents from many sources including those of psychic origin. Salivation is brought about by reflex mechanisms associated with the nuclei of cranial nerves supplying motor fibers to the salivary glands.

4. Thermal. Heat-sensitive neurons are found in the medial reticular for-

mation of the medulla. They may help to remove excessive heat by inhibiting the vasomotor center. Their role may be limited to emergency states such as heat exhaustion when massive vasodilatation occurs in the body.

Motor functions

Brain mechanisms involved in the control of the skeletal musculature exert their effects in large part through the reticular formation. From here, descending pathways transmit either facilitatory or inhibitory impulses into the cord. Their actions may be exerted on the large anterior horn cells supplying motor nerves to the muscles or they may modify the gamma efferent discharges to the muscle spindles. These functions have already been described in Chapter XII.

1. Facilitatory reticular area. The greater part of the reticular area provides a background of excitation for spinal reflexes. Stimulation augments cortically induced movements, enhances skeletal muscle tone, and increases tendon jerk responses. Damage, on the other hand, causes loss of power in the muscles, diminishes tone and reduces the tendon jerks. The hypotonia observed in patients after transection of the cord is due to loss of facilitatory supraspinal influences.
2. Inhibitory reticular area. Reticular-spinal pathways terminating on short interneurons in the cord transmit impulses from the cerebral cortex, basal ganglia, and the cerebellum. Stimulation depresses cortically induced movements, decreases muscle tone, and reduces reflex activity. Damage leads to hypertonia and exaggerated reflexes. The postural changes associated with spasticity are due to the fact that the inhibitory area becomes less active and facilitation predominates.

Sensory function

Brain mechanisms capable of modifying sensory inputs exert their effects through non-specific elements in the reticular formation. In their course through the brain stem, the ascending sensory tracts give off extensive collaterals to the reticular neurons whose axons ramify in a multi-synaptic network. Reticular neurons respond to a wide variety of stimuli; many are spontaneously active and respond to the stimulus by an increase or decrease of firing rate. Their discharges are greatly influenced by afferents converging from different sources, and consequently, the output patterns of reticular neurons have few of the characteristics exhibited by the peripheral nerves. Furthermore, the latencies of the evoked responses are considerably longer than those recorded from the primary sensory tracts at the same level.

Evidently, reticular neurons have poor localizing or discriminating ability and are not concerned with transmitting detailed information. Impulses inducing facilitatory responses are conducted more rapidly through the reticular substance as shown by their fairly short latencies. But their influence is exerted bilaterally and widely dispersed on the diencephalic structures and thalamic nuclei on which they terminate. Such effects no doubt enhance the excitability of the sensory systems and serve as a driving force for powerful stimuli or significant events or alter the perceptual experience of important signals like pain. Impulses inducing inhibitory interactions are generally depressed or blocked altogether or else repeated synaptic delays slow down conduction to higher levels. The reticular neurons can therefore serve as a screening device for weak and unwanted signals and help to eliminate background noise. In these ways the reticular formation imposes a measure of control over the sensory input, acting in a non-specific way on functions such as preference, selectivity, and attention.

Arousal functions

In 1949, Magoun and Moruzzi showed that stimulation of the brain stem reticular formation in a sleeping animal modified the electrical activity of the cerebral cortex. The characteristic pattern of the EEG recorded during sleep (high-voltage slow waves and spindle bursts) was replaced by the low-voltage fast rhythms of the waking state. This change in the EEG pattern was called the "arousal reaction" (Fig. 173). At the same time the sleeping animal became wide awake. It was subsequently demonstrated that strong stimulation of the reticular formation induced a state of excitement or alarm. On the other hand, if the reticular substance was destroyed, while sparing the lemniscal pathways, the animal fell into a deep comatose state and could be aroused only by intense sensory stimulation. Anatomically,

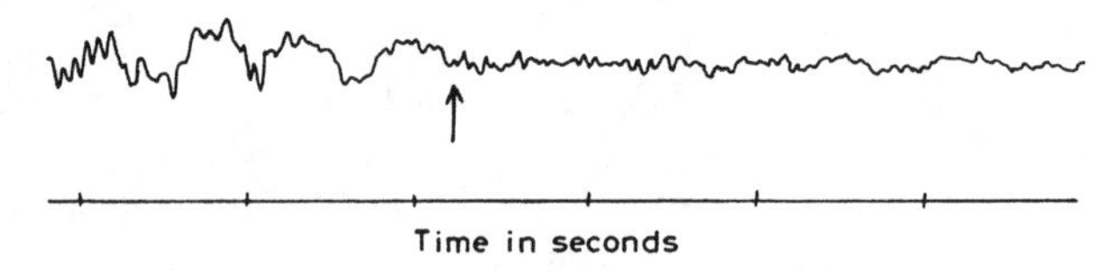

Fig. 173. Arousal reaction. On stimulating the ascending reticular activating system in a sleeping animal, the sleep pattern of the EEG is changed to the rhythm of the waking state. Point of stimulation is marked by arrow.

the pathways involved in performing the arousal function are known as the ascending reticular activating system.

The above findings have been amply confirmed by experiments on dogs and monkeys equipped with chronically implanted electrodes. It appears that normal wakefulness is dependent upon the generation of activity within the reticular formation, which in turn receives such a diffuse input from the cord that almost any sensory stimulus in the body can be effective. The influence exerted by the reticular activating system is transmitted upward to the reticular, mid-line, and intralaminar nuclei of the thalamus, all of which have multiple connections with the specific thalamic nuclei, subthalamus, and hypothalamus. The outflow is completed through diffuse thalamocortical projections on all parts of the cerebral cortex in both cerebral hemispheres. In addition to activation of the reticular system by sensory impulses, there are an exceedingly large number of fibers that pass to it from the motor regions of the brain, for motor activity is usually associated with a high degree of wakefulness. The significance of these relationships will now be considered.

RETICULAR-CEREBELLAR RELATIONSHIPS

Apart from extending its influence upward to the cerebral cortex and downward into the spinal cord, the reticular system has close anatomical and functional relationships with the cerebellum. It is admirably situated in the brain stem to integrate functions and affect excitability levels in all directions.

Functional anatomy

Despite the apparent lack of topographical arrangements within the reticular substance, the projections to and from the cerebellum appear to be fairly well organized. *Afferents* to the cerebellum come from both cerebral and spinal sources: Descending fibers travel in the cortico-pontine tracts to the nucleus tegmenti pontis; from here, relays pass as mossy fibers to the cerebellar vermis. Ascending fibers travel in the spino-reticular tracts to the lateral reticular nucleus; from here, relays pass as mossy fibers to the vermis of the anterior lobe. *Efferents* from the deep nuclei of the cerebellum emanate mainly from the fastigial nucleus; the axons give off many collaterals that subsequently divide into ascending and descending branches.

Cortico-reticular responses

Electrophysiological evidence suggests that activity evoked in the cerebellum following stimulation of the cerebral cortex may be mediated by the brain stem reticular formation. The most notable contributions come from the sensorimotor areas but responses can also be obtained from the orbital cortex, anterior cingulate gyrus, and the hippocampus. Attempts to trace the descending pathways have proved extremely difficult owing to the diffuse nature of the fiber systems converging on the reticular neurons and to the fact that the responses are nearly all bilateral.

The traces shown in Figure 174 are from an experiment in which responses evoked by stimulation of the orbital cortex were recorded simultaneously from the brain stem and the anterior lobe of the cerebellum. The cerbellar response in the upper trace is seen to have early and late components. The early component has a latency range between 5 msec and 8

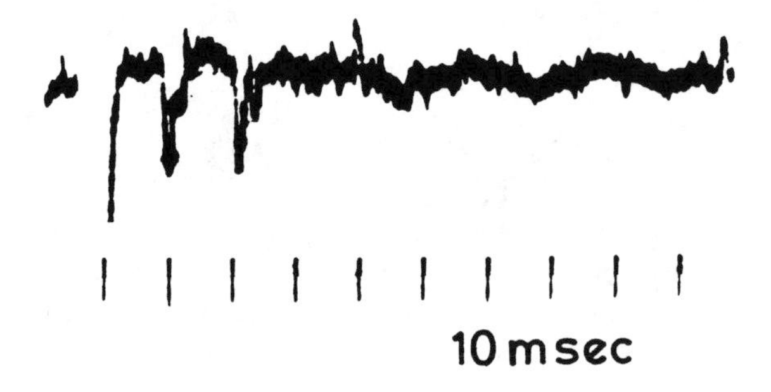

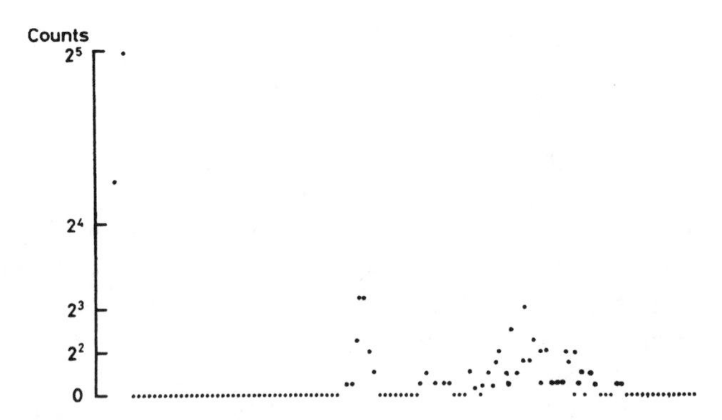

Fig. 174. Potential changes evoked in the cerebellum of the cat following electrical stimulation of the orbital cortex. At top note early and late components recorded from the culmen of the anterior lobe. Time intervals are 10 msec. At bottom post-stimulus time histogram shows the number of events occurring at two distinct peaks of time during 32 successive sweeps. Inter-point time is 160 μsec.

msec—relatively fast conduction times for impulses relayed through the reticular formation. On the other hand, the late component has a latency range between 12 msec and 18 msec—much slower conduction times, suggesting the possibility of relays through the inferior olive. Also in the figure is a post-stimulus histogram showing the occurrence of events at two distinct peaks of time.

Spino-reticular responses

A high percentage of reticular neurons respond to mechanical or electrical stimulation of peripheral nerves from both sides of the body. Many can also be influenced by auditory and visual stimuli. A typical response following stimulation of the splanchnic nerve is illustrated in Figure 175. In the upper trace, a single stimulus delivered to the central cut end of the nerve evokes a brisk multiple-spike discharge. The lower trace records the action potential of the A beta group of afferent fibers. The latency of the discharge often varies with each successive stimulus and with different units sampled. However, if recordings are obtained simultaneously from the anterior lobe of the cerebellum, it is found that responsive cerebellar units have a mean latency of about 15 msec. In contrast, splanchnic-evoked responses mediated by the inferior olive have a mean latency of about 20 msec for the first spike of the discharge. These observations lead to the conclusion that splanchnic afferent impulses are conducted to the cerebellum by two distinct routes, one of which involves the caudal part of the reticular formation.

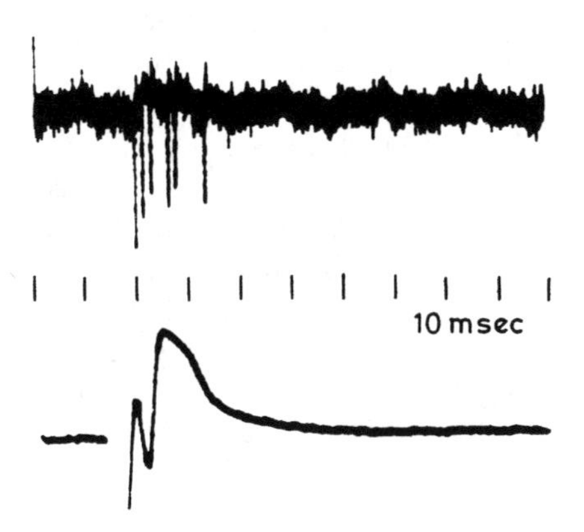

Fig. 175. Reticular unit response to peripheral nerve stimulation. Upper trace reflects typical multispike discharge following a single stimulus. Time intervals are 10 msec. Lower trace demonstrates splanchnic nerve action potential simultaneously recorded.

Cerebellar stimulation

Sherrington was the first to demonstrate that stimulation of the anterior lobe of the cerebellum inhibits the extensor rigidity of the decerebrate animal. It is now known that such influences on extensor tone are mediated by the inhibitory area of the reticular formation. Consequently if the cerebellum is ablated, the rigidity increases. Stimulation of the lateral zones of the anterior lobe increases the tone in the limbs; this effect is mediated by the facilitatory area of the reticular formation. Cerebellar-reticular influences on spinal motoneurons provide the basic mechanisms needed for the effective control of the musculature by the higher centers of the nervous system.

ELECTRICAL ACTIVITY OF THE BRAIN

In 1929, Hans Berger first demonstrated electrical potentials derived from the human brain through electrodes attached to the scalp. The rhythmic oscillations develop best when the eyes are closed and, when suitably amplified, are known as an electroencephalogram. Berger found that the brain waves were altered by any sensory stimulus that had an arousal or attention value. This observation is now familiar as the arrest of the alpha rhythm when the subject is alerted or simply opens his eyes. Attempting to see in a darkened room is almost as effective as a visual stimulus, since attention is the critical factor in the blocking of the alpha rhythm. Adrian postulated that the alpha rhythm was due to a large number of neurons discharging synchronously at the same rate when the subject was mentally relaxed. The rhythm was broken up or desychronized by the arrival of impulses in the visual cortex (Fig. 176). Thus, alpha rhythm is recorded most strongly from scalp leads over the occipital lobes.

Jasper recorded rhythms faster than alpha from more central regions of the head and showed that these too could be blocked by arousal stimuli. In contrast, when the subject fell into a deep sleep, the waves became much slower in rate. It is apparent that the electrical activity of the brain is determined to a great extent by the overall level of excitation that exists between the sleeping and waking states. This level, as we have seen, is dependent upon the functions of the reticular activating system.

Method of recording

A rubber cap is fitted over the head of the subject who sits in a quiet room. Electrodes are placed under the cap in firm contact with the scalp;

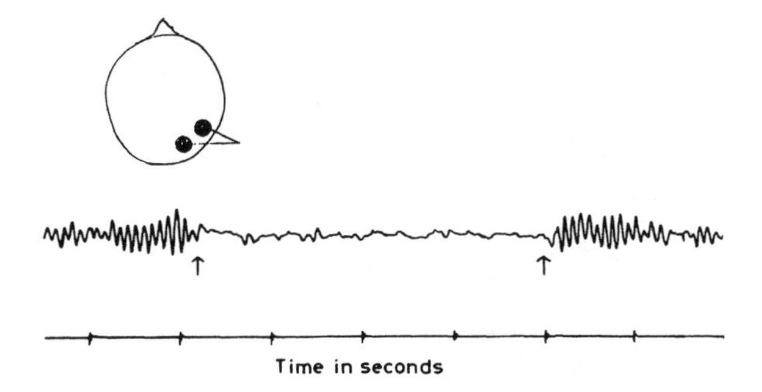

Fig. 176. Effect of opening and closing the eyes on alpha rhythm. Electroencephalographic record was taken from scalp leads as shown in inset. Eyes were opened at first arrow and closed at second arrow. Note replacement of alpha rhythm with faster rhythm of lower amplitude.

these usually consist of silver plates wrapped in gauze and soaked in saline. Each pair of electrodes is connected to an amplifying channel, and one electrode, placed over the mastoid process, is connected to ground.

Clinical electroencephalography employs at least six independent channels for simultaneous recordings from which several combinations of electrode placements can be selected. The recording instrument is usually an ink-writing oscillograph with storage and averaging facilities added. Interpretation of the tracings requires knowledge of the paper speed, localization of the electrodes, calibration of the gain control, and recognition of possible artifacts. The latter may be caused by transient electrical interference, faulty electrode contacts, or muscle activity due to the patient's swallowing, eye blinks, or inability to relax.

Normal brain rhythms

The normal record is made up of different rhythms, which are all illustrated in Figure 177. They are classified as alpha, beta, theta, and delta.

1. Alpha rhythm is a regular series of sinusoidal waves about 50 μv in amplitude and occurring at a frequency between 8 and 12 per second. It is the dominant rhythm in most individuals when they are relaxed and awake. As stated above, it is best recorded from the occipital regions of the scalp and is usually absent from the frontal regions. It originates from the thalamic portion of the reticular activating system and is promptly replaced by faster rhythms when the brain becomes more active.

2. Beta rhythm is smaller in amplitude and occurs at frequencies between 14 and 60 per second. One type of beta rhythm disappears with mental activity, while another type persists. Low-voltage fast activity is normally

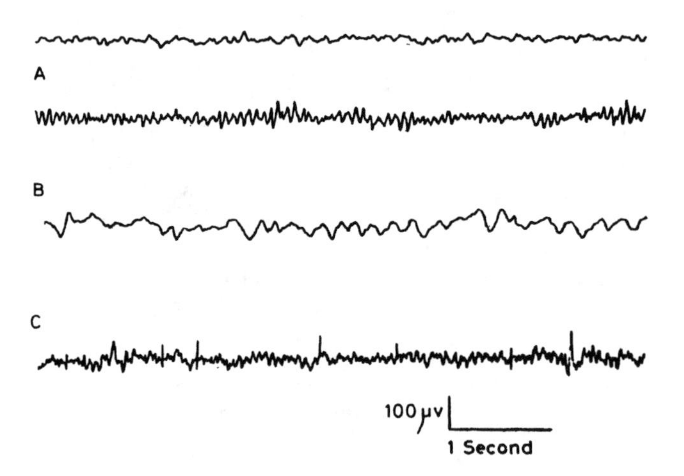

Fig. 177. Electroencephalograms recorded from different subjects. A shows EEG of subject relaxed with eyes closed; upper trace is from frontal region, lower trace from occipital region. B reflects theta rhythm recorded from temporo-parietal region. C demonstrates typical record of tense subject; EEG was recorded from parieto-occipital region. Note appearance of occasional muscle spikes superimposed on the trace.

recorded from the frontal regions, but the rhythm may be encountered in any head region after forced overbreathing, which causes a reduction in the CO_2 level of the blood.

3. Theta rhythm consists of low voltage waves occurring at frequencies between 4 and 7 per second. It is usually recorded from the parietal and temporal regions of the head in subjects, especially children, suffering from emotional stress. When not visible in an ordinary tracing, it can sometimes be demonstrated by an instrumental analyzer which plots a histogram of the wave components.

4. Delta rhythm refers to all brain waves slower than 4 per second and normally appears in deep sleep. The waves may attain an amplitude of 100 µv. Delta rhythm occurs in patients under anesthesia. It is also commonly present in infancy and early childhood, as normal adult patterns are not achieved until the age of puberty.

Abnormal brain rhythms

Electroencephalography is a useful aid in the diagnosis of cerebral disorders. Abnormalities of cortical function may affect the rhythm, amplitude, and pattern of the brain waves. Also, the disturbance may be generalized or focal.

1. Changes in rhythm. Slow or fast rhythms and changes of the rhythm with forced breathing are often associated with certain types of cerebral disorders.

• Generalized slow waves are found in anoxic conditions, hypoglycemia, cerebrovascular disease, and increased intracranial pressure. All these disturbances interfere with normal cerebral metabolism. Generalized fast waves are found in hyperexcitable states and may herald the onset of an epileptic attack. As long as the level of excitability remains below a critical threshold, no attack occurs.

• Focal slow waves may indicate the presence of a cerebral tumor or abscess. The slow waves do not originate from the lesion itself, which is electrically inactive, but from the surrounding cortical tissue whose function is depressed owing to pressure or traction. The site of the lesion can often be determined from the electroencephalogram, as illustrated in Figure 178. The slow waves recorded from opposite ends of the lesion show reversal of polarity to each other, indicating that the origin must lie in the area between the two respective electrodes. Localized fast waves may indicate the existence of an epileptogenic focus, which may also be detected by the method of phase reversal. Once such a focal point is found, surgical excision prevents future epileptic attacks.

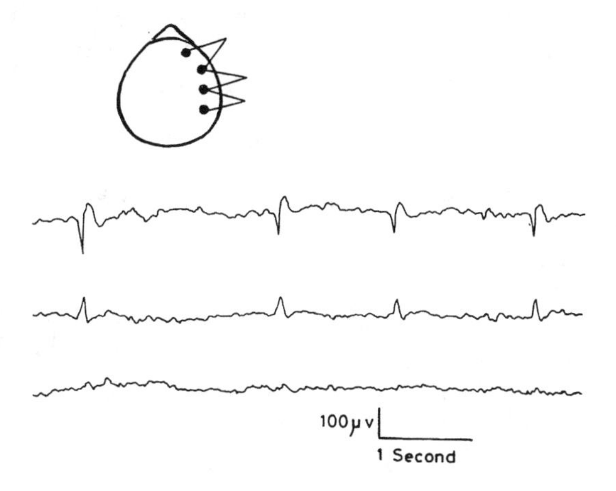

Fig. 178. Localization of a focal lesion in right anterior frontal region showing phase-reversal of slow wave discharges.

2. Changes in amplitude. Low amplitude waves without changes in frequency or waveform have no particular significance. High amplitude waves, usually accompanied by an increase in frequency, develop in epileptic patients prior to the onset of an attack (Fig. 179A). The seizure is caused

by synchronous discharges of hyperexcitable neurons. The electrical disturbances appear first in the region of origin of the attack, but may spread to other regions until the entire cortex becomes involved. The patient loses consciousness during the seizure when the body may be enveloped in generalized tonic convulsions, followed toward the end of the attack by muscular spasms called clonic convulsions. In the post-convulsive state the appearance of high-voltage slow waves is not uncommon. When recordings are made between the attacks, the brain rhythms are often normal in character. However, in some patients high-amplitude fast waves can be induced by forced breathing. This feature is known as a larval or subclinical attack, which soon disappears and does not cause any loss of consciousness.

3. Changes in pattern. Abnormal patterns occurring irregularly or as a continuous rhythm are conventionally described as spikes or sharp waves. The electroencephalogram can be useful to localize abnormal spikes originating from some organic lesion of the brain, such as a scar from a destroyed area following a head injury or from a small tumor. In the condition known as petit mal, the normal rhythm is replaced by a new pattern consisting of a high-voltage flat-topped wave and a spike, repeated at the rate of about 3 per second (Fig. 179B). This pattern can be recorded from all electrode placements, suggesting that the origin of the abnormality is in the reticular activating system. The patient generally has a momentary lapse of consciousness without being aware of it and without any convulsive signs or falling, apart from a few twitches and blinking of the eyes.

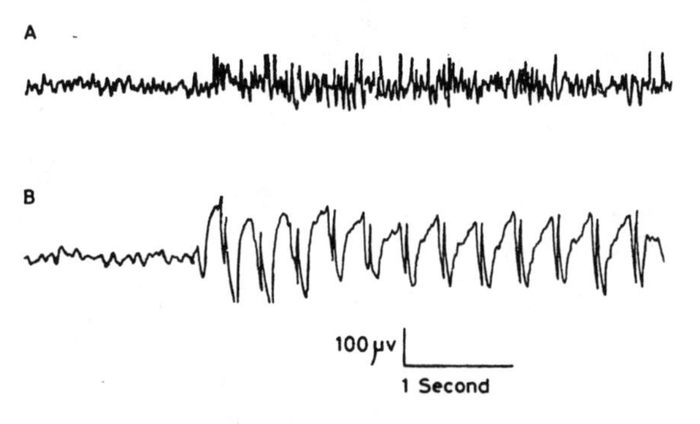

Fig. 179. Electroencephalographic patterns in epilepsy. A illustrates typical high voltage fast waves characteristic of an epileptic discharge. B records classical spike and slow wave pattern seen in a petit mal attack.

CEREBRAL BASIS OF CONSCIOUSNESS

Consciousness may be defined as a state of being aware of one's surroundings. It is not a simple on-off mechanism in terms of being awake or asleep, even though loss of consciousness may occur abruptly and without warning. On the contrary, there are various grades of consciousness ranging from drowsiness to alertness—from minimal responsiveness to the highest pitch of excitement. The faculties may be at their sharpest when the physical body is completely relaxed. When a person is fully conscious his attention may be restricted to the point of exclusion of all other forms of distraction, so that he appears to be unconscious of events in the outside world. Finally, there are a number of states intermediate between drowsiness and sleep, such as hypnotic trances and dream-like conditions, while at the other extreme, the mind can be developed by training to achieve a higher level of perceptual enlightenment. Many disorders of consciousness occur in association with certain psychological disturbances or as a consequence of drug action, epileptic attacks, or brain damage.

Electroencephalographic changes and levels of consciousness

Beginning with wakefulness and proceeding through deep sleep to coma, the electroencephalogram shows the following changes:

1. Alertness. The predominant rhythm consists of fast low amplitude waves and partial desynchronization.
2. Relaxation. Synchronized cortical activity occurs with maximal alpha rhythm.
3. Drowsiness. A reduction of alpha rhythm coincides with occasional low-amplitude slow waves from the occipital region, but desynchronized activity is seen in other regions.
4. Light sleep. Low voltage slow waves are interrupted by short bursts of spindles (12 to 14 per second).
5. Deep sleep. High voltage delta waves occur at a rate of 2 to 3 per second.
6. Coma. Only slow waves are recorded. The condition differs from deep sleep in that the subject cannot be aroused by strong sensory stimulation.

Types of sleep

Two different types of sleep can be identified, slow wave sleep and paradoxical sleep.

1. Slow wave, or synchronized, sleep occurs in four stages. At first, the brain rhythms show characteristic spindle bursts. Then as sleep becomes progressively deeper, the waves become larger in amplitude and slower in frequency. There is a gradual slowing of the heart and respiratory rates and a reduction in skeletal muscle tone. No rapid eye movements (NREM) occur during this restful type of sleep.

2. Paradoxical, or desynchronized, sleep occurs in short bursts when the brain rhythms show low-voltage fast activity similar to the pattern of the waking state. The heart and respiration rates may be increased, but skeletal muscle tone is depressed and the tendon jerks absent. During this type of sleep irregular muscle movements occur including rapid (to and from) eye movements (REM).

Most adults sleep for about 8 hours every night during which the two types of sleep alternate in variable cycles. Most of the sleep is of the slow wave variety, especially if the subject is tired after a hard day. Paradoxical sleep is less restful, occurring in bouts of about 10 to 30 minutes, and is often associated with dreaming. The bouts tend to occur less frequently in older people, who also seem to require fewer hours of sleep.

Mechanisms inducing sleep

Although the reticular system is responsible for the cycles of sleeping and waking and for all grades of consciousness, the two types of sleep have different neurophysiological mechanisms.

1. The neural structures causing slow wave sleep are believed to be in the caudal part of the brain stem reticular formation. The evidence is as follows:

- In animals, recordings from single reticular neurons during the early stage of sleep reveal tonic burst discharges associated with synchronous slow wave patterns in the electroencephalogram. Local cooling of the area produces an arousal response.
- Bilateral electrolyte lesions in the caudal reticular formation lead to the occurrence of desynchronized fast activity in the electroencephalogram.

The conclusion to be drawn from the above evidence is that the caudal part of the brain stem induces the onset of sleep by inhibition of the reticular activating system.

2. The neural structures causing paradoxical sleep are believed to lie in the pontine and mesencephalic reticular formation. The evidence is as follows:

- In animals, complete destruction of the nucleus reticularis pontis caudalis eliminates REM sleep without affecting slow wave sleep.

• Stimulation of the pontine reticular formation activates the bulbo-reticular area in the lower brain stem, which causes a decrease of muscle tone and depression of spinal reflex arcs. The evidence suggests that the rostral part of the brain stem possesses an intrinsic mechanism operating in short cycles of increased and decreased activity. Its ascending influences raise the level of cortical excitability sufficient to cause dreaming and eye movements but without awakening the person; its descending influences produce depression of muscle tone throughout the body.

In summary, natural sleep is brought about by a change of neuronal activity from that operating during the waking state. The change involves two different mechanisms. One causes inhibition of the reticular activating system and the other triggers REM cycles. Both mechanisms maintain the level of cortical excitability below the threshold of consciousness.

Theories of sleep

It is generally accepted that sleep restores the energy of body and mind when a person is overcome by fatigue, while prolonged lack of sleep has exactly the reverse effect. Less is known about the intrinsic mechanisms of falling asleep and waking up to set times, the phenomenon of jet lag, or the reasons some healthy persons require less sleep or more sleep than normal. The following theories have been postulated about the causation of sleep:

1. Stimulus deficiency. Procedures that minimize sensory stimulation favor the onset of natural sleep. Thus a warm bed, comfortable surroundings, a dark and silent room, and muscular relaxation are all useful preparations for restful sleep. It is supposed that a reduction in the flow of excitatory impulses to the sensory cortex also decreases the excitability of the ascending reticular activating system which is essential for maintaining the state of wakefulness. There also exists positive feedback from the cortex to the reticular system that contributes to the excitability of the system.

2. Conditioning. Pavlov suggested that sleep was an inhibitory process involving the entire cortex on the basis that conditioned inhibition could spread from one cortical analyzer to another until all conditioned reflexes became extinct. The lower brain centers were also implicated, as shown by loss of tone in the muscles of the limbs and a fall of blood pressure.

3. Sleep-wake centers. The integrity of the posterior hypothalamus seems to be necessary for maintaining the state of wakefulness. Hess found that stimulation in this region produced excitement and arousal and preparations for defence, while localized ablation left the animal permanently asleep. Hess also believed that there was a center for sleep lateral to the massa intermedia of the thalamus. Nauta located a sleep center in the

anterior hypothalamus, observing that electrical stimulation in this region caused the waking animal to fall asleep.

4. Trigger mechanisms. The existence of trigger mechanisms in the brain stem reticular formation suggests that these structures become active with the onset of sleep. The transition from drowsiness to light sleep is thought to result from inhibition of the reticular activating system and posterior hypothalamus. As sleep becomes progressively deeper, the NREM mechanism is capable of synchronizing cortical rhythms and generating delta waves. A second trigger mechanism in the pontine reticular region discharges in cycles during REM sleep, causing rapid eye movements and muscular twitches as well as inhibition of spinal reflex centers.

Disorders of sleep.

Most disorders of sleep can be placed in one of the following categories—insomnia, narcolepsy, hypersomnia, and parasomnia.

1. Insomnia. True insomnia, or sleeplessness, must be distinguished from the patterns of healthy individuals who are poor sleepers or who require less sleep than normal. The term is applied to delay in falling asleep, failure to maintain sleep, or to poor quality of sleep. Bouts of insomnia occur for no apparent reason, but restlessness is a common experience in travellers who are constantly changing their surroundings. Delay in falling asleep often results when a person is excessively tired, while difficulty in staying asleep is likely at times of mental or physical stress. The quality of sleep may be affected during an illness or in persons suffering from an acute emotional disturbance. The cause of insomnia may be attributed to a failure of the trigger mechanisms that normally induce sleep and maintain it for an appropriate period of time.

2. Narcolepsy. This condition is defined as sudden irresistible attacks of sleep often occurring at inappropriate times. If the subject is resting after a meal, the attack may last several hours; otherwise it may come on during any kind of activity and this sleep may last several minutes. The need to sleep is powerful, but the subject can be easily roused and awakens refreshed. Such dramatic changes in the sleep-wake cycle are believed to be due to a metabolic defect of the brain stem trigger mechanism. Narcolepsy is often associated with other signs of sleep disorder like cataplexy and sleep paralysis.

Cataplexy is marked by a sudden loss of postural tone, which may be limited to a few muscle groups or may cause the patient to fall to the ground where he lies, incapable of movement. The attacks, often precipitated by emotional excitement, last only a few seconds and recovery is instantaneous. There is no loss of consciousness.

Sleep paralysis (night nurse paralysis) is due to a disturbance of the REM sleep mechanism with activation of the inhibitory reticular-spinal pathways. The attack causes a flaccid muscular paralysis and, since full consciousness is retained, it is often accompanied by vivid dreaming, an intense feeling of fear, and frightening hallucinations.

3. Hypersomnia. This is a condition of prolonged periods of NREM sleep or "sleep drunkenness," which may last several hours and does not refresh the individual. In fact, patients often have to be awakened by persistent stimulation and remain confused and disorientated for some time afterward. Two clinical syndromes are commonly described:

• The Kleine-Levin syndrome affects male adolescents who have recurrent sleep attacks associated with marked hunger, vivid waking fantasies, and overeating. Restlessness is a common feature after the attack. The patients tend to be hostile and quarrelsome and suffer from auditory and visual hallucinations. In view of the hyperphagia associated with the somnolence, a hypothalamic dysfunction has been postulated as the cause of the disturbance.

• The Pickwickian syndrome, named after the fat boy in Charles Dickens' *Pickwick Papers,* is characterized by gross obesity and periodic respiration during sleep. Disturbance of respiratory control mechanisms together with hypotonus of the pharyngeal muscles lead to airway obstruction and arousal from sleep. Frequent arousals during the night are compensated by daytime sleeping.

4. Parasomnia. This term is applied to one of many behavioral disturbances that frequently occur during the hours of sleep or in the transition stage between sleep and wakefulness.

• Sleep-talking, meaningless vocalization and teeth-grinding are common manifestations of disturbed sleep in tired and excessively anxious persons.

• Sleep-walking (somnambulism) is more common in children than in adults. It resembles a trance-like state with a low level of awareness; the subject may answer if spoken to, although comprehension is limited. Activities range from sitting up suddenly or walking aimlessly about the bedroom to wandering out of the room as if searching for something. There is usually total amnesia for the incident and no dream is recalled on waking up.

• Sleep terrors occur mainly in young children but they may persist into adult life. They arise during the NREM period of sleep and are ushered in by screaming lasting a few minutes with eventual return to sleep. Nightmares are similar disorders developing in REM sleep. The subject does not awaken during the episode, but appears distressed or terrifed and recalls vivid dreams afterward.

• Bed-wetting (nocturnal enuresis) is the most frequent sleep disorder of childhood, occurring predominantly in NREM sleep. Periodic rises in intravesicular pressure cause contraction of the detrusor muscle and emptying of the bladder. The cause of bed-wetting is unknown, since emotional disturbances in the child or unsatisfactory relationships between child and parents account for only a small percentage of the cases. The sleep cycles of the electroencephalogram are similar to those of normal children.

Transcendental meditation

The practice of transcendental meditation (TM) has become increasingly popular in recent years and its value in alleviating pain in very anxious subjects was briefly indicated in Chapter XVI. The object of meditation is to achieve complete relaxation of the body by breathing exercises and deep rest, which in turn help the individual to attain maximal stability in all physiological systems. The beneficial effects of meditation are shown by the following observations made after about six weeks of training:

1. Metabolic rate and oxygen consumption were decreased.
2. Respiratory rate and heart rate were decreased and blood pressure was lowered.
3. Increased skin resistance was related to diminished anxiety.
4. Reaction times were faster and muscular coordination improved.
5. Brain rhythms showed greater synchronization.

These findings when taken together suggest a state of deep metabolic rest which, it is hoped, can be maintained during the intervals between meditation sessions. Advocates of the method believe that total relaxation can lead to removal of anxieties and stress. The physical well-being promoted in this way causes an expanding awareness and more orderly use of the intellectual faculties. It is also claimed that a higher level of consciousness can be developed with practice in the same way that enlightenment of the mind was sought by the mystical practices and bodily deprivations of Eastern cultures.

Hypnotism

Many attempts have been made to explain the nature of hypnotism as a function of the brain. Hypnotism has been practiced for centuries by primitive peoples for ceremonial purposes and by ancient civilizations for the purpose of healing. Even today there are many who believe in the power of healing by touch or by direct contact between the subject and the hand of the healer. It was a belief of this kind that was used by Mesmer in the eighteenth century to explain the success of his treatment, which he called "animal magnetism." The concept that the phenomenon was subjective in nature, acting through the subject's own mind, was first suggested by James Braid who renamed it "hypnotism" in 1843. Essentially, hypnotism is a relationship between two persons in which the motivations of the subject are released from conscious control and yielded to the will of the operator. Almost any person can be hypnotized. It is not a condition of hysteria, nor is it a modified form of sleep.

1. Neural mechanism. The discovery of the ascending reticular activating system and its role in sleep and wakefulness have given a lead to the possible mechanism involved in producing the hypnotic state. Under normal conditions, a harmonious relationship exists between the cerebral cortex and the reticular system, whereby ascending impulses exert a tonic facilitatory influence on the higher centers, and descending impulses exercise a powerful inhibitory influence on the lower centers. Since volitional control is modified during hypnosis, the brain must be capable of interrupting this relationship without loss of consciousness. Independent action of the reticular system released from cortical control allows free display of all bodily functions within physiological limits. Motor effects can be increased or decreased with resultant changes in posture or simulated paralysis, and sensory effects can give rise to changes resembling parathesia or anesthesia.

2. Dissociation theory. Freud believed that a large part of mental experience was subconscious and remained repressed except when forced into consciousness by dreaming or by the command of the hypnotist. Pavlov maintained that dreaming was the result of inhibition of the lower centers when all kinds of memory traces were liberated from organized cortical control. This would account for the chaotic, unregulated actions in dreams and for their illogical character. Both of these ideas express the view that the hypnotic state is the result of a dissociation of functions between lower and higher centers of nervous activity. Dissociation is brought about by means of suggestion. The induction phase relies on monotony and rhythm to narrow down the sensory channels until the operator becomes the sole channel of communication. When this stage is achieved, the subject loses contact with the rest of the external world and appears to be deprived of his critical faculties and of the inhibitions that are usually imposed on volitional behavior. Thus the hypnotized subject may behave completely out of character and perform amusing, ridiculous, or undignified actions from which he would refrain in the ordinary waking state. Likewise, post-hypnotic behavior can be regarded as a temporary return to the trance-like,

detached state, depending upon some residual memory factory persisting in time until evoked by words previously suggested by the operator.

3. Levels of perception. The physiological changes that take place in hypnosis take the form of excitation and inhibition as in all other central nervous processes. Suggestion in the form of crisp, repetitive commands may act as a stimulus whereby attention is narrowed down in the subconscious field to the exclusion of all other influences. As each ascending influence falls out of action, the level of volitional control is reduced to the point where cortical inhibition becomes ineffective. Hypnosis is therefore a condition of selective excitation of the brain which retains the electroencephalographic patterns of the waking state, but fails to exercise its inhibitory functions that normally hold the subconscious in check. The theory does not give a complete answer to the problem, but it does meet the requirements of those who believe in two levels of perception, conscious and subconscious, and who advocate the idea of dissociation. It also helps to explain the remarkable manifestations of hypnotism on the basis of release from volitional control.

NEUROCHEMICAL ASPECTS OF RETICULAR FUNCTION

It is now recognized that biochemical factors play an essential role in the normal functioning of the reticular formation. Reticular neurons are particularly sensitive to the products of metabolism and circulating hormones, while many of them secrete specific substances that increase or decrease the activity of this area. A balanced state is achieved by interactions involving three principal neurochemical systems—cholinergic, catecholaminergic, and serotonergic. Destructive lesions in the reticular activating system will obviously disrupt the flow of impulses to the cortex and thereby affect the states of sleep and wakefulness. More subtle changes, however, may be brought about by anomalies attributed to the enzyme processes essential to the synthesis or breakdown of these secretory substances.

1. Role of acetylcholine. The reticular formation contains relatively large amounts of acetylcholine and the enzymes necessary for its synthesis or inactivation. When injected into the blood stream, acetylcholine has a strong excitatory effect on the reticular activating system, producing cortical desynchronization and electroencephalographic patterns of arousal. The effects of acetylcholine on single reticular neurons have been studied in the unanesthetized decerebrate cat when some neurons responded with an increased discharge rate and others with a decreased discharge rate. Using multi-barrelled micropipettes, one barrel was used for recording neuronal activity, while the drug and related compounds were released at the tip of

other barrels by the technique of micro-iontophoresis. The results showed conclusively that acetylcholine had both excitatory and inhibitory actions on reticular neurons in the brain stem. A typical inhibitory effect of graded doses of the drug is illustrated in Figure 180.

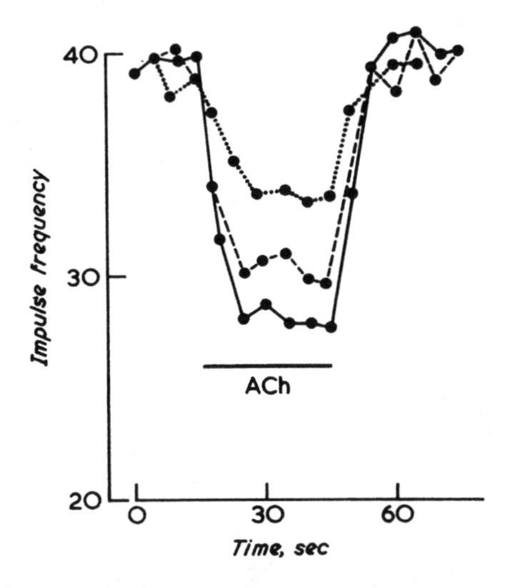

Fig. 180. Effect of graded doses of acetylcholine on a single reticular neuron in the unanesthetized decerebrate cat. Impulse frequency is plotted against time. Each point on the graph represents a count over a period of 5 sec. Note inhibition of the discharge during iontophoresis followed by recovery. The three plots were obtained by using different current strengths to expel the drug. (From Bradley, Dhawan, and Wolstencroft.)

2. Role of noradrenaline. Neurons that contain and release noradrenaline can be visualized by the method of histochemical fluorescence. Such neurons are found in the locus ceruleus, an area in the dorsal part of the pons associated with the paradoxical episodes in sleep. According to Jouvet, noradrenergic pathways have a facilitatory effect on the ascending reticular activating system and therefore contribute to mechanisms maintaining the waking state.

3. Role of serotonin. This important neurotransmitter is derived from tryptophan and is rapidly inactivated by amine oxidase. Drugs like reserpine increase its turnover rate while, on the other hand, its actions are opposed by lysergic acid diethylamide (LSD). Serotonin, also known as 5-hydroxytryptamine (5-HT), has a marked tranquilizing effect on behavior when administered in therapeutic doses. Neurons containing high concentrations of serotonin are found in the raphe area of the reticular formation and in most of the limbic structures such as the hippocampus and amygdaloid nuclei with which it is connected. Stimulation of the raphe area

induces cortical synchronization and electroencephalographic patterns of slow wave sleep in the cat, whereas localized destruction of the raphe causes behavioral arousal. It appears that the serotonin pathways suppress the activity of the noradrenergic pathways and consequently, serotonin depletion releases the reticular activating system from inhibition.

Interactions between the serotonergic and noradrenergic systems seem to be one of the most important mechanisms regulating the sleep-wake cycle. The diagram in Figure 181 illustrates a possible feedback relationship between the raphe and locus ceruleus. The 5-HT system of the raphe inhibits the NA system of the ceruleus during slow wave sleep. Negative feedback from the NA system causes suppression of the 5-HT system during paradoxical sleep. These findings support the concept that antagonistic biochemical mechanisms may be involved in the control of the conscious and unconscious states.

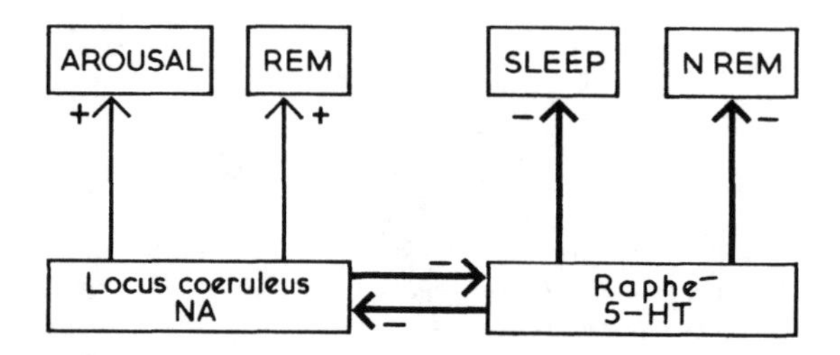

Fig. 181. Scheme of feedback relationship between noradrenergic and serotonergic systems. The NA system of the locus ceruleus controls arousal mechanisms and REM sleep. Thin arrows denote activation. Negative feedback causes suppression of the 5-HT system during REM. The 5-HT system of the raphe suppresses the NA system and controls slow wave sleep (NREM). Thick arrows denote inhibition.

In summary, interactions between brain monaminergic systems offer new hypotheses for the clarification of sleep-wake mechanisms, the degree of attentiveness, and the levels of excitability during specific phases of mental activity.

SELECTED BIBLIOGRAPHY

Adrian, E.D., and Matthews, B.H.C. The Berger rhythm: potential changes from the occipital lobes of man. *Brain* 57:355-385, 1934.

Brodal, A. *The Reticular Formation of the Brain Stem. Anatomical Aspects and Functional Correlations.* Edinburgh: Oliver and Boyd, 1957.

Delgado, J.M.R. *Physical Control of the Mind.* New York: Harper and Row, 1969.

Dement, W.C., and Miller, M.M. An overview of sleep research: past, present and future. In

Hamburg, D.A., and Brodie, H.K.H. (Eds.). *Amer. Handbook of Psychiatry* 6:130-191, 1975.

Eccles, J.C. *Brain and Conscious Experience.* New York: Springer-Verlag, 1966.

Globus, G.G., Maxwell, G., and Savodnik, I. *Consciousness and the Brain.* London: Plenum Press, 1976.

Jouvet, M. Neurophysiology of the states of sleep. *Physiol. Rev.* 47:117-177, 1967.

Moruzzi, G., and Magoun, H.W. Brain stem reticular formation and activation of the EEG. *Electroen. Clin. Neurophysiol.* 1:455-473, 1949.

Wyzinski, P.W., McCarley, R.W., and Hobson, J.A. Discharge properties of pontine reticulospinal neurons during sleep-waking cycle. *J. Neurophysiol.* 41:821-834, 1978.

CHAPTER XXV

Learning and Memory

The capacity to learn must be distinguished from instincts or patterns of behavior laid down before birth. A baby begins to recognize its mother by visual imprinting within a few hours of birth. As growth proceeds, learning takes place by trial and error and then by repetition and practice. At a later stage learning is evolved by the assocation of ideas and events, while the rate of learning can be increased by incentives and rewards. Many environmental factors influence learning in children, some of whom may be advanced or precocious and others late starters. Eventually, all learning is based on experience, which recognizes the existence in the brain of a memory bank or storage system, as well as on the ability to retrieve information by voluntary effort.

The part played by the nervous system in learning and memory has been investigated in several different ways. Behavioral studies came first with the classical work of Pavlovian conditioning followed by an era of experiments with problem boxes and mazes. Second, use was made of neurosurgical techniques in an attempt to localize significant regions by selective ablations. Third, electrophysiological methods contributed to an understanding of the neural pathways and mechanisms involved in the learning process. These included conditioning-testing stimuli, single unit recordings and studies of functional changes at synaptic knobs. Fourth, a biochemical approach was instigated at the suggestion of Hydén that a change in the base sequence of ribonucleic acid (RNA) might represent the synthesis of a memory molecule or engram for reproducing impulse patterns. Thus the synthesis of new RNA molecules in the brain of one animal could facilitate the performance of a task when the substance was transferred to another animal. Fifth, the effects of drug administration on brain functions provided an opportunity for studying intracellular processes in learning and memory. In particular, extensive work has been reported on drugs that interfere with the metabolism, storage, and uptake of chemical substances involved in synaptic transmission. Sixth, clinical studies of amnesia and other disorders of memory, such as confabulation and forgetfulness, have indicated the extent to which learning processes and memory are related to the emotional life of the individual.

The ability to learn even a simple task requires that some change take place in the nervous system which is not an inborn reflex. The change may take the form of a transient membrane displacement influencing the character and availability of transmitter; more permanent effects may result from the formation of new protein linkages between synapses. Therefore, the possession of a brain is not essential for learning since even the most primitive life forms are able to learn, as evidenced by their accepting or rejecting food, attacking or retreating, and modifying their behavior according to circumstances. Most animals can learn by training, but usually require the incentive of reward or the fear of punishment to perform more difficult tasks. Higher animals solve their problems by resorting to their intellectual faculties which, in the human being, have the added advantages of language, knowledge, and rational thought. The mechanisms by which the brain performs intellectual operations are very complex indeed. The human brain makes use of previous experience to think things over, to make logical deductions, and to calculate the probable consequences before committing itself to action.

BEHAVIORAL STUDIES

Most behavioral studies of learning fall into two main classes: Type I is generally known as Pavlovian conditioning in which the animal is trained to discriminate between different signals; Type II is exemplified by the maze in which learning is accomplished by trial and error through one or more cues.

Type I—Pavlovian conditioning

In 1906, Pavlov described the conditioned reflex in dogs and the factors that govern it. A conditioned reflex depends upon associating two experiences so that when one is presented again, the other is remembered. For example, the sight of a whip causes an animal to obey since it remembers the punishment of a previous occasion. Pavlov used the salivary reflex whereby he could measure the degree of response in terms of drops of saliva. The parotid duct was cannulated in a preliminary operation, and the animal was placed in a quiet room with its body supported by a harness. Food and other stimuli were presented by the observer in an adjoining room. The introduction of food into the mouth elicited a reflex flow of saliva whereas other stimuli–for example, the sound of a bell–had no such effect. Pavlov called the former an unconditioned stimulus and the latter a neutral stimulus–i.e., no flow of saliva. When, however, the two stimuli were combined–the bell was rung while the animal was taking food, and the procedure repeated a few times–the neutral stimulus acquired new properties. The sound of the bell alone elicited a flow of saliva. Pavlov called the bell in this experiment a conditioned stimulus and the response a conditioned reflex. The following is a summary of the principal findings based on Pavlov's experiments:

1. Any type of stimulus can be used to elicit a conditioned reflex–e.g., tactile, auditory, visual, or mild noxious stimuli.

2. Secondary and tertiary stimuli can given positive conditioned responses. Thus, if an animal is conditioned to the sound of a bell and a neutral stimulus–e.g., a flash of light–is presented during ringing of the bell, the light alone becomes a conditioned stimulus. Tertiary conditioned reflexes can be established if the light is now combined with another neutral stimulus–e.g., a rotating disc. Note that neither the flash of light nor the rotating disc is applied during the administration of food.

3. Time can be incorporated to establish short and long trace responses. Thus, if food is given some time after ringing the bell, a flow of saliva is elicited only after the same interval of time.

4. All conditioned reflexes need reinforcement or are soon forgotten. Thus, if ringing of the bell is never followed by food, salivary flow is reduced until the reflex ceases altogether. This progressive decrease in the conditioned response is called "extinction."

5. Repeated conditioning stimuli without reinforcement set up a negative conditioned reflex in which the salivary response ceases but returns after a period of rest. Pavlov suggested that the negative reflex was due to a state of cortical inhibition. If the procedure is repeated without periods of rest, the inhibitory process may involve the whole cortex and even spread to the subcortex when the animal becomes drowsy and falls asleep.

6. Stimuli of a similar nature can be differentiated. If a positive conditioned response is established to a tone of a given frequency and reinforced by food while a slightly different tone is not reinforced by food, the animal rapidly learns to discriminate between the two tones. Pavlov used the term "acoustic analyzer" to describe the part of the cortex involved in the inhibitory differentiation of sounds. Similarly, the abilities to detect small differences of temperature or luminosity were referred respectively to cutaneous and visual analyzers.

7. The stimulus responsible for teaching the animal to discriminate is

called a conditioned inhibitor. Pavlov considered learning to be a property of all the cortical analyzers. If one analyzer was removed, the appropriate conditioning capacity and power of differentiation were lost.

Present-day methods of experimental conditioning employ closed-circuit television, with the operator remaining unobserved, and discrimination tasks are presented by remote control. The electrical activity of the brain is continuously monitored, and stimulation procedures carried out through stereotactically positioned intracerebral electrodes.

Type II–Instrumental conditioning

The animal is required to perform some motor function like pressing a lever or escaping from a shock in order to obtain a reward. Many devices have been developed to test for discriminative ability, rate of learning, capacity for retention and decision-making, and to correlate these performances with specific cerebral structures.

1. The maze. Lashley favored the use of mazes in studying the effects of brain lesions on learning and retention. The apparatus is easy to construct and has the advantage that it can be varied in difficulty. A small animal can find its way through a simple maze after a few trials; the same cue is used each time as a habit. Complicated maze patterns may require the animal to use several cues in making the correct choice.

2. Problem-boxes. In this technique, the animal learns to press a lever or to turn a wheel in order to obtain food. Skinner developed a problem-box with automatic recording of the lever movements in response to a light signal or other conditioning stimulus. Pigeons can be trained to peck a key, rats can be taught to push a bar or platform, while monkeys can develop much manipulative skill with their hands to find the correct solution to a problem.

3. Avoidance conditioning. Most animals can be trained to make decisions by rewarding a correct choice and punishing incorrect ones. Avoidance conditioning is a relatively simple method of training in which the animal lifts its paw or moves away to avoid an electric shock. For instance, a suitable voltage is applied to a metal shelf containing food and the animal is placed on a platform above it. The circuit is completed when the animal steps on the shelf. After a few trials the animal learns not to step off the platform. Likewise, if one rung of a metal ladder is connected to a voltage source, the animal soon learns to step over the rung to avoid the shock.

4. Brain stimulation. Reference has already been made to the use of the chemitrode system for stimulating the brain by remote control.

- Delgado implanted electrodes in selected regions of the brain and trained animals to rotate a wheel to prevent the shock. When paired with a conditioning stimulus–e.g., a flickering light–such animals operated the wheel, whereas untrained animals gave no response to the light.
- Animals can be taught to shock themselves for a reward by turning a wheel despite inflicting pain on themselves. Self-stimulation can become an obsession as the animals spend most of their time at the wheel.
- Prolonged stimulation of the brain has apparently no deleterious effect and may indeed improve learning ability. On the other hand, electroconvulsive shock induces long-lasting neurochemical changes which may adversely affect the memory factor in learning. This aspect of brain function has important clinical implications for electroshock treatment in psychiatric practice.

NEUROSURGICAL STUDIES

Many efforts have been made to determine the parts of the nervous system involved in learning and to substantiate the contention that learning is impaired by damage to specific anatomical structures. Early experiments were based on ablation studies combined with classical conditioning or avoidance conditioning, but there was little to suggest that any function of learning was restricted to a localized region of the brain. Discriminative abilities were naturally impaired following lesions of the sensory cortex or visual cortex, and manipulative skills were diminished following lesions of the motor cortex. Such deficits, however, could not be attributed to interference with the learning process. More recently attention has been directed toward the role of the hippocampus in learning and memory since both in animals and in man large hippocampal lesions result in loss of learning capacity and memory storage. On present evidence, the facts relating hippocampal function to learning must be viewed as an open question.

Spinal cord

It is doubtful whether the spinal cord alone is able to learn anything. Tactile receptors in primitive animals provide information about food, but the decision to attack or retreat cannot be made by simple reflex mechanisms. Thus an octopus responds to a crab by seizing it. If the brain of an octopus is removed, it will not attack. If the brain is intact and the octopus is given an electric shock, it learns to leave the crab alone.

Cerebellum

It is unlikely that the cerebellum is important for learning an instrumental task. Damage to the cerebellum does not affect the accuracy of a previously learned maze habit; when the animal is deprived of proprioceptive cues, it will perform successfully by the aid of other cues.

Reticular system

There is some basis for assuming that extensive lesions of the reticular activating system interfere with emotionally motivated learning, since the arousal factor is significant in the fixation of the memory trace. Thus events are remembered best when they have either a pleasant or painful association. The existence of reward and punishment centers in the hypothalamus may also be relevant to reticular function in motivational learning.

Cerebral cortex

Ablation studies

Lashley trained rats to learn a maze habit and then observed the effects of cortical ablations on retention or on their ability to relearn the habit. He showed that all parts of the cortex were involved equally and only the size of the lesion, not its locality, affected the rate of learning. Lashley concluded that the cerebral cortex had a non-specific role or "mass action" in learning and retention and that the greater the amount of cortex removed, the more limited was the range of performance in subsequent training.

The prefrontal cortex is essential for delayed conditioned responses when the animal is required to wait some minutes before making a choice. Where the performance is based on the animal's previous experience, there is often considerable impairment of memory after prefrontal lesions.

The behavioral effects of temporal lobe lesions in monkeys were first reported by Klüver and Bucy, and from then on it became clear that one of the dominant features was the loss of recent memory. Penfield observed that ablation of the hippocampus in human beings deprived the patients of the capacity to recognize and compare incoming signals with information already stored from a previous experience.

Transfer between cerebral hemispheres

Pavlov found that if a dog were conditioned to a tactile stimulus from one side of the body, stimulation of a symmetrical point on the opposite side of the body would elicit a flow of saliva. He then showed that the response was abolished after section of the corpus callosum.

The transfer of information from one hemisphere to the other has been demonstrated in cats trained to perform a visual discrimination test with one eye blindfolded. The animals were first subjected to an operation in which the optic chiasm was sectioned; this permitted the isolated training of one half of the brain. Nevertheless, the animals were able to perform just as well with the other untrained eye. The information obtained visually during training must have been transmitted through the corpus callosum to the other hemisphere, for if the corpus callosum was divided, the animals showed no memory at all for what had been learned.

ELECTROPHYSIOLOGICAL STUDIES

Information processing is the first essential step in the learning mechanism. Information sent from the receptors in the form of coded signals is transmitted to selected areas of the brain for detailed analysis. The brain reacts to specific qualities of information by generating new signals, rejecting others, and transforming complex neuronal circuits into pattern detectors. The electrical activity involved in all these processes has been investigated from many angles, ranging from simultaneous recordings from large populations of cells to recording the excitability changes in the presynaptic terminals and the synapses themselves. From a practical point of view it seems reasonable to suggest that some local electrical change may take place in the nervous system when learning occurs.

Alpha block

In this experiment a light is used as an unconditioned stimulus to produce blocking of alpha rhythm. When the sound of a bell precedes the light by a short interval, the sound acquires conditioning properties. A tactile stimulus paired with a visual stimulus can also cause blocking of alpha rhythm after a few trials. If the conditioning stimulus is never reinforced, the response becomes extinct. If two different sounds are used, only the reinforced sound blocks the alpha rhythm, an example of differential conditioning.

Evoked potentials

Auditory conditioning

In recording from the auditory cortex of the cat, it is found that the evoked responses to click stimulation are extremely variable in amplitude, phase, and distribution. When, however, the click is combined with a conditioning stimulus—e.g., a weak shock to the paw—the evoked response increases in size and becomes more regular. At the same time a withdrawal reflex may develop in response to click stimulation alone. It is thought that the conditioning procedure stabilizes the evoked potential by facilitatory interaction occurring between pathways connecting the two sensory inputs.

Learning during sleep

Claims that learning is possible during sleep are based on the fact that potential changes may be evoked in the cerebral cortex on stimulation of a sensory nerve. Thus tape recordings played while the subject is asleep are believed to lay down traces that the subject can remember on waking. This suggests that learning can take place, provided that a fair measure of cortical activity is still present. Learning cannot occur during the deep stages of slow wave sleep, but it may be feasible during paradoxical sleep and in the transition stage between sleep and wakefulness.

Reverberating circuits

One of the suggestions that have been made for short-term memory is a reverberating circuit in the brain. This is based on the demonstration by Lorente de Nó of the closed internuncial chain, whereby collaterals permit recirculation of impulses through the chain. There are, in fact, many closed loops in the reticular formation and hippocampus which, no doubt, serve this type of storage mechanisms. The repetitive impulse activity is supposed to produce synaptic modifications resulting in a temporary trace on which more permanent structural changes can be built.

Synaptic properties

Several ideas have been put forward to explain the learning mechanism in terms of permanent changes in the synapses. Among the most important of these are the following:

Growth theory

While there is no evidence that the central nervous system can produce new cells, the possibility that learning may promote the formation of new synapses seems attractive. The theory proposes that structural growth may occur by protoplasmic expansion of existing dendrites to form more complex branching or that growth may occur by proliferation of dendritic spines. In either case the result would be an increase in the number of functional contacts available to the axon terminals. Electron-microscopic methods are being employed to determine whether enriched or deprived experience can modify dendritic structure. Some investigators believe that they can observe changes in presynaptic terminals that have been subjected to prolonged activity. Thus rats exposed to light for various periods show an increase in the number of synaptic contacts in the deeper layers of the visual cortex as compared with a control litter kept in the dark. The development of new links between neurons would make possible the flow of impulses between networks that were not previously connected.

Post-tetanic potentiation

The amount of transmitter liberated at a presynaptic terminal depends to some extent on previous conditioning. Tetanic stimulation of a fiber causes a subsequently increased excitability and larger size of the test response. The potentiation is due to mobilization of synaptic vesicles leading to more effective transmission. Eccles postulated that the conditioning stimulus may in fact cause an increase in the size of the synaptic knobs, giving a wider area of contact for transmitter action on the subsynaptic membrane. As a corollary, diminished synaptic efficiency following a long period of disuse may be due to shrinkage in the size of the synaptic knobs and consequent reduction of available transmitter. Such alterations in synaptic structure could be the basis for short-term memory.

Idea of plasticity

The classical conditioning procedure depends upon the establishment of a functional relationship between two sensory inputs previously unrelated. When conditioning takes place, certain cells common to the two input pathways undergo modification of a temporary or lasting character. This suggests that the structural pattern of the nervous system is not rigidly fixed. The synapses, in particular, are considered the most likely structures to register the changes since their distribution over the soma-dendritic surface permits overall control of the cell's activity. A theoretical consideration of the neuronal circuits employed in a conditioned reflex is illustrated in Figure 182. The theory postulates the existence of a convergence center or common target for impulses derived from unconditioned and conditioned sources. An unconditional stimulus (US) induces a change in the cells of the convergence center and also evokes a specific reflex response (R_1). When paired with a conditioned stimulus (CS) over a number of trials, the same reflex is evoked in response to the conditioned stimulus alone. Obviously, some modification must take place in synaptic transmission to increase the excitatory potency of the convergence center.

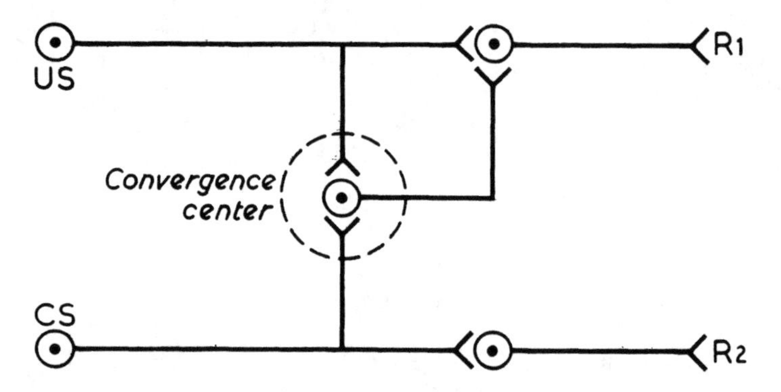

Fig. 182. Basic circuit employed in a conditioned reflex. The unconditioned stimulus (US) excites cells in the convergence center and also elicits the reflex response R1. Likewise, the conditioned stimulus (CS) excites the convergence center and elicits the reflex response R2. When paired over a number of trials, the conditioned stimulus alone elicits the response R1. Theoretically, increased synaptic efficiency at the site of convergence is brought about by combined presentation of the two stimuli.

NEUROCHEMICAL STUDIES

Perhaps the most fruitful approach to the problem of learning and memory has come from the field of molecular biology and the concept that genetic information is encoded in molecules of desoxyribonucleic acid (DNA). Halstead was the first to propose that the nucleoproteins in nerve cells transform themselves from random to organized patterns as a result of learning, and Fessard suggested that any modification of neural function must be sustained by self-reproducing macromolecules of nucleic acid or proteins. Research along these lines has continued in the hope of defining the biochemical systems involved in learning and memory related processes. Learning ability and memory storage may in fact depend upon two distinct changes, one being an electrical event of a temporary nature and easily disrupted by a variety of agents such as anesthesia, electroshock, or trauma, and the other representing a more stable and permanent change in the chemistry of the nervous system.

Events in the learning process

The mechanisms contributing to the acquisition of a learned task may be divided into three phases:

1. An initial short-term memory consists of specific electrical events set up by environmental information processing.

2. In a period of consolidation, changes occur in protein metabolism, and enzyme activity at central synapses leading to the formation of new macromolecules. These are believed to carry a type of chemical code essential for the storage of new information.

3. Long-term memory. Many parts of the nervous system may be involved in the storage of a permanent memory trace. Present ideas tend to favor the macromolecular theory of cerebral protein synthesis in which new peptides are formed by alterations in the relative amount and sequence of side chains–for example, by modification in the base ratios of the RNA molecule.

Evidence for molecular basis of memory

The discovery that DNA and RNA can act as codes for synaptic transmission has led to the theory that these substances are responsible for learning abilities and for transferring short-term memories into permanent store. Evidence will be described for studies on environmental influences in learning, experiments on chemical changes in the nervous system, and experiments showing behavioral effects of chemical transfer from trained to untrained animals.

Environmental influences

1. Observations made on rats reared in "enriched" or in "impoverished" environments show a marked difference in learning and problem-solving abilities. In these experiments environmental stimulation was provided by placing a variety of objects in the cage such as wheels, boxes, and ladders. Later, when the brains were examined, the "enriched" rats were found to

have an increase of weight and depth of cortical tissue accompanied by an increase of total RNA content. In contrast, rats kept in an "impoverished" environment showed a reverse situation for brain weight and chemistry as well as for learning abilities.

2. Mice kept in prolonged isolation develop strong aggressive behavior associated with diminished learning ability. Analysis of the brain chemistry revealed a variety of enzyme changes relevant to neurotransmission and neuronal RNA levels. In particular, a decrease in turnover of brain serotonin was a constant finding during deprivation, which could account for the disturbed learning of such animals.

Chemical changes

1. All nerve cells contain ribonucleic acid (RNA) which is continually breaking down and being resynthesized. Hydén trained rats to climb a wire in order to obtain food and observed that after a few days of training the proportion of RNA in neural tissue increased significantly. Other rats were trained to use a forepaw to obtain the food and neurons in the motor cortex were analyzed for their RNA base compositions. The contralateral cortex was considered to be trained, while the ipsilateral cortex served as a control. The neurons from the trained cortex contained higher levels of RNA with more adenine and less cytosine, as compared with the neurons from the untrained side.

2. Corning and John trained planarian worms to perform a conditioned response and then cut the worms in half. The parts allowed to regenerate in pond water retained their conditioning ability, but those regenerated in a solution containing ribonuclease did not. It was assumed that RNA was the memory molecule.

3. Several workers have reported on the use of radioactive isotopes injected intravenously or directly into the brain. Animals were given shock-avoidance training and subsequently examined for uptake of the precursors of cerebral RNA. The results indicated that the rate of RNA synthesis was higher in the brain of the trained animals than in untrained control animals.

4. Certain substances injected into an animal's brain interfere with the performance of a previously learned task. It is thought that such substances may disrupt memory by limiting the availability of materials with which memory traces are built. There are a number of experiments reporting the use of drugs which inhibit cerebral protein synthesis, indicating that new proteins are necessary for the neuronal events requiring learning to be consolidated into a more permanent form. Effective inhibitors of RNA synthesis are serotonin, puromycin, and cycloheximide. Thus mice injected

with puromycin appear to lose all memory of a previously learned discriminative task.

The above experiments suggest that changes in the base ratios of the RNA molecule may represent a possible chemical mechanism for long-term storage of the memory trace. The changes occurring in the synaptic structures during learning are permanent changes that modify transmitter actions and thus regulate the flow of information patterns between neuronal networks.

Transfer of learning

There is good evidence that chemical substances formed during training can be transferred to another animal, which then acquires the learned response. The first transfer experiment, reported by McConnel, was conducted with planaria trained to a conditioning task. When these worms were fed to untrained ones, the latter learned faster than did worms in a control group. The validity of the experiment, however, remained in question since it was considered unlikely that planaria were able to learn at all. Accordingly, more convincing evidence was obtained by comparable experiments performed on vertebrates:

1. Rats were trained to perform various tasks by classical conditioning with visual or auditory stimuli. When extracts from the brains were injected into another group of rats, transfer of information occurred: The recipient group made fewer errors than the control group.
2. Mice were trained to carry out specific tasks with Skinner boxes and were sacrificed at the end of training. RNA extracts of brain tissue were given intraperitoneally to recipient mice whose performance was significantly more accurate than those in a group receiving control extracts. This is an example of transfer of task-specific information.
3. Most rodents show a strong preference for moving from a light box to a dark box and staying in the dark. If the animals are shocked in the dark box, they learn to avoid it. After sacrificing the animals, prepared brain extracts are injected into untrained mice. The recipients show a strong tendency to avoid the dark box as compared with mice in a group receiving only untrained extract.
4. Goldfish have been used quite extensively to study transfer effects. Training is carried out by delivering shocks to one compartment of the vessel when, after a few days, the fish learn to escape. Extracts prepared from the brains of the trained fish are injected into untrained goldfish. The recipients are found to avoid the shock compartment much sooner than control fish.

The chemical nature of the transfer factor has been the subject of much speculation. Ungar believed that a peptide called *scotophobin* was responsible for the effects of dark avoidance. Other workers have confirmed that the active component is a combination of small peptides of RNA which vary in amino acid content and sequence. The mechanism of action of these peptides is still unknown.

PHARMACOLOGICAL STUDIES

The administration of drugs to laboratory animals has given useful clues to the biological processes involved in learning and memory. There is also reason to believe that drugs may improve learning performance in man and aid such disturbances as loss of memory, especially in elderly individuals.

Enhancement of learning

Any information drawn on the enhancement of learning by drug therapy must be viewed with caution since it is often difficult to distinguish the drug effects from purely motivational influences on performance.

1. Moderate doses of dextro-amphetamine usually facilitate learning of a wide variety of tasks. Amphetamine is a well-known stimulant of the reticular system and would therefore be expected to enhance central alerting mechanisms in a non-specific manner. However, reticular pathways converging on the hippocampus may induce a synaptic potentiation and an increase of RNA which in turn acts on memory consolidation processes.
2. Low doses of lysergic acid (LSD) improve performance in avoidance conditioning and discrimination tasks. The drug may act as an antagonist to serotonin, which is known to decrease the rate of learning. High doses of lysergic acid have the opposite effects on learning ability.
3. Cholinergic drugs exert an extremely potent effect on learning mechanisms. Essman showed that there was an increase in turnover of acetylcholine in the brain of mice reared together in a large cage, as compared with mice reared in isolation. Anticholinesterase facilitates retention after training, while anticholinesterase inhibitors have the opposite effect.
4. The daily injection of pentylene-tetrazol increases learning ability in mice. The effects are believed to be due to enhanced RNA synthesis, which facilitates the consolidation process. Picrotoxin and strychnine in suitable doses have comparable actions in many species.

Little is known about the mechanisms underlying pharmacological enhancement of learning and memory. One possibility is that the drugs affect the enzyme systems involved in RNA synthesis. Another more likely theory is that the durgs simply modify neurotransmission in learning circuits, especially in structures mediating arousal or motivational functions.

Memory consolidation

The process of consolidation is a function of the nervous system whereby information acquired through learning is stored as a permanent feature of memory. It may also be defined as a period of time during which events are transformed into memory traces that may span an entire life. As stated earlier, short-term memory is dependent upon the electrical events acquired from information processing; it is limited in capacity and soon forgotten. Long-term memory, on the other hand, has a great capacity and is durable in character. Since it can be recalled into consciousness, it must somehow become consolidated into the neuronal circuits. It is assumed that the process of consolidation represents a specific macromolecular change involving the synthesis of new proteins. Hence, any factor that interferes with such events will also disrupt the memory trace before it has become firmly established.

Many pharmacological agents have been shown to influence the memory trace in animals to the detriment of their learning rate and performance. The following examples will serve to indicate that impairment can affect all stages of the process–registration, fixation, and retrieval:

1. Actinomycin D injected into a rat's brain interferes with the performance of a previously learned discriminative task. Since the drug inhibits the synthesis of RNA, it is thought that it may disrupt the consolidation process to cause impairment of retention.
2. Puromycin injected into a mouse's brain immediately before learning a task causes the animal to make many more errors than animals in a control group. Puromycin was thought to inhibit cerebral protein synthesis but it now seems more likely that the drug acts by interference in the later stages of the memory process.
3. Serotonin injected into a mouse's brain immediately after learning an avoidance conditioning response results in a significant loss of retention as compared with a control group injected with saline. Since serotonin reduces the turnover rate of RNA, it is thought that an excess of serotonin may interfere with the acquisition of retention by modifying the biochemical processes critical to memory consolidation.

In summary, certain pharmacological agents acting on the nervous system may modify the rate of learning, the storage of traces, and memory retrieval. Such agents exert their influence mainly by inhibiting RNA synthesis and other protein molecules relevant to synaptic transmission. Therefore, drugs causing a change in the chemical structure at synaptic sites may contribute to the mechanism of memory disruption. Other agents act in a non-specific way to influence the motivational aspects of learning, possibly through limbic and hypothalamic mechanisms, while for several drugs commonly used in behavioral studies, it is impossible to distinguish between the effects on learning and the effects on performance.

CLINICAL STUDIES

Observations from clinical material provide an opportunity to relate abnormalities of memory function to a variety of pathological disturbances affecting the brain. It is usual to divide memory function into a number of stages–registration, storage, retrieval, and utilization. Any of these stages can be altered by disease, leaving the other aspects of memory intact.

Character of memory defects

1. Amnesia, or loss of memory, may involve both short-term and long-term events. Anterograde amnesia is an impaired ability to acquire new information or skills. Retrograde amnesia is an impaired ability to recall past events.
2. Confabulation is the product of false memories, fantasies, and inventiveness, which the patient believes to be real. It occurs when there is a failure to recollect events accurately and resort is made to fabrication of past experiences.
3. Disturbances of time and space represent defects in registration and storage. Thus, a patient may be unable to place historical events in their correct sequence of time. Or, with impairment of topographic memory, a patient may be so absent-minded as to forget the purpose of his visit to another place; he may even have difficulty in remembering the situation of his own home.
4. Hypomnesia is characterized by the inability to recall names or faces. Most people experience this difficulty to a lesser or greater extent and usually write everything down so as not to forget an appointment. Elderly people sometimes repeat stories they have already related on a previous occasion.
5. Hypermnesia is a persistence of thoughts that torments the patient or causes him to pour out an inexhaustible flood of memories, often completely trivial ones that one usually forgets.

Forgetting

Little is known about the physiological mechanism of forgetting except that it may be an active process as well as a defect of registration. Dreams are quickly forgotten since they are never registered in the fully conscious state unless they are written down immediately on awakening. What is experienced under hypnosis also escape registration and so cannot be retrieved. On the other hand, unpleasant memories may be deliberately repressed, especially those with a strong emotional content. Forgetting then becomes a failure of voluntary recall since "lost" memories may be retrieved by sudden shock, hypnosis, or by psychoanalysis.

Clinical disorders of memory

Amnesia is a manifestation of a number of diseases in which memory is deranged out of proportion to other intellectual functions of the brain. Where, however, there is evidence of diffuse cerebral degeneration, memory disturbance may be merely one facet of a widespread impairment of the intellect. The following types of amnesia are commonly described:

Transient amnesia

A brief period of amnesia occurs in many conditions, followed as a rule by complete recovery, although the patient has no recollection of the episode.

1. An example of transient amnesia is seen in epilepsy. Patients with temporal lobe epilepsy have some form of memory impairment and often have no memory of the actual seizure. Petit mal is accompanied by transient memory loss lasting only a few seconds.

2. Cerebral anoxia caused by drugs, alcoholic poisoning, or by acute toxic states is generally accompanied by confusion and disturbed consciousness. It can be induced experimentally by breathing a low O_2 mixture from a Douglas bag connected to a soda-lime chamber for absorption of CO_2. The mental performance of the subject deteriorates as the hypoxia increases. The experiment simulates the fall in oxygen tension that occurs on ascending to a high altitude. The same effects are also encountered in decompression chambers and can arise when artificial breathing circuits, such as an anesthetic apparatus, are used.

3. Transient global amnesia occurs in elderly persons. During the attack

there is no loss of consciousness, but defects of memory for recent events may persist for several hours.

Traumatic amnesia

Head injury from any cause, irrespective of brain damage, may interfere with memory processes. The patient may not remember events just prior to the accident and after apparent recovery there may be further periods of amnesia.

1. Concussion refers to a temporary suspension of cerebral function following a heavy blow to the head. The resultant amnesia is thought to be due to sudden disruption of the metabolic processes required to achieve a stable memory trace.

2. Amnesia following physical damage to the brain may persist when other cerebral functions appear intact. Recent memories are more affected than long-standing ones. Hemorrhage in the temporal lobe may cause permanent memory defects especially when the lesion involves the hippocampus.

3. Cerebral electroshock has a profound effect on short-term memory and the memory consolidation process. In experiments on animals, electroshock causes retrograde amnesia for a condition avoidance response. Neurochemical analysis of the brain after a single convulsive shock reveals an inhibition of cerebral protein synthesis as compared with the protein levels in unshocked animals. The experiments support the view that electroshock induces a retrograde amnesia by disrupting molecular events necessary for memory fixation and storage. Patients who receive shock therapy in psychiatric practice have impaired memory for events prior to the treatment and also have difficulty in remembering new events for a short period afterward.

Korsakoff syndrome

Features of this syndrome include retrograde amnesia and confabulation. The patient is usually alert and attentive, but memory defect prevents him from assimilating new information and skills. The condition often occurs against a background of chronic alcoholism, or it may result after a severe infection of the brain causing degenerative changes in the region of the third ventricle.

Hysterical amnesia

This is more a personality disturbance than a disorder of memory. Registration and storage are unaffected, but there is an unconscious resistance to remembering particularly unpleasant episodes. Hysterical amnesia tends to occur in stressful situations or domestic upheavals from which recovery is usually complete.

CONCLUDING REMARKS

It will be evident from the above account that multidisciplinary studies on learning and related memory processes have contributed a large amount of material from which an increased understanding of the problem is now emerging. It appears that two different aspects of brain function must be recognized:

In the first place, the brain requires certain basic drives or incentives to learn. The stimulus of an enriched environment is important in order to sustain interest and concentration. Also, conscious effort is needed to retrieve information after learning has occurred. These are primarily non-specific functions linked to intrinsic neural networks and activating mechanisms underlying motivational behavior. Such functions are readily disrupted by deprivation, lack of incentive, or distraction since learning becomes less efficient and errors creep into performance when the individual is tired, confused, or emotionally agitated.

Second, the brain serves as a source of reference or storehouse of information which is relatively stable under normal conditions of health, but is very susceptible to injury and disease. There is probably no particular region of the brain where newly formed memories are stored, for these are derived initially from electrical events spread over extensive projection fields. Likewise, it is extremely unlikely that the memory consolidation process is established in any one locality, despite the fact that the temporal lobes and hippocampal formations appear to be especially implicated in memory function.

The relationship between long-term memory and cerebral protein synthesis suggests that changes in synaptic structure in the circuits involved in learning may constitute the foundations upon which permanent memory traces are built. An essential part of the physiology of memory may therefore include chemical processes required for the development of a relatively fragile memory into a more stable and permanent one. Any factors that interfere with these processes or substances that inhibit protein synthesis also tend to impair learning and retention. The possible relevance of these observations to memory disruption in amnestic states cannot be ignored. Information storage has long been the subject of much speculation and remains an attractive problem for research in the future.

SELECTED BIBLIOGRAPHY

Burns, B.D. *The Mammalian Cerebral Cortex.* London: Edward Arnold, 1958.

Corning, W.C., and John, E.R. Effect of ribonuclease on retention of response in regenerated planarians. *Science* 134:1363-1365, 1961.

Doty, R.W. Conditioned reflexes elicited by electrical stimulation of the brain in macaques. *J. Neurophysiol.* 28:623-640, 1965.

Essman, W.B. Some neurochemical correlates of altered memory consolidation. *Trans. N.Y. Acad. Sci.* 32:948-973, 1970.

Essman, W.B., and Nakajima, S. (Eds.) *Current Biochemical Approaches to Learning and Memory.* New York: Halsted Press, 1973.

Gaito, J. DNA and RNA as memory molecules. *Psychol. Rev.* 70:471-480, 1963.

Grossman, S.P., and Mountford, H. Learning and extinction during chemically induced disturbance of hippocampal functions. *Amer. J. Physiol.* 207:1387-1393, 1964.

John, E.R. *Mechanisms of Memory.* New York: Academic Press, 1967.

Landauer, T.K. Two hypotheses concerning the biochemical basis of memory. *Psychol. Rev.* 71:167-179, 1964.

Lashley, K.S. Mass action in cerebral function. *Science* 73:245-254, 1931.

Morrell, F. Electrophysiological contributions to the neural basis of learning. *Physiol. Rev.* 41:443-494, 1961.

Pavlov, I.P. *Conditioned Reflexes.* New York: Dover Press, 1960.

Rosenblatt, F., Farrow, J.T., and Herblin, W.F. Transfer of conditional responses from trained rats to untrained rats by means of a brain extract. *Nature* 209:46-48, 1966.

Russell, I.S., and Ochs, S. Localization of memory traces in one cerebral hemisphere and transfer to the other hemisphere. *Brain* 86:37-54, 1963.

Zinkin, S., and Miller, A.J. Recovery of memory after amnesia induced by electro-convulsive shock. *Science* 155:102-104, 1967.

Index